METHODS OF SOIL ANALYSIS

METHODS OF SOIL ANALYSIS

Contributors :
Jeffrey G. Arnold,
Elizabeth B. Haney, *et al.*

KOROS PRESS LIMITED
London, UK

Methods of Soil Analysis
Contributors : Jeffrey G. Arnold *and* Elizabeth B. Haney, *et al.*

Published by Koros Press Limited

www.korospress.com

United Kingdom

Copyright 2017

Printed in 2017 for Sale in the Indian Subcontinent

The information in this book has been obtained from highly regarded resources. The copyrights for individual articles remain with the authors, as indicated. All chapters are distributed under the terms of the Creative Commons Attribution License, which permit unrestricted use, distribution, and reproduction in any medium, provided the original author and source are credited.

Notice

Contributors, whose names have been given on the book cover, are not associated with the Publisher. The editors and the Publisher have attempted to trace the copyright holders of all material reproduced in this publication and apologise to copyright holders if permission has not been obtained. If any copyright holder has not been acknowledged, please write to us so we may rectify.

Reasonable efforts have been made to publish reliable data. The views articulated in the chapters are those of the individual contributors, and not necessarily those of the editors or the Publisher. Editors and/or the Publisher are not responsible for the accuracy of the information in the published chapters or consequences from their use. The Publisher accepts no responsibility for any damage or grievance to individual(s) or property arising out of the use of any material(s), instruction(s), methods or thoughts in the book.

No part of this publication maybe reproduced, stored in a retrieval system or transmitted in any form or by any means, electronic, mechanical, photocopying, recording, scanning or otherwise without prior written permission of the publisher.

Methods of Soil Analysis

ISBN: 978-1-78163-559-9

British Library Cataloguing in Publication Data
A CIP record for this book is available from the British Library

Exclusively distributed by CBS Publishers & Distributors Pvt. Ltd. Sales & Distribution Rights only for India, Pakistan, Bangladesh, Sri Lanka, Nepal and Bhutan.This book is not to be sold outside these territories.

PREFACE

Soil as a natural resource is a complex body made up of interacting mineral, organic, water, and air components with both biotic and abiotic features. From a practical viewpoint, soil is nonrenewable; thus, its characterization is of prime importance in regard to conservation strategies. Soil science, the study of soil as a natural resource, is interdisciplinary in character and has important fields of study in pedology, agriculture, plant science, forestry, engineering, geology, geography, and biology. The need to describe and understand soil material requires the continued development of suitable analytical methods to characterize soil chemical, physical, and biological properties.

Overall, the book aims to establish a middle ground between the so called "cook-book" approach and the comprehensive, in-depth type of manual. In general, methods which allow some measure of standardization have been selected based on their commonness of use and ease of duplication, and their facility for accuracy and speed. Where possible, the range, limitations and potential for each method is characterized. Within any one volume all- inclusive coverage of any one method is impractical, thus sufficient references are supplied to inform the reader about the complexity and application of the methods and provide a source for further reading.

This handbook is a reference guide for selecting and carrying out numerous methods of soil analysis. It is written in accordance with analytical standards and quality control approaches. It covers a large body of technical information including protocols, tables, formulae, spectrum models, chromatograms and additional analytical diagrams. The approaches are diverse, from the simplest tests to the most sophisticated determination methods.

This page left intentionally blank.

Contents

Preface	*v*
1. Soil Genesis	**1-7**
Introduction	1
Soil Genesis and Development: Rocks, Minerals, and Soils	2
Concepts of Soil Genesis	5
2. Soil Organisms	**8-33**
Introduction	8
Soil Organisms: The Recyclers and Soil Builders	13
Important Soil Organisms	15
Soil Organisms Affect Plants	17
Nitrogen-Fixing Associations	27
3. Soil Microbiology	**34-59**
Characterisation of Soil Micro-organisms	34
Scope and Importance of Soil Microbiology	40
Characterisation of Soil Microbial	43
Soil Micro-organism and Inhabiting	49
Definition of Soil Microbiology and Soil in View of Microbiology	53
Soil Micro-organisms in Biodegradation of Pesticides and Herbicides	56
4. Organic Constituents of Plants	**60-95**
Carbon	61
Carbonic Acid	61
Hydrogen	62
Nitrogen	62
Nitric Acid	62
Ammonia	63
Oxygen	63
Source of the Inorganic Constituents of Plants	75

The Constituents of Plants	77
Oily or Fatty Matters	81
Chemical Changes During Plant Growth	85
The Inorganic Constituents of Plants	89

5. SOIL TAXONOMY — 96-119

Introduction	96
Arid Soils (Aridisol)	97
Soil Taxonomy — Classifying Soils and the Soil Orders	98
Universal Soil Classification System	104
Applying Soil Taxonomy	108
Procedures for Amending Soil Taxonomy	111
Category of the Soil Series	112

6. SOIL CLASSIFICATION — 120-157

Introduction	120
Soil Description and Classification	122
Soil Classification and Characterization	129
Classification of Soils	141
United States Soil Classification System	146
Soil Classification for the Needs of Regional Planning	154

7. SOIL SURVEY — 158-180

Soil and Soil Survey	158
Soil Survey : Examination and Description of Soils	168
Soil Survey Information and Hydric Soils	172
Using Soil Survey to Identify Hydric Soils	178

8. SOIL FERTILITY — 181-214

Introduction	181
Split Applications and Nitrogen Source	182
Maintenance of Soil Fertility	184
Soil Fertility and Productivity	189
Factors of Fertilizer Application	190
Principles of Soil Manuring	193
Soil Fertility Programmes	201
Options for Improving Soil Fertility	206
Facilitated Learning in Soil Fertility Management	210

9. COMPONENTS OF SOIL — 215-231

Soil	215

Soil Humus	220
Soil Micro-organism	221
10. MANAGEMENT EFFECTS ON SOIL BIOTA	**232-243**
Grasing of Soil Biota	233
Cultivation	234
Fertilizers	235
Inorganic Fertilizers	235
P and S Fertilizers	236
N Fertilizers	237
Lime	237
Organic Fertilizers	237
Plant Residue Retention	239
11. ABIOTIC SOIL COMPONENTS	**244-265**
Chemical Input on Abiotic Soil Characteristics	244
Organic Chemicals and Metals	248
Microcosm Tests Including those on Soil Micro-flora	248
Chemical Composition of Soil	257
Risk of Chemicals for Soil Organisms *(Prognosis)*	260
12. AEROBIC MICROBIAL OF SOILS	**266-273**
Micro-organisms	268
Substrates	268
Ph Control and Risks of Contamination	269
Biomass Measurement of Soil	270
13. ORGANIC PASTURE SYSTEMS OF SOIL	**274-286**
Pastures	274
Pest and Disease Management	276
Long-term Field Trials	281
Research Findings Relating to the Bio-Dynamic Preparations	283
14. ANALYSIS METHODS FOR THE DETERMINATION OF ANTHROPOGENIC ADDITIONS OF P TO AGRICULTURAL SOILS	**287-300**
Richard L. Haney, Virginia L. Jin, Mari-Vaughn V. Johnson, Elizabeth B. Haney, R. Daren Harmel, Jeffrey G. Arnold, Michael J. White	
Introduction	288
Materials and Methods	290
Results and Discussion	292
References	297

This page left intentionally blank.

LIST OF CONTRIBUTORS

Richard L. Haney
United States Department of Agriculture-Agriculture Research Service, Grassland, Soil and Water Research Laboratory, Temple, TX, USA
Email: rick.haney@ars.usda.gov

Virginia L. Jin
United States Department of Agriculture-Agriculture Research Service, Agroecosystem Management Research Laboratory, University of Nebraska, East Campus, Lincoln, NE, USA

Mari-Vaughn V. Johnson
United States Department of Agriculture-Natural Resource Conservation Service, Temple, TX, USA

Elizabeth B. Haney
Texas A&M University, Texas AgriLife Research & Extension Center, Temple, TX, USA

R. Daren Harmel
United States Department of Agriculture-Agriculture Research Service, Grassland, Soil and Water Research Laboratory, Temple, TX, USA

Jeffrey G. Arnold
United States Department of Agriculture-Agriculture Research Service, Grassland, Soil and Water Research Laboratory, Temple, TX, USA

Michael J. White
United States Department of Agriculture-Agriculture Research Service, Grassland, Soil and Water Research Laboratory, Temple, TX, USA

This page left intentionally blank.

CITATION

CHAPTER 1

Haney, R.L., Jin, V.L., Johnson, M.-V.V., Haney, E.B., Harmel, R.D., Arnold, J.G. and White, M.J. (2015) Analysis Methods for the Determination of Anthropogenic Additions of P to Agricultural Soils. Open Journal of Soil Science, 5, 59-68. http://dx.doi.org/10.4236/ojss.2015.52007.

This page left intentionally blank.

Chapter 1

Soil Genesis

INTRODUCTION

Soil primarily had its beginning from rock together with animal and vegetable decay, if you can imagine long stretches or periods of time when great rock masses were crumbling and breaking up. Heat, water action, and friction were largely responsible for this. By friction here is meant the rubbing and grinding of rock mass against rock mass. Think of the huge rocks, a perfect chaos of them, bumping, scraping, settling against one another.

Then, too, there were great changes in temperature. First everything was heated to a high temperature, then gradually became cool. Just think of the cracking, the crumbling, the upheavals, that such changes must have caused! You know some of the effects in winter of sudden freezes and thaws. But the little examples of bursting water pipes and broken pitchers are as nothing to what was happening in the world during those days.

The water and the gases in the atmosphere helped along this crumbling work. From all this action of rubbing, which action we call mechanical, it is easy enough to understand how sand was formed. This represents one of the great divisions of soil sandy soil. The sea shores are great masses of pure sand. If soil were nothing but broken rock masses then indeed it would be very poor and unproductive. But the early forms of animal and vegetable life decaying became a part of the rock mass and a better soil resulted. So the soils we speak of as sandy soils have mixed with the sand other matter, sometimes clay, sometimes vegetable matter or humus, and often animal waste. Clay brings us right to another class of soils clayey soils. It happens that certain portions of rock masses became dissolved when water trickled over them and heat was plenty and abundant.

This dissolution took place largely because there is in the air a certain gas called carbon dioxide or carbonic acid gas. This gas attacks and changes certain substances in rocks. Sometimes you see great rocks with portions sticking up looking as if they had been eaten away. Carbonic acid did this. It changed this eaten part into something else which we call clay. A change like this is not mechanical but chemical. The difference in the two kinds of change is just this : in the one case of sand, where a mechanical change went on, you still have just what you started with, save that the size of the mass is smaller. You started with a big rock, and ended with little particles of sand. But you had no different kind of rock in the end.

Mechanical action might be illustrated with a piece of lump sugar. Let the sugar represent a big mass of rock. Break up the sugar, and even the smallest bit is sugar. It is just so with the rock mass; but in the case of a chemical change you start with one thing and end with another. You started with a big mass of rock which had in it a portion that became changed by the acid acting on it.

It ended in being an entirely different thing which we call clay. So in the case of chemical change a certain something is started with and in the end we have an entirely different thing. The clay soils are often called mud soils because of the amount of water used in their formation. The third sort of soil which we farm people have to deal with is lime soil.

Remember we are thinking of soils from the farm point of view. This soil of course ordinarily was formed from limestone. Just as soon as one thing is mentioned about which we know nothing, another comes up of which we are just as ignorant. And so a whole chain of questions follows. Now you are probably saying within yourselves, how was limestone first formed ? At one time ages ago the lower animal and plant forms picked from the water particles of lime. With the lime they formed skeletons or houses about themselves as protection from larger animals. Coral is representative of this class of skeleton-forming animal. As the animal died the skeleton remained. Great masses of this living matter pressed all together, after ages, formed limestone.

Some limestones are still in such shape that the shelly formation is still visible. Marble, another limestone, is somewhat crystalline in character. Another well-known limestone is chalk. Perhaps you'd like to know a way of always being able to tell limestone. Drop a little of this acid on some lime. Then drop some on this chalk and on the marble, too. The same bubbling takes place.

So lime must be in these three structures. One does not have to buy a special acid for this work, for even the household acids like vinegar will cause the same result. Then these are the three types of soil with which the farmer has to deal, and which we wish to understand. For one may learn to know his garden soil by studying it, just as one learns a lesson by study.

SOIL GENESIS AND DEVELOPMENT : ROCKS, MINERALS, AND SOILS

Most soil parent materials were rocks at some time in their history. The minerals in rocks contribute to soil fertility and other soil properties long after the original rock is gone. Consequently, it is a valuable skill to be able to identify a few broad categories of rock. Geologists classify rocks into igneous, sedimentary and metamorphic rocks, according to their origins. In this lesson you will learn how to identify these major rock types and about some common rock forming minerals.

MINERALS AND ROCKS

It is important to understand from the beginning what is the difference between minerals and rocks.

Simple, but useful definitions, are :
- Mineral — a naturally-occurring solid material with a limited range of chemical composition and a specific structure;
- Rock — a naturally-occurring solid material composed of one or more minerals.

Mineral particles may be very small (not visible without magnification), but still have the same limitations on composition and specific structure as a hand-sised specimen.

Minerals in soil are important for several reasons :
- They provide volume and mass to the soil;
- As they weather, they supply elements that are required to grow plants;
- As they weather, they provide the materials to form other minerals.

It is beyond our ability to describe or learn about the approximately 3000 minerals in nature. Luckily, only a few of those minerals are important in soil and those are the ones we will describe here.

First, we need to make a few more definitions :
- Primary mineral — a mineral which forms under conditions different from those at the surface of the earth, often by solidification from magma.
- Secondary mineral — a mineral which forms at or near the surface of the earth, often from weathering products of primary minerals.

The table below provides brief descriptions of minerals found in soils. You can link back to the table below from the rock and mineral sections that follow.

Note that this classification will be useful for us to complete this lesson, but more exact categories are necessary if you want to learn more about minerals and the rocks they form. Also, the stability of a mineral is relative to the other minerals in the list and to a particular environment. Forsterite (an olivine), for example, is described as unstable because it will not persist in a surface soil as long as, for example, orthoclase (a feldspar). Forsterite may still persist in a soil for thousands of years, though.

Common Mineral Groups: Elemental Composition and Relative Stability				
Mineral Group	Specific Example(s)	Primary or Secondary	Principal Elemental Composition	Stability at Earth's Surface
feldspars	orthoclase, plagioclase	primary	Na, Ca, K, Si, Al, O	stable
nesosilicates	forsterite (olivine)	primary	Mg, Fe, Al, Si, O	unstable
chain silicates	diopside (pyroxene) hornblende (amphibole)	primary	Mg, Fe, Si, O	unstable
network silicates	quartz	primary	Si, O	stable
layer silicates (clay minerals)	montmorillonite, biotite, muscovite	primary or secondary	Si, Al, O, K, Mg, Fe	stable
carbonates	calcite	secondary	Ca, Mg, C, O	unstable

Igneous Rocks

Igneous rocks are formed through crystallisation from melt. All igneous rocks, with the exception of volcanic glass (obsidian), are made up of interlocking crystals. In rocks that cool slowly, deep below the surface of the earth, these crystals can be quite large and visible to the naked eye. Common coarse-crystalline rocks that cooled slowly in a magma chamber are granite or gabbro. In rocks that cool rapidly from lava after a volcanic eruption, these crystals may be too small to be visible with the naked eye, but a hand lens or microscope easily reveals the crystalline nature of these rocks. Basalt is an example of a fine-crystalline rock that cooled rapidly at the surface of the earth. Igneous rocks can be thought of as the ultimate parent material for most soils. The igneous rock may have been ground up and transported innumerable times between when it first solidified and what we see today.

Sedimentary Rocks

Sedimentary rocks are formed via the breakdown and redeposition of other materials, often older rocks. Sedimentary rocks are classified by their dominant particle size (sand, silt, or clay) and their mineral composition. Classic sedimentary rocks (*i.e.*, sandstone or shale) are composed of minerals, grains and rock fragments derived from older rocks; while biogenic sedimentary rocks are made up of shells, their recrystallised remnants (*i.e.*, limestone) or plant fragments (*i.e.*, coal). In general, the environment in which particles are deposited determines their grain size and mineral composition. On a beach or near the shoreline, chemical weathering and physical weathering are continuously breaking down mineral grains. Near-shore sediment tends to be well sorted sand, consisting mainly of quartz, which is very resistant to all types of weathering. Farther offshore, where water currents slow down, smaller particles can drop out of the water and form siltstones or shales. Sedimentary rocks make up the source material for more soils than igneous rocks do, because sedimentary rocks are more common at the earth's surface.

Metamorphic Rocks

Metamorphic rocks are recycled rocks that have been subjected to varying degrees of pressure and temperature. As a result of all this extreme heating and pressure, old minerals become unstable and new ones begin to form. In many cases, the resulting rocks display a preferred orientation of minerals, as new minerals form perpendicular to the maximum pressure exerted on the rock. Some important metamorphic rocks types are gneiss ('nice'), schist ('shist'), and slate. Metamorphic rocks are fairly uncommon at the earth's surface so they usually do not contribute greatly to forming soils. Metamorphic rocks also weather slowly because of their hardness.

CONCEPTS OF SOIL GENESIS

One of the earliest land evaluation systems that incorporated a soil classification was established during the Vao dynasty (2357-2261 B.C.) in China. Soils were graded into nine classes, based on their productivity. It has been suggested that property taxes were based on the size of the individual land holding and soil productivity.

In former times (< 1600 A.C.), soil was solely considered as a medium for plant growth. Knowledge of soil behaviour and crop growth was passed from generation to generation gained by observation. For example, in the Middle Ages it was well know that manure applied to soils improved crop growth. There was a close relationship between plant and animal production. For instance, the 'Plaggen cultivation' was practiced for a long time in Europe, which left 'Plaggen soils' : The top of grassland was peeled off and used as litter in the stables. This material mixed with manure was applied to arable land to improve crop production. In 1840, the German chemist Justus von Liebig initiated a revolution in soil science and agriculture. He proved that plants assimilate mineral nutrients from the soil and proposed the use of mineral fertilizers to fortify deficient soils. Crop production was increased tremendously using mineral fertilizers. Another effect was the shift from extensive to intensive techniques in agriculture, which influenced soils. Thaer (1853) published a classification that combined texture as a primary sub-division with further sub-divisions based on agricultural suitability and productivity. Several classifications based largely on geologic origin of soil material were also proposed in the 19th century.

From the 1660s onwards, various members of The Royal Society of London proposed schemes of soil classification that incorporated elements of a natural or scientific approach in their criteria. From this period on, the disciplines of agricultural chemistry (with a strong focus on soil fertility), geography, and geology provided a broad but somewhat fragmented background from which pedology emerged as a separate discipline in the late 19th century more or less independently in Russia (Dokuchaev and colleagues) and in the United States (Hilgard and colleagues). In 1883, Dokuchaev carried out a comprehensive field study in Russia, where he described the occurance of different soils thoroughly using soil morphologic features. Due to his observations in the field he hypothesised that different environmental conditions result in the development of different soils. He defined soil as an independent natural evolutionary body formed under the influence of five factors, of which he considered vegetation and climate the most important. Dokuchaev is generally credited with formalising the concept of the 'five soil forming factors', which provides a scheme for study of soils as natural phenomena. The soil classification developed by Dokuchaev and his colleagues (Glinka, Neustruyev) was based on the soil forming factors > soil forming processes -> and diagnostic horizons / soil properties. The focus in his soil classification approach was in soil genesis, therefore the classification system is called 'genetic'.

In the United States, Hilgard (1892) emphasised the relationship between soils and climate, which is known as the climatic zone concept. Coffey (1912), produced the first soil classification system for the United States based on the soil genesis principles of Dokuchaev and Glinka. Marbut (1951) introduced the concepts of Coffey into soil survey programmes in the U.S. carried out by the USDA (United States Department of Agriculture). Between 1912 and the 1960 the soil classification in the U.S. used a genetic approach. Jenny (1941) put together a detailed description of the five soil forming factors responsible for the development of different soils. In 1959, Simonson stressed that many genetic processes are simultaneously and / or sequentially active in any soil. Hence, a soil classification based on principles of soil genesis would not be favourable. Other soil scientists in the U.S. like Smith (director of Soil Survey Investigations for the USDA Soil Conservation Service) agreed, that soil genesis is very important for soil classification, but that genesis itself cannot be used as a basis for soil classification because the genetic processes can rarely be quantified or actually observed in the field.

Present Perspective in the U.S.

The soil classification based on external environmental factors and an assumed genesis as differentiating characteristics led to dissatisfaction among

U.S. soil scientists in the 1950's and 1960's. It was stressed that there is a lot of uncertainty involved dealing with soil forming factors and changing environmental conditions in a landscape. Many processes go on in any soil, often offsetting one another. Hence, it is often difficult to identify the processes in soils because most soils are polygenetic. Furthermore, it is difficult to assess the relative importance of each soil forming factor contributing to the development of a distinct soil class.

There was much concern of not being able to classify certain soils with adequate agreement because of uncertainties or disagreement concerning their genesis. The knowledge about processes was and still is limited, because we do not completely understand soil processes. It became obvious that a more 'objective' soil classification scheme should be developed to replace the 'subjective' one. *As a consequence a completely new system of soil classification was developed :* The U.S. Soil Taxonomy. It is termed a 'natural' soil classification system, which attempts to organise the division of soils from a more holistic appraisal of soil attributes. The pedon is used as the sampling unit for soil classification and mapping in schemes employing the U.S. Soil Taxonomy. Diagnostic horizons are used to define different soil taxa. Since that time, the Keys to Soil Taxonomy have undergone six published revisions. In this course the 6th edition of Keys to Soil Taxonomy is used.

The principles of Soil Taxonomy are :
- Classify soils on basis of properties
- Soil properties should be readily observable and / or measurable
- Soil properties should either affect soil genesis or result from soil genesis.

Soil processes are not refused in the Soil Taxonomy but the focus is on soil properties rather than on processes. The rationale behind this is that processes are important to produce soil properties and that those are used for classification, *i.e.* genesis 'lies behind' the soil taxa. Soil processes provide a framework for understanding the Soil Taxonomy, for example, the soil forming process of decomposition and humification forms an A horizons. The soil properties recognisable in the field are a dark colour of the A horizon with a granular soil structure. The soil properties selected should be observable or measurable, though instruments may be required for observation or measurement. Properties that can be measured quantitatively are to be preferred to those that can be determined only qualitatively.

Chapter 2

SOIL ORGANISMS

INTRODUCTION

Scientists have devised various schemes for characterising and classifying soil organisms in order to be able to cope with their great diversity. A part from the conventional taxonomic approach, there are classification schemes based on body size, function (decompo-sition, etc.), and the role of organisms from an anthropocentric point of view. The last-mentioned of these systems recognises 'productive' biota (such as crop plants), 'destructive' biota (pests, pathogens, and weeds) and 'resource' biota (species that contribute to soil processes such as decomposition but that do not produce a harvestable product). Under body size, there are three main groups: the microbiota (<100 µm diameter), the mesobiota (100 µm to 2 mm diameter), and the macrobiota (>2 mm diameter). Micro-organisms are the most abundant members of the soil biota. They include species responsible for nutrient mineralisation and cycling, antagonists (biological control agents against plant pests and diseases), species that produce substances capable of modifying plant growth, and species that form mutually beneficial (symbiotic) relationships with plant roots. This last group includes mycorrhizal fungi, various actinomycetes, and some bacteria. Within the soil biota, the most important groups of both destructive and resource organisms are the bacteria, fungi, nematodes, arthropods (such as mites and insects), earthworms (mostly beneficial), and weeds.

Mites

Although these are among the most common soil-dwelling mesofauna, they tend to be restricted to the leaf litter and surface layers. They are more diverse than any other single group of soil arthropods (including the insects), and this diversity is reflected in their feeding habits and life history. In general, they are more important in their role as resource biota than as destructive biota. Some species feed on fungal spores, while numerous predatory species attack nematodes, other mites, insect eggs, and larvae. Other species feed on plant debris, dung, or carrion and are important members of the decomposer community. Although some mite species feed on living plants and others are parasitic on livestock, in general these potential pests are only a minor component of the mite fauna.

Insects

The number of taxa of soil-dwelling insects is relatively small compared with the diversity that marks other major groups of soil organisms. With many species, only part of the life cycle (egg, larva and/or pupa) is spent in the soil, although some taxa are associated with specific soil types. Root-feeding species can reduce crop yields by reducing a plant's ability to absorb water and nutrients and may cause further losses by facilitating the entry of soil-borne pathogens. In contrast, predatory and parasitic insects can contribute to the biological control of both invertebrate pests and plant pathogens. Some Collembola, for example, show a marked preference for feeding on the spores of fungal pathogens, including Rhizoctonia solani and Fusarium oxysporum.

Species that feed on organic matter (detritivores) may be important members of the decomposer community. Some insects have ambivalent roles. For example, although termites are usually viewed as pests, in arid tropical soils with low earthworm populations the burrowing activities of termites can also help decompose organic matter and improve soil structure and porosity. Termites and ants are usually the dominant components of the soil insect biomass. As such, they are probably more important than all other insects in their effects on soil structure. They can also be significant pests. In parts of Malawi, for example, it has been estimated that the two most destructive termites (Pseudacanthotermes militaris and Macrotermes michaelseni) damage crops on 72 per cent and 49 per cent of all small-holdings, respectively. In Africa as a whole, the sub-family Macrotermitinae is considered the most damaging group, affecting a wide range of crops, including tree and pasture species.

Their success as pests has been attributed to their ability to survive cultivation, feed on both living and dead plant material, and cultivate saprophytic fungi that serve as food in the dry season. Other important root-feeding insects include the 'white grub' larvae of various scarab beetles (Scarabeidae), weevils (Curculionidae), wireworms (Elateridae), and false wireworms (Tenebrionidae). Some Hemiptera, Orthoptera, and larval Lepidoptera (*e.g.*, cutworms) are likewise important in parts of the world. The significance of different species as pests is related to their host range. Those capable of feeding on a wide variety of plants, so called polyphagous species, are generally the most difficult to control. The distribution and damage potential of many soil-dwelling insects is also strongly influenced by soil moisture and rainfall patterns, with the more successful species being able to survive periodic drought. As noted earlier, damage by root-feeding insects can result in secondary infections by plant pathogens. Root crops are particularly susceptible to this type of damage, since the harvested part of the plant is directly affected. Potatoes, for example, soon become infected with bacteria and fungi if damaged by larvae of the potato tuber moth (Phthorimaea operculella), resulting in the rapid rotting of the tuber. The same is true of sweet potato tubers damaged by larvae of the sweet potato weevil (Cylas formicarius). In groundnuts, the level of aflatoxin (caused by the fungus Aspergillus flavus) can increase if the pods are damaged by termites, since the latter can transmit fungal spores.

Bacteria

The most abundant members of the vast community of soil organisms are bacteria. Per gram of soil, they can reach densities of one billion (one thousand million) individuals — and an estimated 20 000 to 40 000 species. The magnitude of bacterial bio-diversity has only recently been revealed through molecular techniques that can differentiate between hard-to-culture taxa. Bacteria play important roles in many soil processes, including the cycling of nitrogen, carbon, and phosphorus, and the degradation of pesticides and other potential pollutants. They can multiply rapidly under favourable conditions, although very high growth rates are generally limited by the availability of nutrients.

Bacterial populations are typically greater and more diverse close to plant roots. It has been estimated that 5 to 10 per cent of root surfaces may be occupied by bacteria. Various endophytic bacteria (*e.g.*, species of Azotobacter, Acetobacter, and Azospirillium) not only fix nitrogen but also stimulate the production of root hairs. This increases the host plant's capacity to take up water and nutrients and compensate to some extent for grasing by root-feeding pests. Various bacteria that colonise seed coats and plant roots produce compounds capable of affecting plant growth. These growth-promoting rhizobacteria (including species of Pseudomonas, Bacillus, Serratia and Arthrobacter) can also increase plant resistance to pests and diseases through a variety of mechanisms, including the production of structural materials that strengthen root tissues and antibiotic metabolites that directly affect other elements of the soil biota.

In some cases, these beneficial bacteria displace potential pathogens by competing for nutrients. Or, they may stimulate increased synthesis of defensive compounds by the plant itself, resulting in so-called induced resistance or systemic acquired resistance. Rhizobium etli, for example, induces systemic resistance in potato roots to the potato cyst nematode (Globodera pallida). Some rhizobacteria are able to suppress weed growth while leaving crops unaffected. The relationship between bacterial endophytes and their host plants is a subject of increasing interest to those seeking better ways to manage the soil biota. Plant roots themselves produce a number of compounds that affect microbial populations in their vicinity (the so-called rhizosphere). These compounds may act as attractants, repellents, biocides, or biostats (compounds that inhibit microbial growth), and as such offer new possibilities for the manipulation of both pathogens and beneficial species.

Fungi

As with the bacteria, the great diversity of fungi remains poorly documented. It has been estimated that only about 5 per cent of fungi have so far been described.

Fungi play as important role in soil processes as do bacteria, but tend to be more abundant in slightly acid soils. They vary widely in size, preferred habitat, and mode of life. Fungal plant pathogens (*e.g.*, some species of Fusarium and Verticillium) can cause diseases such as root rots and vascular wilts that are significant problems in many parts of the world. In the early 1990s, for example, the increasing prevalence of fungal root rots severely restricted the viability of bean crops in parts of Kenya. Many soil-borne fungal pathogens (*e.g.*, Rhizoctonia solani and Pythium spp.) are capable of infecting a range of plant genera.

Furthermore, plants infected with fungal diseases may be more vulnerable to attack by soil-dwelling insects. In Malawi, for example, it was found that groundnuts were more vulnerable to attack by termites if they were infected by fungal pathogens such as F. solani. However, some fungi form symbiotic relationships with plant roots, enhancing the plant's ability to take up nutrients. There is some evidence that such relationships can help less competitive plants to become established in pasture systems. Many species of soil fungi are saprophytic (*i.e.*, grow on dead organic matter), while others are parasitic on animals or plants. Some are important antagonists or biological control agents of soil-borne pests or diseases (*e.g.*, species of Dactylaria and Arthobotrys for nematodes, Beauveria and Metarhizium for insect pests, and Trichoderma and Coniothyrium for plant-pathogenic fungi).

Nematodes

Nematodes (roundworms or eelworms) are the most abundant microfauna in the soil and are particularly numerous in the top 5 cm. In the top 2 cm of soil, their numbers may exceed two billion per hectare. The number of nematode species worldwide has been estimated at 80 000 to 100 000 species. The majority of soil-dwelling nematodes feed on bacteria and fungi, but many prey on other nematodes, protozoa and rotifers, and some species are parasitic on insect pests. More than 2000 species are parasites of higher plants. These include cyst and root-knot nematodes, such as certain species of Heterodera, Globodera, Cactodera, and Meloidogyne), and 'migratory' species such as Pratylenchus spp., Ditylenchus destructor, and Scutellonema bradys.

Their feeding lowers crop yields by disrupting water and nutrient uptake or by decreasing fruit or tuber quality or size; it can also allow fungal and bacterial pathogens to gain access to damaged roots, causing secondary infections. The presence of root-knot nematodes, for example, is known to increase the incidence of root rots and fusarium wilts in a wide range of crops. Plant parasitic nematodes tend to be more damaging pests in the tropics than in temperate zones. This is because their population growth rate is favoured by warm, humid climates and the longer growing seasons in these regions, which allows for more reproductive cycles per year. Nematode pests are associated with almost all crops grown in the tropics and cause losses of millions of dollars each year.

It has been estimated that, worldwide, plant-parasitic nematodes annually reduce agricultural production by about 12 per cent. Losses are particularly severe in developing countries due to insufficient expertise for species identification, inadequate quantification of the pest problem, and a limited range of management options. Worldwide, root-knot and cyst nematodes are especially important pests and infestations by them continue to undermine national efforts to improve food security and alleviate poverty. Losses to individual crops often go unnoticed or are attributed to other causes. This is because most symptoms of nematode damage such as chlorosis (yellowing), patchy growth stunting, and wilting in hot weather, are easily confused with nutrient deficiencies or sometimes with bacterial or fungal diseases. Compared with the plant-parasitic nematodes, much less is known of the biology and ecology of other nematodes. This is particularly true of the bacteria-feeding nematodes, which are generally considered to be beneficial or harmless. These species, often concentrated in the root zone of higher plants, play an important role in soil nutrient cycling and can help distribute bacteria that promote plant growth. Although nematodes are not capable of widespread dispersal on their own, in agricultural systems they may be spread via contaminated machinery, land levelling, irrigation, and soil erosion.

Earthworms

The Greek philosopher Aristotle referred to earthworms as the "intestines of the earth". The description is apt not only because these invertebrates physically resemble digestive organs but also because they perform something of a digestive function when they break down plant residues in the soil and recycle nutrients. Experiments with maise, for example, have shown that plant residues can degrade 30 per cent faster in surface soil containing earthworms than in soil without them. Earthworms perform a variety of other ecosystem services as well. For instance, they enhance soil porosity thereby reducing rainwater run-off and allowing for infiltration of moisture to plant roots. They also play a vital role in water purification, agrochemical detoxification, and maintenance of soil stability.

When it comes to sheer biomass, earthworms usually account for the bulk of invertebrates that make their home in the soil. Along with ants and termites, they have a special distinction as 'ecological engineers'. This is due to their ability to excavate remarkably large amounts of soil and to create organomineral structures through their casts (excrement). To date about 3700 species of earthworms have been scientifically described. However, the actual number is estimated to be at least twice that, with the overall knowledge gap being significantly greater for tropical earthworm species than species in temperate zones. Although many aspects of earthworms' soil engineering role have been documented, interactions with other members of the soil biota, and how those affect plant growth and health, positively or negatively, are much less well understood.

Weeds

In the tropics, crop losses due to competition with weeds are on the order of 25 per cent, with weed removal often the most labourintensive task in small-holder systems. In most areas, the weed population centers on about 20 particularly troublesome species. The individual taxa may be indigenous or introduced, but they usually share a few key traits, especially rapid vegetative growth, high fecundity, and persistence in the soil seed bank. Weed species that resemble the crop in their early stages (*e.g.*, grass weeds in cereal systems) are particularly difficult to deal with when hand weeding is the main control technique. Besides competing with the crop for water and nutrients, weeds can also act as alternative hosts for pest nematodes and plant pathogens. And by boosting the humidity around the base of crop plants, weeds also increase the likelihood of infections by pathogens such as R. solani and Sclerotium rolfsii. Some weed species are parasitic on certain staple food crops, in some cases causing total crop failure.

The role of weeds in relation to soil-dwelling insect pests has not received a lot of scientific attention, although in some cases weeds are known to act as alternative hosts, maintaining the insect pest between crop cycles. Wild Ipomoea species, for example, can support the sweet potato weevil (Cylas formicarius) between crops, and in Zimbabwe, the groundnut plant hopper (Hilda patruelis) can survive the dry season on a variety of different weeds, allowing it to invade groundnuts as soon as they emerge. But weeds may also play a constructive role by supporting higher numbers of soil-dwelling predators that serve as pest control agents.

SOIL ORGANISMS : THE RECYCLERS AND SOIL BUILDERS

Soil organisms - many are microscopic workers spending lifetimes endlessly recycling dead plant materials and maintaining the fertility of the soil. In areas where their life cycles are not damaged soils naturally conserve themselves or actually build. These include uncut forests, soils under ponds and lakes, prairies and meadows with permanent plant cover and to some extent no till and mulched fields and gardens.

- "We know more about the movement of celestial bodies than about the soil underfoot." –Leonardo Da Vinci, circa 1500's

We've managed to figure out how to measure aspects of the soil's chemistry so we think we know all about the soil. I know when I do a little test and find out my soil's pH is 8 I know a whole bunch. But, equally, if not more important, are the soil organisms living within the mineral stew. The problem with soil organisms is that most of them are invisible to us. The most important, the bacteria and fungi, are microscopic and we think that they might be a problem. I mean have you ever considered loving your bacteria — or your fungi. Plus they just aren't all that cute. They're blind — which makes absolute sense given that they live in the dark — so they don't have eyes. They aren't pretty colours — most variations on the black and white theme. Basically you have to be someone like Grissom (from CSI) to love them.

Soil Organisms

Even the smallest amount of information about soil organisms can threaten your world view — and with luck entice you to love and care for these small beings. Or at least it has done so for me. I'm a committed organic type gardener and have been for some time, but, I have to admit I didn't really understand what was going on in that soil underfoot my feet touch everyday. I still don't really — it's too complex and brilliant for me to totally fathom. But thanks to a few soil nerds who've been kind enough to share their work with the world in books I've been partly enlightened. Here is the information in a nutshell. Soil, or at least healthy soil, is teeming with life. This life lives in an extremely symbiotic and mutualistic way unfailing offering their services to help the whole system work.

In a thimble full of soil — about a gram weight — you can expect to find
- 100 million to 1 billion bacteria
- Several yards to several miles of fungi depending on whether you have agricultural or forest soil
- Several thousand protozoa and up to several hundred thousand in forest soils
- From ten to several hundred nematodes.

This is all in a thimble full of soil. And this doesn't begin to count the larger mites, spiders and earthworms who together form a complex called the Soil Food Web. I suspect that we would love soil organisms a lot more if we could see them. But the fact is we can't see most of them and to a large extent out of sight is out of mind. Remember, in a natural system you don't need to be adding a bunch of fertilizers, pesticides, herbicides and fungicides every year to keep things growing and healthy. Natural systems maintain themselves thank you very much and they do this with the work of the soil food web.

First the Bacteria and Fungi — foundations of the web :
- Hold nutrients in the root zone for your plants
- Build soil structure so that water both drains and holds better
- helps prevent soil erosion
- Produces plant stimulating hormones
- Fixes nitrogen from nitrogen gas to plant usable nitrogen
- makes phosphorus available in a plant usable form
- Open the soil so that plant roots can penetrate more deeply
- Compete with bad guy bacteria and fungi thus keeping them in check and helping plants fight disease or stay disease free.

The other soil organisms, the protozoa, nematodes and other larger members of the soil food web eat the bacteria and fungi and in doing so release the nutrients the plants need when they need them. Pretty amasing really.

Care and Feeding of Soil Organisms

For a moment think of your soil organisms as very cute and lovable puppies. Naturally you want to give them a comfortable and safe home, good food, and the air and water they need. Guess what — soil organisms have similar needs. Unfortunately they aren't cuddly and you can't play with them, for the most part you can't even see them, so you might not notice that you're killing them with either your gardening or farming methods or sheer neglect.

Habitat — most of the soil creatures live in the top few inches of the soil. When we turn the soil over with a rototiller or a shovel we can bury our babies in a place where they no longer can function. Fungi have the worst time with our cultivating methods. Their filaments get chopped up and the long straws they were forming to syphon nutrients to their plant partners get severed and big parts cease to function. Many farmers have adopted no till methods in their fields to improve soils and this will serve to give the soil food web a chance to recover.

Air — when we walk on our gardens we compact the soil under our feet.

This is many times worse when we drive over land with a tractor. What it does is eliminate some of the pores that were holding oxygen — always handy for the beneficial soil organism. As oxygen goes down the anaerobic bacteria and fungi go up and the bad news is these are the guys that cause plant diseases.

Feed Good Food — like us the soil organisms need good food. Bacteria tend to like the green foods and sugars and fungi the brown foods like woods and tough leaves. Compost, being a blend of green and brown foods, is the perfect solution — and it has the added benefit of increasing the numbers of bacteria, fungi, protozoa and nematodes at the same time.

Avoid Bad Food — In the past — and even now — we tend to feed the garden chemical fertilizer. This changes the habitat fast — too fast for our soil pets to move or adapt so they can die. Essentially we then put our plants on life support - something akin to an IV drip instead of real food. In doing so we kill the chance of real food coming their way through the network of the soil food web. Plus we of course have bills to pay — fertilizer bills.

Safety — here is where the — cides come in. When the biology of the soil gets out of whack we tend to have a bunch of problems crop up — insect pests run amuck, fungi and bacteria shift from help mode to destroy mode and weeds take over. So naturally we reach for pesticides, herbicides and fungicides. They are effective killers but like big bombs they tend to kill a lot more than we intended. We lose our soil food web and all it's benefits once again.

IMPORTANT SOIL ORGANISMS

Each trophic level of the soil food web is comprised of several types of organisms. This section will describe some of the most prevalent and important soil organisms; however, it is important to note that there are many other types of organisms that exist in the soil.

Earthworms

Earthworms are very important soil inhabitants. While not required for a soil to be healthy, they are an indicator of soil quality. The burrowing and ingestion of soil materials by earthworms helps to physically mix the soil, decompose plant materials, and cycle nutrients. The casts of earthworms can also improve soil structure by enhancing aggregation. Earthworm activity is usually low in Florida's sandy soils.

Arthropods

Arthropods are animals with an exoskeleton, such as insects, crustaceans, and arachnids. In the soil, arthropods are responsible for physically shredding organic materials, stimulating microbial activity, and cycling nutrients. Soil arthropods are categorised based on their function in the soil. Large arthropods like millipedes and termites are shredders that chew up dead plant materials. Herbivores like mole crickets feed on plant roots. Shredders and herbivores can become pests when their populations are high. Predator arthropods, including centipedes and spiders, help to control the populations of other soil arthropods. Some arthropods eat soil fungi and bacteria.

Nematodes

Nematodes are non-segmented worms, most that live in soil are microscopic. Many nematodes feed on bacteria, fungi, algae, or are predators of microscopic animals. Other nematodes are parasites of animals or plants. In the soil, nematodes assist with nutrient cycling, are a source of food for other soil organisms, and can suppress or cause disease. While root-feeding nematodes can cause plant disease, fungal-feeding, bacterial-feeding, and predatory nematodes often work to suppress plant diseases or pests.

Soil Protozoa

Protozoa are single-celled, mobile organisms that are important for nutrient cycling and regulating populations of bacteria in the soil. Protozoa, which are larger than bacteria, are often abundant in soils. Protozoa populations can reach up to 1 million per teaspoon of soil in highly fertile soils. When they feed on bacteria or other protozoa, these organisms often release excess nitrogen in a form that can be used by plants. Protozoa depend on soil moisture to feed and for movement. Therefore, soil moisture influences the type and size of protozoa in the soil. Soil protozoa are often most active near plant roots.

Soil Fungi

Fungi are a diverse group of Micro-organisms that are abundant in the soil. Most fungi (*e.g.*, molds and mushrooms) are multi-cellular and grow in long strands called hyphae, but a few species like yeast are single-celled. Fungi play an important role in the nutrient cycling, organic residue decomposition, and disease suppression. Decomposers, mutualists and pathogens are the three main groups of soil fungi. The decomposers are responsible for the breakdown of dead organic material. Mutualist fungi will colonise plant roots and provide nutrients to the plant in exchange for carbon (an energy source). The pathogens cause disease or death when they colonise and feed on living organisms like plant roots or nematodes.

Bacteria

Bacteria are microscopic, single-celled organisms that are abundant in soils. In fact, a teaspoon of soil can contain up to 20,000 different species of bacteria. In the soil, bacteria are important for nutrient cycling, disease suppression, and even pollutant cleanup. Soil bacteria can also improve soil structure by enhancing the formation of aggregates. Similar to soil fungi, soil bacteria are categorised as decomposers, mutualists and pathogens. There is also a fourth type of bacteria called chemoautotrophs. Decomposer species are responsible for the breakdown of organic residues. Some mutualist bacteria species are capable of converting nitrogen in the atmosphere to a form that is available to plants. These nitrogen-fixing bacteria colonise the roots of plants called legumes (*e.g.*, clover, soybean, alfalfa). Chemoautotrophs obtain energy from nitrogen, sulfur, or iron compounds rather than from carbon. These bacteria are often capable of decomposing harmful chemicals like oil and pesticides that find their way into the soil.

Factors Affecting Soil Organism Populations

The type and number of organisms present the soil environment is directly related to the availability of food. However, other factors like climate, vegetation, and soil pH also impact the type and quantity of organisms present. For example, forests usually have more diverse soil organism populations than disturbed areas (*e.g.*, cultivated fields, urban areas). Highly disturbed soils, such as those found in urban areas, typically have smaller numbers of soil organisms than forest or grassland soils. Also, soil organisms are usually more active in areas with a warm, moist climate than in areas with cold or dry climates. Typically, soil Micro-organisms (*e.g.*, bacteria and fungi) account for most of the biological activity in soils.

SOIL ORGANISMS AFFECT PLANTS

Soil organisms interact with plants in diverse ways. Some organisms increase plant growth, while others decrease it. Some alter root structure but others have little effect. Micro-organisms can also have important influences on the way that roots branch and grow in soil. These structural modifications affect the uptake of water and nutrients by plants. Plant roots are of course the main structures through which soil organisms and plants interact. Roots provide a concentrated supply of carbon and energy for many heterotrophic soil Micro-organisms. In turn, soil animals that feed on fungi and bacteria multiply in response to the higher food supply around roots. Some organisms can invade plant roots. Living roots are particularly important for many fungi, especially fungal pathogens and mycorrhizal fungi. Most root pathogens form associations with specific plants. Other pathogens colonise roots of a wide range of plants. Arbuscular mycorrhizal fungi generally form non-specific associations with many plant species. Ectomycorrhizal fungi are more specific in the types of roots they colonise. Before this section considers the effects of soil biota on plant roots, it first examines how plant roots shape the environment of soil organisms.

Plant Roots and Soil Biota

The rhizosphere is the region of soil that surrounds, and is influenced by, plant roots. A major influence in the rhizosphere is the release of organic molecules into the soil from roots. Although the exudates and mucilaginous materials released by roots diffuse through the soil, they always occur in their highest concentrations adjacent to the root surface. Because soil organisms are attracted to these exudates as a source of food, the abundance of soil organisms also increases close to roots. This is a dynamic relationship. The rhizospheres of old and young roots provide very different habitats for soil organisms. As roots age, they release different types and quantities of carbon substrates, which in turn affect soil Micro-organisms. The rhizosphere also differs from the remainder of the soil in other ways. For example, the amount of inorganic nutrients can be either higher or lower in the vicinity of roots compared with that in the soil some distance away from roots. The concentration or depletion of nutrients around roots depends on physiological processes associated with root function, including nutrient uptake by plants. In addition, roots release hydrogen or bicarbonate ions into the soil, causing the pH in the rhizosphere to decrease or increase. Roots also consume or release oxygen. These contrasting effects, in addition to the considerable exudation of carbon compounds from roots into the soil and the compression of soil as roots grow in length and breadth creates a dynamic environment for soil organisms. Some nutrients move towards roots in soil water by the process of mass flow. Other nutrients are less mobile because they are adsorbed onto soil particles. The availability of adsorbed nutrients such as phosphorus and copper to plants depends on the extent to which roots explore the soil and intercept these ions. This in turn affects the extent of the rhizosphere and the spatial distribution and activity of soil Micro-organisms.

Water use by plants adds to the dynamic nature of the rhizosphere. Daily cycles of water uptake and loss by roots cause roots to expand and shrink. This adds to physical changes induced by seasonal cycles of wetting and drying or freesing and thawing. These daily and seasonal cycles alter the soil habitat, which influences the growth and activity of soil organisms. As a root develops, the number of organisms near the root surface increases rapidly. For many plants there is not a clear distinction between the root surface and the soil. Roots are dynamic, growing and casting off cells and navigating soil cavities and rough surfaces. As a consequence, the root surface can be highly irregular.

What is the Origin of Root Exudates?

A variety of organic compounds released from roots provide a source of energy and carbon for soil organisms. In a study of exudation from wheat roots, the plant was labelled with 14C. This technique allows the carbon released from the plant to be followed precisely. The greater the amount of 14C released into the rhizosphere of the wheat plant, the greater the quantity found in the microbial biomass. The quantity of the compounds released from roots is not the same along the length of a root. Older roots can be more 'leaky' than younger roots, although in this experiment there was no difference, at least for one line of wheat. The release of relatively large quantities of organic carbon into soil under normal conditions stimulates microbial activity and provides a source of energy that can accelerate the decomposition of organic matter.

How do Roots Affect Soil Organisms?

Roots provide an important habitat for bacteria, fungi and very small soil animals. The number of organisms can be about 500 times more in the rhizosphere than in the bulk soil. The surface of a root is not completely covered with bacteria and fungi, except in special circumstances, such as where roots are colonised by an ectomycorrhizal fungus. Usually less than 10 per cent of the root surface is colonised. Bacteria accumulate in crevices along the edges of outer root cells or where the root surface is damaged.

Some organisms form loose associations with roots and simply depend on the plant releasing molecules that they can use. Other organisms have more specific interactions with roots. For example, root nodule bacteria can stimulate the root to form additional cell tissue that proliferates to form a nodule. Root nodule bacteria live and multiply inside these root cells. When roots begin to disintegrate with age or through disease, there are more chances for opportunistic bacteria and fungi to enter through crevices between the surface cell layers or dead root cells.

Roots indirectly affect soil animals because they respond to changes in the abundance of bacteria and fungi associated with roots. The types of organisms present in the rhizosphere and in the bulk soil change over time (Lussenhop and Fogel 1991). These changes are associated with the intensity of grasing by soil animals and the abundance of other bacteria and fungi. Furthermore, there is a wide range in numbers of different types of animals. Overall, there is great diversity in type and number of organisms in the vicinity of roots.

How are Roots Affected by Soil Organisms?

Micro-organisms can indirectly increase or decrease root growth. For example, organisms influence the movement of carbon from shoots to roots. This can alter root growth as well as stimulate the release of carbon into the rhizosphere. Some organisms increase root growth by producing compounds that function as root growth hormones. Organisms may decrease root growth by creating a significant drain on carbon reserves in the plant. This reduces the quantity of carbon in the root that would otherwise be used to form more roots.

The effects of soil organisms on roots can be either indirect or direct. For example, earthworms indirectly promote root growth because their burrows provide channels for roots to grow more deeply into the soil profile. In contrast, some rhizosphere bacteria and fungi impede root growth. Roots may grow more extensively if such Micro-organisms are absent than if they are present. Ectomycorrhizal fungi also reduce root growth by colonising the surface of roots and stunting their growth. In addition to the thickening of the root, the roots often branch in response to hormones produced by the fungus.

Arbuscular mycorrhizal fungi can reduce root growth. In a study of maise, Micro-organisms decreased both root hair length and root hair number. However, the extent of the decrease depended on whether arbuscular mycorrhizal fungi were present or not. In the presence of these fungi, both root growth and root hair number were considerably impeded in comparison to when these fungi were absent.

Generalisations about the effects of soil organisms on root growth are difficult to make because different species of plant and different types and combinations of soil organisms interact in different ways. Bacteria in the genus Azospirillum are interesting examples of Micro-organisms that have the potential to influence root growth. Species of Azospirillum commonly live in the rhizosphere of grasses. Some bacteria in the genus Azospirillum increase root growth by contributing nitrogen to the plant and others stimulate root growth by hormone production. Strains of species of Azospirillum vary in their capacities to fix atmospheric nitrogen or produce hormones. Nitrogen fixing strains of Azospirillum are not expected to increase root growth if nitrogen levels in the soil are already adequate for the particular plant species; if root growth is affected, other mechanisms must be involved.

Changes in root structure alter the capacity of root systems to absorb nutrients and water. Principally, this is related to the way that roots are distributed in soil. However, close associations between roots and Micro-organisms enhance nutrient uptake by plants if organisms have access to nutrients to which the plant does not. An example of this is the association between roots and nitrogen-fixing bacteria. Another example is the association between various types of mycorrhizal fungi and roots that allow root systems access to nutrients well beyond the rhizosphere.

Root pathogens commonly interfere with nutrient and water uptake by damaging the root surface and root cortex or by blocking the xylem vessels in the vascular system inside the root. Damage to the cortex or the vascular system will reduce nutrient and water transport between roots and shoots. Soil animals such as nematodes may also damage roots. Parasitic nematodes induce gall formation on some plants and animals such as springtails graze on small roots and damage their surface. Disturbances of this kind create opportunities for a variety of fungi and bacteria to enter roots that would not otherwise be able to do so.

Mycorrhizal Associations

The majority of plant species form mycorrhizal associations. Mycorrhiza means 'fungus root' and refers to several specialised associations between fungi and roots. There are four major types of mycorrhizas : ectomycorrhizas, arbuscular mycorrhizas (also called vesicular arbuscular mycorrhizas), ericoid mycorrhizas and orchid mycorrhizas. These associations are generally beneficial to both the plant and the fungus, with nutrient exchange between the partners. However, under some conditions, at least for arbuscular mycorrhizal, there is no beneficial effect and plant growth can even be reduced.

Orchid mycorrhizal associations are very different from the other three types. The minute orchid seed has no reserves of carbon and is totally dependent on the fungus during the early stages of its growth. Almost all plant species form one of the four types of mycorrhizal associations. Each plant species usually forms only one type of mycorrhizal association. In the less common situation in which more than one type of mycorrhiza is formed on the same plant, the different mycorrhizas usually occur on different roots or are present at different times during the life cycle of the plant.

One type of mycorrhiza is usually dominant in each ecosystem. For example, plants that form ericoid mycorrhizas often dominate heathland ecosystems. Trees that form ectomycorrhizal associations commonly dominate forests in Europe and North America. Tropical forests may contain more equal proportions of plant species that form arbuscular and ectomycorrhizal associations. The type of mycorrhiza present generally depends on the diversity of plant species that occur there. The eucalypt trees of Australian forests predominantly form ectomycorrhizal or arbuscular mycorrhizal associations. For example, in a jarrah forest in south-west Australia, seventy two per cent of the 93 species of plant studied formed either or both of these two mycorrhizas. The 25 per cent found with no mycorrhizal association were mostly members of the family Proteaceae, which does not form mycorrhizas but they can form specialised roots 'proteoid' roots which facilitate nutrient uptake.

Most annual agricultural plants form arbuscular mycorrhizas. Ectomycorrhizal associations are less common in disturbed ecosystems and are more frequently observed on perennial plants. Horticultural species form a variety of mycorrhizal associations, depending on whether they are ornamentals (all types of mycorrhizas are represented in this group) or vegetable and orchard plant species (most of these form arbuscular mycorrhizas).

Arbuscular mycorrhizas are the most common type of mycorrhiza and they are formed by a fungal group that occurs in most soils. The distinctive fungi that form arbuscular mycorrhizas have co-evolved with plants over millions of years. However, worldwide, there are relatively few species of arbuscular mycorrhizal fungi in comparison with other groups of soil organisms. Fewer than 200 species have been described, although many others occur which have not yet been formally named and described. These 200 species all occur within a relatively small number of genera including : Glomus, Acaulospora, Archaeospora, Gigaspora and Scutellospora. The genus of each fungus can be identified from the characteristics of their spores and molecular analysis is now also used to distinguish among species and genera. The most characteristic feature of arbuscular mycorrhizal fungi is the arbuscule. Arbuscules occur within cells of the plant and are key sites for nutrient exchange between the two partners of the association. The distinctive vesicles formed inside roots by some, but not all of these fungi contain high concentrations of lipids and appear to be storage structures.

Biology of Arbuscular Mycorrhizal Fungi

The roots of most plants are colonised by several species of arbuscular mycorrhizal fungi at the same time. There is little evidence of specificity between particular plants and these fungi, although some plants may be colonised to a greater extent by some of them. The relative abundance of different arbuscular mycorrhizal fungi within roots depends on soil conditions, root development and characteristics of the fungi such as the rapidity of spore germination and growth of hyphae.

Arbuscular mycorrhizal fungi are dependent on the plants they colonise for their survival. The fungi are "biotrophic" which means that they are unable to complete their live cycle without a plant. Because of this unusual characteristic of arbuscular mycorrhizal fungi, it is difficult to grow large quantities of arbuscular mycorrhizal fungi on artificial media in the laboratory without a plant. When newly formed hyphae of AM fungi intercept a root, the fungus enters with no apparent alteration to the root surface. Species, and even strains, of the same fungus colonise roots to different extents, but generally the more hyphae there are in the soil the greater the extent of mycorrhizal root formed. Arbuscular mycorrhizal fungi can survive for long periods in the soil as spores or as pieces of hyphae, even when the soil is dry or frozen. Most species produce spores that have relatively thick walls that are resistant to desiccation and are not easily eaten by organisms in the soil or colonised by pathogens. They have been shown to contain bacteria that appear to be natural inhabitants. Spores of arbuscular mycorrhizal fungi germinate when soil conditions become suitable, providing they are not in a dormant phase. Pieces of hyphae can survive in dead root pieces or directly in the soil.

Effects of Arbuscular Mycorrhizal Fungi on Plants

An important impact of AM fungi on plants is the alleviation of phosphate deficiency. Arbuscular mycorrhizal fungi form a network of hyphae in soil surrounding roots, and in doing so, effectively extend the root system. This can increase uptake of phosphorus from soil into roots. For example, some plants that are well colonised by arbuscular mycorrhizal fungi can grow just as well with only half the amount of phosphorus available in soil as that required by plants that do not have these fungi within their roots.

The extent to which a plant benefits from its arbuscular mycorrhizal fungi depends on the availability of nutrients and water. In terms of phosphorus nutrition, only plants growing where there is an inadequate supply of phosphorus will receive any benefit from having arbuscular mycorrhizas (this also requires that water and other nutrients are not limiting to growth). In soils where there is very little available phosphorus or where adequate phosphorus is available, there is generally no direct nutritional benefit of arbuscular mycorrhizas. Plant root architecture also influences the ability of a plant to supply itself with phosphorus, and therefore its dependence on an association with arbuscular mycorrhizal fungi. Plants such as grasses have fibrous root systems and are capable of thoroughly exploring soil for phosphorus. As a result, they do not benefit nutritionally from mycorrhizas to the same extent as a plant such as clover which has a coarser root system.

This effect of arbuscular mycorrhizal fungi on plants through increased phosphorus uptake has usually been studied only for young plants or plants grown in controlled conditions. The contribution of arbuscular mycorrhizal fungi to the growth of older plants, especially those growing in field soils, is not well understood. Obviously the relationship between the way the symbiosis functions and the nutritional status of the plant is complex and the plant does not always benefit. For example, under some conditions, arbuscular mycorrhizal fungi reduce plant growth, possibly because they take a large amount of carbon from the plant. Alternatively, Scutellospora calospora can increase plant growth and phosphate uptake into plants when the level of phosphorus in soil is extremely deficient. But, it can also decrease plant growth when the quantity of phosphorus in the soil is almost sufficient for the plant to reach its maximum potential growth. At higher levels of available soil phosphorus, the fungus appears to remove excessive carbon from the plant to an extent that is detrimental to the growth of the plant.

The abundance of arbuscular mycorrhizal fungi within plant roots changes the supply of phosphorus. Very low levels of phosphorus can be insufficient for the growth of the plant as well as the fungus, keeping fungal numbers low. As more phosphorus becomes available, both the plant and the fungus increase in growth. There can be an increase in the length of root colonised by arbuscular mycorrhizal fungi and a corresponding decrease in the proportion of the root colonised. This occurs if the fungus does not grow fast enough to keep up with the growth of the roots. In soil where phosphorus is present at an adequate level for maximum plant growth, the quantity of arbuscular mycorrhizal fungi in roots decreases substantially. This may either be due to a detrimental effect of excessive amounts of phosphorus on the fungi in the soil or to inadequate carbon supply in the roots for fungal growth.

Clearly, it is not always practical to classify a fungus as being either effective or ineffective at enhancing plant growth. The effectiveness of fungi can be altered by changing the amount of fungus in the soil, by changing the amount of phosphorus in the soil or by altering soil conditions that affect the growth of the fungus. In addition, strains of the same fungus may respond differently, so the diversity of fungi present is also significant.

Phosphorus is not the only nutrient taken up by the hyphae of arbuscular mycorrhizal fungi. For example, the supply of Cu (copper) and Zn (zinc) to some plants is improved in the presence of arbuscular mycorrhizal fungi. However, arbuscular mycorrhizal fungi do not increase the supply of nitrogen to plants. The reason that AM fungi enhance the uptake of some nutrients and not others is related to the way that nutrients interact with soil particles. Nutrients (such as nitrate) that move with water towards plant roots in most soils are not likely to become more available to a plant if it has arbuscular mycorrhizas. In contrast, P, Cu and Zn become adsorbed to soil particles and diffuse through a soil rather than move with water towards roots. Roots need to explore the soil to intercept these nutrients that move slowly by diffusion. By acting as an extension of the root system, the hyphal network of arbuscular mycorrhizal fungi increases the likelihood that roots have access to nutrients that diffuse slowly in soil.

Even if no nutrient benefit is gained from the presence of arbuscular mycorrhizal fungi, there can be a benefit to the physical state of soil if hyphae help to stabilise soil aggregates. Hyphae of arbuscular mycorrhizal fungi may also protect roots from disease. There is some evidence that arbuscular mycorrhizal fungi assist plants in water uptake, especially under dry conditions. Demonstration and interpretation of such benefits is confounded because the fungi also take up phosphorus. Enhanced growth may be due to increased phosphorus uptake as well as to water uptake if the soil is phosphate deficient. Therefore, increased water uptake may be an additional benefit in some circumstances, but the effects related to water and phosphorus are difficult to separate.

Within soil, arbuscular mycorrhizal fungi form extensive networks of hyphae that connect the roots of many plant species into an integrated system. This provides a pathway for nutrient exchange between different plant species. The extent to which this occurs, and its importance for plant growth needs further investigation. One particularly interesting area is the role of these networks in aiding seedling growth. Seedlings emerging from a soil containing a network of arbuscular mycorrhizal hyphae may gain access to nutrients from existing plants and grow better than might be expected if they are able to connect into this network. In addition, arbuscular mycorrhizal fungi may increase the chance of survival of seedlings in ecosystems where plants compete for nutrients and light. This is especially important for slower growing plants. Therefore, successful establishment of arbuscular mycorrhizas may help maintain or increase plant species diversity in a natural ecosystem.

Ectomycorrhizal Fungi

Ectomycorrhizal fungi are the dominant mycorrhiza in many natural ecosystems, even though they form associations with a smaller number of plant species than do AM fungi. Plants that form ectomycorrhizas include well-known trees and many smaller perennial plants.

The main features of an ectomycorrhizal association are the hyphal mantle on the root surface, the Hartig net formed between the cells of the root closest to the root surface and the extension of hyphae into the soil. The hyphae of ectomycorrhizal fungi do not enter root cells in the same way as AM fungi. Rather, the Hartig net is formed when the ectomycorrhizal fungi enter roots between epidermal cells and spread out to form a network of hyphae through the root. Ectomycorrhizal fungi are usually clearly visible on the surface of roots. They promote root branching and restrict root extension which results in a great diversity in architecture of ectomycorrhizas. Roots become covered with hyphae to varying degrees. Ectomycorrhizas are formed by many different species of fungi, in contrast to the relatively small number of species that form arbuscular mycorrhizas. Most belong to families of the basidiomycetes and a number to families of the ascomycetes. Some are clearly visible in forests and other reasonably undisturbed ecosystems because they produce mushrooms and similar fungal spore-containing structures above the soil surface. Others form their reproductive structures below the soil surface. Small mammals dig up and eat these structures. Truffles are a well-known example of fungus-produced underground fruiting structures.

Some ectomycorrhizal fungi form associations with many different plant species, while others form associations with fewer plants. For example, most species of Amanita have an intermediate or broad host range, whereas most of species of Suillus have a narrow host range. In contrast, 25 per cent of species of Russula have a narrow host range and 30 per cent have a broad host range. Compared to AM fungi, many species of ectomycorrhizal fungi can grow in artificial, laboratory media, making them easier to study, although they often grow very slowly. Others are either difficult to grow or have not yet been grown at all under laboratory conditions.

Biology of Ectomycorrhizal Fungi

It is not easy to define a simple life cycle for ectomycorrhizal fungi because many genera of fungi are involved. Ectomycorrhizal fungi are commonly associated with perennial host plants that are relatively long-lived. Therefore, ectomycorrhizal fungi within the roots of living plants have the potential to survive for long periods.

Ecological successions of ectomycorrhizal fungi occur on the same root systems leading to a gradual change in dominance of fungi (Mason *et. al.* 1984). Some plants form more than one type of ectomycorrhizal association, either on different roots of the same plant or at different times. In this study, fruiting structures of the fungus Laccaria sp. generally occurred further from the tree base than those of Lactarius. However, the presence of these structures is not a direct measure of the extent of mycorrhizal fungus present in soil or in roots. Recent molecular studies have highlighted the lack of a quantitative relationship between fungi in roots and the number of fruiting structures formed above the ground. New roots usually become colonised by the hyphae of ectomycorrhizal fungi attached to existing mycorrhizal roots on the same or on a nearby plant. The spores that are formed by these fungi do not appear to be very important for initiation of new mycorrhizas. Various degrees of specificity exist between ectomycorrhizal fungi and host plants. A process of molecular recognition takes place prior to colonisation of the root by the hyphae. Colonisation of the root by the fungus proceeds only if the fungus and root are compatible. Recognition of compatibility between a root and a fungus involves molecular signalling using genetic information in both the fungus and plant.

Effects of Ectomycorrhizal Fungi on Plants

It is more difficult to estimate the contribution of ectomycorrhizal fungi to plant growth than for AM fungi, except at early stages in plant development. Studies on seedlings clearly illustrate the contribution of ectomycorrhizal fungi in the early stages of tree development, but extrapolation of these benefits to older trees is not reliable. Ectomycorrhizal associations can improve the phosphate status of plants in soils that are phosphate deficient. Under some circumstances, ectomycorrhizal fungi make nitrogen more available to plants and increase plant resistance to drought and disease. These benefits can lead to a more robust plant that has a greater chance of survival. There is substantial evidence for benefits of ectomycorrhizal fungi in nitrogen nutrition of plants where nitrogen is deficient. Some ectomycorrhizal fungi can use simple or more complex forms of organic nitrogen. For example, all three fungi in this study were effective at using glutamine as a nitrogen source, but none could use histidine, which is a very complex form of nitrogen. In contrast, Elaphomyces was able to use nitrogen from arginine, but the other two fungi were not. These differences are likely to be important in soils with low levels of inorganic nitrogen and high levels of organic matter. The nitrogen that is taken up by the ectomycorrhizal fungi is passed onto their plant hosts, improving what would otherwise be relatively poor growing conditions for the plant. Ectomycorrhizal fungi also appear to directly help plants avoid disease and survive periods of drought. Where the hyphal mantle covers the root, it provides protection from pathogenic fungi, which can not penetrate the hyphal layer. In addition to this, due to the presence of the ectomycorrhizal fungi changes occur in root physiology that minimises the susceptibility of the root to infection by pathogenic fungi or bacteria. The hyphal networks around mycorrhizal roots enables water to be taken up from parts of the soil that the roots cannot reach. Protection from both disease and drought are contributions that ectomycorrhizal fungi may make to various degrees at different stages in the growth cycle of the host plant. The contributions depend on the type of fungus and the soil conditions.

Ericoid Mycorrhizas

Ericoid mycorrhizas are the specialised mycorrhizal associations formed with members of the plant families Ericaceae (Smith and Read 2008). These mycorrhizas have a different morphology from the mycorrhizas already mentioned and play an important role for their host plants (Perotto *et. al.* 2002). Ericoid mycorrhizal fungi colonise the individual cells of very fine hair-like roots directly from the soil, forming coils of hyphae within the root cells.

The hyphal connections to the soil coalesce around the roots as a fine network. The hair roots colonised by ericoid mycorrhizal fungi are short lived. Therefore, ericoid fungi need to continuously re-establish the mycorrhizal associations as the roots grow. There appears to be a high degree of specificity between the fungi that form ericoid mycorrhizas and their host plants. Details of the nature of the recognition process between the fungus and plant are well understood.

Some ericoid mycorrhizal fungi grow in pure culture, and analysis of these cultures shows that ericoid fungi are very different from each other. This was demonstrated by comparing the electrophoretic pattern of enzymes that decompose pectin after they were separated in a gel. The migration of enzymes in the gel indicated differences in the molecular weight of pectic enzymes from some fungi. This is indicative of differences in genetic characteristics among the fungi.

There is evidence that ericoid mycorrhizal fungi have certain characteristics similar to wood degrading fungi that may allow them to degrade complex components of organic matter in the soil. In comparison to a selection of ectomycorrhizal fungi, two ericoid mycorrhizal fungi had greater potential to degrade lignin and soluble phenolic molecules. This conclusion was based on the response of the fungi in four bio-chemical tests that measured characteristics of fungi important in degradation of lignin and other complex plant molecules.

Effects of Ericoid Mycorrhizal Fungi on Host Plants

Ericoid plants occur in nutrient poor soils and the mycorrhizal associations that they form are likely to be important in providing them with nutrients from the organic matter that would otherwise remain unavailable to the plant. In particular, nitrogen appears to be released from organic matter by the activity of ericoid mycorrhizal fungi. There is also a possibility that organic phosphorus is mineralised from organic matter by these fungi. Although ericoid mycorrhizal fungi have characteristics that are very different from those of both AM and ectomycorrhizal fungi, all three types of mycorrhizas coexist in many natural ecosystems.

This illustrates how plant-microbial associations respond to the nutrient resources in a soil. The morphology and physiology of orchid mycorrhizas are very different from those of the other three types of mycorrhizas described above. Orchid seeds are highly dependent on fungi. During the early stages of seedling growth, the fungi supply carbon to the growing orchid seedling.

This is the reverse of what occurs in other mycorrhizas. Some orchid mycorrhizal fungi have been isolated and grown in culture for many years and many belong to the genus Rhizoctonia or to related genera.

NITROGEN-FIXING ASSOCIATIONS

Nitrogen gas (N_2) constitutes nearly 80 per cent of the atmosphere, but in this form nitrogen is not available to plants. Several groups of soil bacteria convert nitrogen gas into ammonium (NH_4^+), which is a form of nitrogen that can be used directly by plants. This transformation is catalysed by the bacterial enzyme nitrogenase. Very few bacteria have the capacity to activate the nitrogenase enzyme system, and those that do occur as free living bacteria in soil, as loose associations with root surfaces or within highly specialised, symbiotic associations with plants.

Bacteria in the genus Azospirillum live in a loose association with root surfaces of grasses and other plants. This contrasts with the highly specific associations that form between cyanobacteria and fungi in lichens. Highly specific nitrogen-fixing associations are also formed between some tree species and bacteria (actinomycetes) in the genus Frankia. The more primitive gymnosperms such as cycads also form characteristic associations with cyanobacteria. The most widely studied of all nitrogen fixing associations are formed between legumes and bacteria from the family Rhizobiaceae.

The biological transformation of N2 gas to ammonium is equivalent to the industrial transformation of N2 that requires excessive application of heat and pressure. This highlights the uniqueness of biological nitrogen fixation and its practical benefit. Nitrogen-fixing bacteria that live in loose associations with roots are called 'associative' nitrogen fixers. Associative nitrogen fixing bacteria grow on the surface of roots and may also colonise the outer layers of a root by entering between epidermal cells.

A common genus of associative nitrogen fixing bacteria is Azospirillum. Species of Azospirillum are not all equally effective at converting atmospheric nitrogen to ammonium. There is conflicting evidence about the extent to which nitrogen fixed by these bacteria is available to plants. Even for the species of Azospirillum that are very effective at fixing atmospheric nitrogen, the nitrogen fixed is not immediately available to the plant. It is not until the bacteria die and are mineralised that the nitrogen becomes available to the plants.

Nitrogen fixed by Azospirillum is not immediately available to plants because of the complex processes of ammonium transfer through the bacterial cell wall. Ammonium excess to the requirements of the bacterium can be transported to the outside surface of the cell wall. However, another process can transfer the ammonium back into the cell, slowing the transfer of ammonium to the plant. Some mutant strains of Azospirillum are effective at permanently transferring ammonium from the bacteria.

The movement of molecules into and out of bacteria and fungi are complex processes. In the case of Azospirillum, an understanding of these processes shows why nitrogen is not immediately available to plants from bacteria that fix nitrogen in the root zone, at least while the bacteria are alive. The benefits of associative nitrogen fixers have been estimated in agricultural soils. The quantity of nitrogen fixed varies greatly. In very nitrogen deficient soils, the quantities of nitrogen fixed are not usually sufficient to overcome nitrogen deficiency in agricultural plants.

Legumes and Nitrogen Fixing Bacteria Function

Legumes are important plants in both natural as well as managed ecosystems. The majority of legumes form symbiotic associations with nitrogen-fixing bacteria. As a consequence of these associations, distinctive structures (called nodules) form on the roots of legumes. The bacteria multiply within the nodule cells.

Many kilograms of nitrogen per hectare can be converted every year by agriculture plants including crop (*e.g.* soybean) and pasture (*e.g.* clover) plants because of their associations with nitrogen-fixing bacteria. These associations are of great importance as a substitute for nitrogen fertilizer. Nitrogen fixing symbioses are also common in natural ecosystems.

Bacteria that form associations with legumes live in the soil as part of the microbial community. However, they do not fix nitrogen directly in the soil. Nitrogen fixation only occurs when bacteria form a nodule on a legume root. When an appropriate legume is present, only one in a million bacteria are necessary to form a nodule. In forming a nodule, the bacterium undergoes a highly specific interaction with the plant, involving several stages and complex signals between the bacterium and plant.

Associations between particular bacteria and their legume host plants are usually highly specific. For example, the bacteria that fix nitrogen in association with species of clover cannot fix nitrogen in association with species of acacia. Nevertheless, the same bacteria can form associations with different clover species and occasionally different bacteria will infect the same plant. However, nodules generally contain only one strain of bacterium that multiplies to fill the central core of nodule cells.

Although there is considerable specificity among root nodule bacteria and legumes, it is interesting that the same bacteria form nitrogen-fixing associations with lupin (Lupinus spp.), a crop legume and serradella (Ornithopus spp.), a pasture legume. Because the architecture of both the roots and shoots of these two plants is very different, the bacteria which colonises both are evidently highly adaptable.

The shape and size of nodules varies considerably, with each determined by the plant, not the bacteria. In some plants, nodules can continue to grow throughout the life of the plant, as for example with clover. Where this occurs on perennial species, layers of new nodule tissue are added annually. In contrast, nodules formed on other legumes have a limit to their growth and are termed determinant nodules. Soybean is an example of a legume that forms determinate nodules.

The bacteria that form symbiotic associations with legumes differ in their ability to fix nitrogen. The effectiveness of these bacteria can also be influenced by soil conditions. Therefore, when planting agricultural legumes it is necessary to select the most effective bacteria that will do best in the environment. This is also true when planting tree species for revegetation and forestry projects.

The bacteria that fix the most nitrogen when in association with a legume are not necessarily those that form the most nodules. Conversely, it is not uncommon to find nodules formed on agricultural plants by bacteria that are ineffective at fixing nitrogen. In soils where there are different strains of bacteria that can nodulate the same legume, many factors influence the success or failure of nodulation. Therefore, the full potential for nitrogen fixation is not always achieved, even if highly effective strains of bacteria occur in the soil.

Diversity among Root Nodule Bacteria

There is considerable diversity among root nodule bacteria (McInnes et al. 2004). Bacteria that form nodules on legumes differ in the rate at which they multiply on artificial laboratory media.

Acacias form nodules with groups of bacteria with very different characteristics and this can affect the detection of bacteria on growth media if the rapidly growing bacteria overgrow those that grow more slowly.

There is a difference in the DNA structure among root nodule bacteria. In some, but not all, DNA is present as circular segments, separate from the main chromosome. These segments, called plasmids, contain the genetic information required to initiate symbiotic nitrogen fixation. Plasmids can replicate separately from the chromosome and transfer from one bacterial cell to another.

Genetic Basis for Nodulation and Nitrogen Fixation

The location of genes on the DNA that code for different stages in the nodulation process illustrates complex interactions that take place between legumes and root nodule bacteria during nodulation. Separate regions of bacterial DNA are responsible for recognition of the legume root surface, initiation of root hair curling, initiation of the infection thread formation, and root cell multiplication to form the nodule.

Bacteria present in the rhizosphere stimulate the plant to produce molecules called flavanoids which are released in root exudates. In turn, these molecules can, to varying degrees activate the nodulation genes in the bacteria. For successful nodulation to occur, the flavanoid molecules need to be present in sufficient quantities to induce a response in the bacteria. Some plants are better able to do this for some bacteria. In this example, the maximum activity of flavanoids produced by Medicago murex occurred at pH 6.0 whereas flavanoids produced by Medicago littoralis were most active at a higher pH. This response partly explains why nodulation changes with soil pH for some legumes. Even though the bacteria may grow well at a certain soil pH, the lack of production of exudates that will initiate nodule formation may hinder their ability to form a nodule.

Legumes in natural ecosystems grow in soil containing indigenous, diverse communities of root nodule bacteria. Similarly, highly managed agricultural systems may contain many different strains of bacteria that have been introduced with agricultural legumes. High diversities of root nodule bacteria can result from a lengthy history of growing different legume species on the land, multiple introductions of commercial inoculant root nodule bacteria, and ongoing genetic diversification of bacteria in soil.

An analysis of the range of strains of root nodule bacteria that occur in Oregon pastures illustrates the level of genetic diversity that is possible. In this study, the bacteria were characterised using serological methods with four distinct serogroups identified and one most abundant in the surface 10 cm of the soil profile. Subsequent investigations using DNA analysis confirm the diversity of strains of rhizobia in soils. When bacteria selected for a particular legume are introduced into a soil, they have to survive as part of the community of soil bacteria. Otherwise, the only way they will be maintained in the soil is by regular re-introductions. The newly introduced bacteria also need to compete with existing strains of similar bacteria for nodulation sites on the legume root. For example, the strain of bacterium that was one of the original inoculants for soybean in the USA was subsequently found to be less effective at fixing nitrogen than some strains introduced later. The original bacterium was very competitive at forming nodules on soybean in comparison to the more recently identified effective strains. It is difficult for a newer bacterium to outcompete older bacteria that have already colonised the soil and are highly competitive at forming nodules.

This is clearly a major problem for the introduction of strains that are more effective at fixing nitrogen. A similar obstacle occurs when two clover species are grown in the same pasture. Nitrogen fixation occurs most effectively for each with different strains of Rhizobium leguminosarum biovar trifolii. However, it is not possible to ensure that each clover species is nodulated by the most appropriate strain of bacterium. The result could be that neither clover species fixes nitrogen according to its potential.

Bacteria that form legume symbioses can survive in a soil in the absence of suitable plants. Once the roots of a suitable plant reappear, the bacteria can re-initiate the symbiotic association. Some bacteria can survive without a plant for longer than others can. Even if nodule formation does not occur as soon as the plant reappears, the plant can stimulate the growth of the bacteria and make it more likely that nodulation takes place when other soil conditions are suitable.

Non-legume Plant Species form Nitrogen-fixing Associations

Frankia is a genus of bacteria (actinomycetes) that has the capacity to fix nitrogen gas in association with several tree species. Examples of such trees include the genera Casuarina and Allocasuarina in the family Casuarinaceae and Alnus in the family Betulaceae.

The associations formed by species of Frankia are generally host-plant specific, but the same bacteria can nodulate closely related plant species. Frankia associations with plants range from those that are highly effective to those that are relatively ineffective at fixing nitrogen.

Nitrogen deficiency in Casuarina can be remedied by inoculation with strains of Frankia that are effective at fixing nitrogen (Reddell and Bowen 1985). Two of the five sources of Frankia studied were effective at increasing growth of C. equisitifolia. Another was highly effective with C. cunninghamiana.

The association between Frankia and roots of their host plant is highly specific. The infection process is initiated through molecular signalling between the bacteria and root as for legume symbioses. When there is a compatible combination of bacteria and plant, molecular communication leads to the formation of a nodule on the plant root that develops a colony of bacteria within it. The bacteria stimulate the plant to form nodule tissue. Nodules can live for a long time, with a new layer of actively growing root cells and bacteria added each year. The bacteria fix nitrogen gas that diffuses through the walls of the nodule cells; the fixed nitrogen is transferred into the root cells and from there into the rest of the plant through the vascular system. A variety of other associations also occur between nitrogen-fixing organisms and plants. For example, Macrozamia riedlei, a member of the family Cycadaceae, forms an association with nitrogen fixing cyanobacteria. This plant forms highly branched roots that contain the cyanobacteria.

Soil-Borne Plant Pathogens

Soils contain diverse communities of microscopic organisms that are capable of damaging plants. A detrimental interaction between a soil organism and a plant is often highly specific. For example, a fungus that causes root-rot of wheat may have no effect on the roots of another plant growing in the same soil. Highly specialised interactions between soil organisms and plants can kill seedlings and even adult trees.

Many organisms target younger plants but others appear as problems at later stages in the life of the plant. Other pathogens are able to cause disease in many different plant species. The soil organisms that have the potential to be plant pathogens include fungi, bacteria, viruses, nematodes and protozoa. Some pathogens of the above ground parts of plants (leaves, stems) survive in the soil at various stages in their life cycles. Therefore, a soil phase of a plant pathogen may be important, even if the organism does not infect roots.

In spite of the potential for severe damage to be inflicted on plants by soil organisms, most plants do not display serious symptoms of disease. Disease usually occurs when conditions are particularly unfavourable, or when a soil organism is accidentally introduced where a highly susceptible plant species occurs. Intensive production in agriculture, horticulture or forestry increases the opportunities for diseases to develop compared with undisturbed natural ecosystems.

Planting of similar plant species together in mono-culture increases the probability of disease outbreak. In contrast, the damage caused by the fungus Phytophthora cinnamomi in many different plant species in diverse natural ecosystems in Australia demonstrates the damage that can be caused by a pathogen that infects the roots of many unrelated plants. Some plant pathogens depend on their host plant for survival and are unable to complete their life cycle without infecting their host plant.

Biotrophic organisms of this type are often difficult to grow in laboratory media. Disease-causing Micro-organisms and soil animals are a natural component of the soil community. The organisms are normally present in relatively low numbers. An outbreak of disease commonly follows either an increase in the abundance of the pathogen or a change in the susceptibility of the host to the pathogen.

Important questions are :
- Why do outbreaks of disease occur?
- What are the soil conditions that allow disease to develop or that suppress disease?
- What are the soil conditions that suppress disease?
- How do management practices influence the soil as a habitat for plant pathogens?
- How do plant pathogens survive in the absence of their specific host plant?

Plant Disease

Plant disease is the result of a complex interaction between the host plant, the pathogen and the environmental conditions including those in the soil. Soil conditions that influence the development of disease are pH, moisture, oxygen, and nutrients. In addition, other soil organisms interact with the pathogen. The level of this interaction is also influenced by soil conditions, which affects the growth and activity of the non-pathogenic organisms. Consequently, whenever the soil environment changes, there is potential for either a positive or negative influence on soil-borne plant pathogens.

Plant disease reduces plant survival and productivity, but the symptoms of plant disease are often difficult to recognise. One reason for this is that they are not necessarily the same in plants of different ages. The amount of a potential pathogen in a soil and the environmental conditions can also alter the symptoms displayed by a plant. Furthermore, symptoms may be related to infection by more than one organism. Interactions between organisms can produce symptoms that are not the same as those that occur when each type of organism is present alone. Similar symptoms can be caused by different pathogens and furthermore, the symptoms of some pathogens resemble those of various nutrient deficiencies. Therefore, it is not always easy to identify a pathogen responsible for a disease simply from the symptoms on the plant. A great deal of experience and knowledge of the range of symptoms produced by different pathogens and knowledge of symptoms of nutrient deficiency is essential for an accurate identification of most pathogens.

Chapter 3

SOIL MICROBIOLOGY

Soil microbes are a diverse group of species that are critical to many of the functions associated with wetlands and to the decomposition of organic matter. They contribute to the development of the morphological features that are used as hydric soil indicators. These contributions are indirect through the development of anaerobic conditions and direct through the reduction of iron (Fe) and manganese (Mn).

CHARACTERISATION OF SOIL MICRO-ORGANISMS

Micro-organisms can be classified metabolically on the basis of their sources for carbon (C) and energy. Chemoheterotrophs derive C and energy from organic sources. The majority of soil microbes are saprophytes, chemoheterotrophs that subsist on dead plant and animal remains and humic substances, and, therefore, are critical to the C cycle. Photoautotrophs are photosynthetic, assimilating C as carbon dioxide (CO_2) and deriving energy from sunlight. They also are critical to the C cycle. Chemoautotrophs assimilate C as CO_2 and derive energy from inorganic compounds. They are important to nutrient cycles and the formation of redoximorphic features.

Micro-organisms can also be categorised on their gaseous oxygen (O_2) requirement for respiration. Obligate aerobes derive energy only through aerobic respiration which is characterised by its use of O_2 as a terminal electron acceptor. Obligate aerobes can not function under anaerobic conditions. Obligate anaerobes can not function under aerobic conditions. They derive energy via anaerobic respiration in which inorganic ions other than O_2 are used as terminal electron acceptors, or through fermentation in which internally-generated compounds are used as terminal electron acceptors. Facultative anaerobes are capable of both aerobic and anaerobic respiration. In the presence of O_2 they rely on aerobic respiration which is much more efficient (produces more energy) than anaerobic respiration or fermentation. Since aerobic respiration is more efficient than anaerobic respiration it results in faster rates of organic matter decomposition.

Fungi are the principal agents for the decomposition of organic matter, especially complex molecules such as cellulose, hemicellulose, and lignin. Soil fungi are obligate aerobes (primarily) and facultative anaerobes. They usually represent the greatest percentage of microbial biomass in a soil at 450 to 4,500 lb./ac. (500 to 5,000 kg/ha). Most fungi produce filamentous growth (hyphae). The total filamentous body is referred to as a mycelium. An individual mycelium may extend several yards (meters) in the soil. Mushrooms and toadstools are reproductive structures for many of the fungal species. Hyphae promote aggregation by binding soil particles together. Some of the fungi infect the roots of higher plants to form a mutually-beneficial relationship or 'symbiosis' referred to as 'mycorrhizae'. The mycorrhizae improve the plant uptake of nutrients, especially phosphate, ammonium, and zinc, and are considered to be essential to the survival of gymnosperms.

Bacteria are the most prevalent soil micro-organisms, numbering in the millions per ounce of topsoil (108 to 109 per gram). Soil bacteria include obligate aerobes, facultative anaerobes, and obligate anaerobes. They are critical to nutrient cycling and the formation of redoximorphic features. Actinomycetes are filamentous bacteria that are much smaller than fungi. The musty smell of freshly plowed soil is due to compounds produced by actinomycetes. Most are saprophytic and are important to the decomposition of soil organic matter. However, most are aerobic, do not grow well in saturated soils, and are acid intolerant. Therefore, their contribution to microbial processes in hydric soils of the Mid-Atlantic Region is less than that of other bacteria and fungi.

Both terrestrial and aquatic algae occur in wetlands. Terrestrial species tend to be single-celled and smaller than the aquatic species. All algae are photosynthetic and, therefore, a source of C inputs. Many algae (and some bacteria) 'fix nitrogen' (N), converting dinitrogen gas (N_2) to ammonia (NH_3). Algal blooms, a rapid proliferation of algae in water, is a sign of eutrophication or nutrient loading of surface waters. Since many of the algae fix N, algal blooms are usually in response to phosphorous (P) pollution. Algae may also bloom at the surface of soils and may impart a greenish colouration to the soil a day after a rainstorm.

Anaerobiosis

Microbes are inherent to the development of hydric soils. They contribute to the development of anaerobic conditions, and are critical to the formation of redoximorphic features. Soils become anaerobic when O_2 use by plant roots and microbes exceeds O_2 diffusion into the soil from the atmosphere. Saturation decreases diffusion rates by four orders of magnitudes. The length of time after saturation required for a soil to become anaerobic depends on soil temperature, available organic C levels, and the O_2 content of the water. High temperatures and a ready supply of available C lead to higher respiration rates by microbes.

Under optimum conditions (high temperature, high levels of available C, and low O_2 concentrations in the water), soil can become anaerobic one day after saturation. Research (unpublished) conducted by the author in seasonally-saturated wetlands on the Delmarva Coastal Plain revealed, in general, a two week delay between the onset of saturation in late winter or early spring and the development of reducing conditions. Only the wettest soils are continuously saturated (peraquic moisture regime).

The majority of hydric soils have a wet period and a dry period in most years, but are wet enough close to the surface to have an aquic moisture regime. Spatial variability is also common as O_2 levels are usually higher near the surface and alternating oxidised and reduced zones may be associated with structural units (ped interiors versus ped faces). Even saturated zones have aerobic microsites. Therefore, both aerobic and anaerobic respiration can occur concurrently in hydric soils.

Carbon Cycle

One characteristic associated with hydric soils that developed under extensive periods of anaerobiosis is a high organic matter content. A number of the hydric soil indicators are a function of organic matter accumulation, and Histosols form in wetlands with frequent and extensive periods of saturation to the soil surface. Soil organic matter is basically plant and animal remains in various stages of decomposition.

Accumulation of soil organic matter occurs when additions exceed decomposition. Therefore, accumulation is favoured by systems with high rates of biomass production and conditions unfavourable to decomposition. Both characteristics are common to wetlands. The level of organic matter in a hydric soil is a function of the hydroperiod, the seasonal pattern of water table fluctuations in a wetland. The longer the water table is close to the soil surface, the longer the period of anaerobiosis. Since most organic materials are deposited either at or near to the soil surface as roots or leaves, hydric soils that develop anaerobiosis close to the surface will accumulate more organic matter.

The rate of organic matter accumulation is also influenced by vegetation type. Grassland soils tend to have higher levels of organic matter than forest soils. Forests produce more organic matter than grasslands. However, the primary source of organic matter in forests is the leaf litter which is readily decomposed, both because of the composition of leaves and because the litter decomposes aerobically on the soil surface. The primary source of organic material in grasslands is root material. Not only is root tissue more resistant to decomposition than leaves, but root decomposition occurs below ground where anaerobic conditions are more prevalent.

Organic matter decomposition in hydric soils during a wet cycle is limited by O_2 as anaerobic respiration is less efficient than aerobic respiration and some of the products of fermentation (*e.g.* ethanol) are inhibitory to many microbes. Gambrell and Patrick (1978) reported that decomposition rates for organic residues in wet soils could be as low as 13 per cent of the decomposition rates of the same residues in aerated soils. For hydric soils that are not continuously saturated, organic matter levels are inversely proportional to the number of wet/dry cycles per growing season. Under aerobic conditions, decomposition rates are limited by the availability of a nutrient, usually N.

One benefit of decomposition is that nutrients required for plant growth are converted from organic forms, which are unavailable to plants, to inorganic or mineral forms, which are available to plants. This process is referred to as 'mineralisation'. The reverse process, immobilisation, ties up nutrients. Both processes occur simultaneously. Whether net mineralisation (mineralisation minus immobilisation) occurs depends on the nutrient composition of the organic material that is decomposing. For example, decomposition of organic tissues with a C : N ratio of greater than 15 or 20 to 1 will result in net immobilisation of N as most of the N will become incorporated in the biomass of the microbial decomposers. However, microbial death and decay will mineralise some of this N. Decomposition of residues with low C : N ratios results in net mineralisation of N as N levels are in excess of the microbial demand.

Redox Reactions

Oxidation-reduction (redox) reactions are characterised by the transfer of electrons from one compound to another. In each redox reaction, one compound is reduced (receives electrons) and one compound is oxidised (donates electrons). Microbially-mediated redox reactions produce many of the morphological features associated with hydric soils and are the driving force behind bio-geochemical cycling. During respiration microbes oxidise organic compounds and transfer the donated electrons through a series of bio-chemical steps. Under aerobic conditions the last compound to be reduced during respiration (terminal electron acceptor) is O_2. Once the soil is depleted of O_2, microbes will use the following compounds as terminal electron acceptors : nitrate (NO_3-), manganic manganese ($Mn4+$), ferric iron ($Fe3+$), sulfate ($SO42-$), and CO_2. Redox potential or Eh is a measure of the tendency of a system to donate electrons. The switch from one terminal electron acceptor to another occurs in an orderly sequence within predictable Eh ranges. They are approximate values as they are pH and temperature dependent, and influenced by other soil water constituents. For more on Eh and its measurement, Monitoring Hydric Soils. The availability of other terminal electron acceptors can inhibit the reduction of specific compounds. For example, the presence of $SO42-$ inhibits CO_2 reduction. The presence of $NO3-$ inhibits $Fe3+$ reduction, but not $Mn4+$ reduction.

Denitrification is the reduction of $NO3-$ (as a terminal electron acceptor) to gaseous N compounds (NO, N2O, N2). Biological denitrification is limited to facultative anaerobic bacteria. It is the process that is primarily responsible for the removal of dissolved N in created wetlands. A related process is dissimilatory nitrate reduction in which bacteria convert $NO3-$ to ammonium ($NH4+$) which is then released from the cell. This process is favoured by continuous anaerobiosis and a high ratio of available C to $NO3-$. At high pH's this can lead to loss of N as

NH_3. A select few species of chemoaoutotrophic bacteria obtain energy by oxidising NH_3 to nitrite (NO_2)

Table : Oxidised and Reduced Forms of Several Elements and Approximate Redox Potentials for Transformation.

Element	Oxidized form	Reduced form	Approx. redox pot. for transformation (mV)
Nitrogen	NO_3^- (nitrate)	N_2O, N_2, NH_4^+	250
Manganese	Mn^{4+} (manganic)	Mn^{2+} (manganous)	225
Iron	Fe^{3+} (ferric)	Fe^{2+} (ferrous)	+100 to -100
Sulfur	SO_4^- (sulfate)	S^- (sulfide)	-100 to -200
Carbon	CO_2 (carbon dioxide)	CH_4 (methane)	Below -200

Iron and Mn are oxidised and reduced both chemically and by microbes. Reduced forms are soluble and, therefore, mobile in the soil solution. Oxidised forms are insoluble and readily precipitate out of solution. Alternate cycles of aerobic and anaerobic conditions result in segregation of oxidised Fe that produces redox concentrations and depletions. Long wet cycles produce the gleyed matrix associated with reduced Fe. The dominant forms of Fe and Mn in the soil depend on Eh and pH. Low Eh and low pH favour Fe and Mn reduction. Ferrous (Fe2+) and manganous (Mn2+) forms are oxidised to Fe3+ and Mn4+ by chemoautotrophic bacteria under aerobic conditions. Many bacteria and some fungi reduce Fe3+ and Mn4+ during anaerobic respiration.

Common to wetlands rich in sulfur (S), such as tidal marshes, is S reduction in which bacteria use sulfate (SO42-) as a terminal electron acceptor producing elemental sulfur (S°) or hydrogen sulfide (H2S) which has the characteristic odour of rotten eggs. Sulfur-oxidising bacteria generate energy through the oxidation of sulfides and S°. Although most are obligate anaerobes, at least one species can also use NO_3^- as a terminal electron acceptor.

Under the most reduced conditions, methanogenic bacteria use CO_2 as a terminal electron acceptor producing methane (CH4) or swamp gas. Another group of bacteria, the methanotrophs, use CH4 as their energy source and oxidise it to CO_2. Methanotrophs are obligate aerobes. In hydric soils they will be active just above the aerobic/anaerobic interface. A given soil can serve as both a CH4 sink and a CH4 source depending on redox potential. Direct CH4 emissions from wetland soils account for one third of the global CH4 emissions to the atmosphere. However, in certain situations, methanotrophs can consume up to 90 per cent of the CH4 generated in a hydric soil before it can reach the atmosphere. Many herbaceous wetland plants contain arenchyma, a spongy tissue found in the pith that readily transports O_2 to the roots. In wetlands dominated by vegetation with arenchyma up to 90 per cent of the CH4 generated may reach the atmosphere via the arenchyma and, therefore, bypass the zone of methanotrophs.

Microbial Activity

The formation of redoximorphic features is mediated by microbial processes. Therefore, the rate at which they develop is a function of the intensity of microbial activity which, in turn, is influenced by soil O_2 level, soil pH, temperature, toxic compounds, and the availability of organic C. Under anaerobic conditions, bacteria are more active than fungi and contribute more to anaerobic decomposition of organic matter and to the formation of redoximorphic features. The majority of bacteria display the greatest growth rates at soil pH's of 6 to 8; while fungi are more competitive at lower pH's (4 to 6). Redoximorphic features are less likely to form at either very high or very low pH's. A temperature of 41°F (5°C) is considered to be "biological zero" as microbial activity slows considerably or stops at lower temperatures. The 41°F threshold is used in growing season calculations for wetland jurisdictional determinations. It should be noted that the concept of biological zero, especially the 41oF threshold, is flawed, as it has been well documented that microbial activity occurs at much lower temperatures. Microbial activity increases with increasing temperatures up to 95 to 105°F (35 to 40°C). It should be emphasised that the formation of redoximorphic features requires Mn and Fe in forms that can be reduced and an active microbial population. Problematic hydric soils occur when one of these two requirements is missing.

Both the quantity and quality of organic C affect microbial activity. Low organic C is usually the ultimate limiting factor for microbial activity in the soil. Organic compounds such as simple carbohydrates (*e.g.* sugars) and amino acids are more readily degraded (more readily available energy source) than complex carbohydrates such as cellulose, hemicellulose, and lignin (cell wall components of higher plants). Redoximorphic features may fail to form on floodplains because moving water readily removes amino acids and simple carbohydrates, which are water soluble. Microbial populations vary with respect to total numbers and species composition according to soil depth and distance from plant roots. This is primarily in response to a gradient of available C. This has major ramifications with respect to the formation of redoximorphic features.

The "rhizosphere" refers to the zone of soil close to and impacted by plant roots. The "rhizoplane" is the surface of plant roots. Microbial numbers are substantially higher in the rhizosphere than in bulk soil and are inversely proportional to distance from the roots. The highest microbial numbers, by far, are found on the rhizoplane. Plant roots supply most of the C that drives microbial activity in soils. Up to 90 per cent of fine roots may die and decompose annually in forest soils. In addition, dead root cap cells slough off to supply organic C and exudates from live roots include readily available C sources (sugars, organic acids), a readily available source of N (amino acids), and growth promoting (and sometimes inhibiting) compounds. The organic acids can increase the solubility of Fe and Mn by lowering the pH and by acting as chelating agents. Therefore, the formation of redoximorphic features may only be associated with root channels. For plants with arenchyma, the O_2 flux is frequently so strong that the roots pump O_2 into the soil producing an oxidised rhizosphere. In such a situation there will be aerobic microbial activity in the rhizosphere and anaerobic microbial activity outside of the rhizosphere.

Microbial activity in soils varies spatially and temporally. Microbial activity drops in response to temperature extremes, soil moisture extremes, or when nutrients are deficient. Under unfavourable conditions for microbial growth, microbes become dormant. Although entering and exiting dormancy results in high mortality rates, enough individuals survive to re-colonise the soil. For example, soil bacteria can increase in numbers by a hundred fold when a dry soil is wetted if no other factors are limiting. Therefore, microbial activity in soil is not limited by microbial numbers, but rather by unfavourable environmental conditions.

SCOPE AND IMPORTANCE OF SOIL MICROBIOLOGY

Living organisms both plant and animal types constitute an important component of soil. Though these organisms form only a fraction (less than one per cent) of the total soil mass, but they play important role in supporting plant communities on the earth surface. While studying the scope and importance of soil microbiology, soil-plant-animal ecosystem as such must be taken into account.

Therefore, the scope and importance of soil microbiology, can be understood in better way by studying aspects like :

- *Soil as a living system* : Soil inhabit diverse group of living organisms, both micro-flora (fungi, bacteria, algae and actinomycetes) and micro-fauna (protozoa, nematodes, earthworms, moles, ants). The density of living organisms in soil is very high *i.e.* as much as billions / gm of soil, usually density of organisms is less in cultivated soil than uncultivated / virgin land and population decreases with soil acidity. Top soil, the surface layer contains greater number of micro-organisms because it is well supplied with Oxygen and nutrients. Lower layer / sub-soil is depleted with Oxygen and nutrients hence it contains fewer organisms. Soil ecosystem comprises of organisms which are both, autotrophs (Algae, BOA) and heterotrophs (fungi, bacteria). Autotrophs use inorganic carbon from CO_2 and are "primary producers" of organic matter, whereas heterotrophs use organic carbon and are decomposers/consumers.

- *Soil microbes and plant growth* : Micro-organisms being minute and microscopic, they are universally present in soil, water and air. Besides supporting the growth of various biological systems, soil and soil microbes serve as a best medium for plant growth. Soil fauna and flora convert complex organic nutrients into simpler inorganic forms which are readily absorbed by the plant for growth. Further, they produce variety of substances like IAA, gibberellins, antibiotics etc. which directly or indirectly promote the plant growth.

- *Soil microbes and soil structure* : Soil structure is dependent on stable aggregates of soil particles-Soil organisms play important role in soil aggregation. Constituents of soil are *viz.* organic matter, polysaccharides, lignins and gums, synthesised by soil microbes plays important role in cementing / binding of soil particles. Further, cells and mycelial strands of fungi and actinomycetes, Vormicasts from earthworm is also found to play important role in soil aggregation. Different soil micro-organisms, having soil aggregation/soil binding properties are graded in the order as fungi > actinomycetes > gum producing bacteria > yeasts. Examples are : Fungi like Rhizopus, Mucor, Chaetomium, Fusarium, Cladasporium, Rhizoctonia, Aspergillus, Trichoderma and Bacteria like Azotobacter, Rhizobium, Bacillus and Xanthomonas.

- *Soil microbes and organic matter decomposition :* The organic matter serves not only as a source of food for micro-organisms but also supplies energy for the vital processes of metabolism that are characteristics of living beings. Micro-organisms such as fungi, actinomycetes, bacteria, protozoa etc., and macro organisms such as earthworms, termites, insects etc. plays important role in the process of decomposition of organic matter and release of plant nutrients in soil. Thus, organic matter added to the soil is converted by oxidative decomposition to simpler nutrients / substances for plant growth and the residue is transformed into humus. Organic matter/ substances include cellulose, lignins and proteins (in cell wall of plants), glycogen (animal tissues), proteins and fats (plants, animals). Cellulose is degraded by bacteria, especially those of genus Cytophaga and other genera (Bacillus, Pseudomonas, Cellulomonas, and Vibrio Achromobacter) and fungal genera (Aspergillus, Penicilliun, Trichoderma, Chactomium, Curvularia). Lignins and proteins are partially digested by fungi, protozoa and nematodes. Proteins are degraded to individual amino acids mainly by fungi, actinomycetes and Clostridium. Under unaerobic conditions of water-logged soils, methane are main carbon containing product which is produced by the bacterial genera (strict anaerobes) Methanococcus, Methanobacterium and Methanosardna.

- *Soil microbes and humus formation :* Humus is the organic residue in the soil resulting from decomposition of plant and animal residues in soil, or it is the highly complex organic residual matter in soil which is not readily degraded by micro-organism, or it is the soft brown/dark coloured amorphous substance composed of residual organic matter along with dead micro-organisms.

- *Soil microbes and cycling of elements :* Life on earth is dependent on cycling of elements from their organic / elemental state to inorganic compounds, then to organic compounds and back to their elemental states. The biogeochemical process through which organic compounds are broken down to inorganic compounds or their constituent elements is known "Mineralisation", or microbial conversion of complex organic compounds into simple inorganic compounds and their constituent elements is known as mineralisation. Soil microbes plays important role in the bio-chemical cycling of elements in the biosphere where the essential elements (C, P, S,

N and Iron etc.) undergo chemical transformations. Through the process of mineralisation organic carbon, nitrogen, phosphorus, Sulphur, Iron etc. are made available for reuse by plants.

- *Soil microbes and biological N_2 fixation* : Conversion of atmospheric nitrogen in to ammonia and nitrate by micro-organisms is known as biological nitrogen fixation. Fixation of atmospheric nitrogen is essential because of the reasons : Fixed nitrogen is lost through the process of nitrogen cycle through denitrification. Demand for fixed nitrogen by the biosphere always exceed sits availability. The amount of nitrogen fixed chemically and lightning process is very less (*i.e.* 0.5 per cent) as compared to biologically fixed nitrogen Nitrogenous fertilizers contribute only 25 per cent of the total world requirement while biological nitrogen fixation contributes about 60 per cent of the earth's fixed nitrogen Manufacture of nitrogenous fertilizers by "Haber" process is costly and time consuming. The numbers of soil micro-organisms carry out the process of biological nitrogen fixation at normal atmospheric pressure (1 atmosphere) and temp (around 20 °C). Two groups of micro-organisms are involved in the process of BNF. A. Non-symbiotic (free living) and B. Symbiotic (Associative)

- *Non-symbiotic (free living)* : Depending upon the presence or absence of oxygen, non symbiotic N_2 fixation prokaryotic organisms may be aerobic heterotrophs (Azotobacter, Pseudomonas, Achromobacter) or aerobic autotrophs (Nostoc, Anabena, Calothrix, BGA) and anaerobic heterotrophs (Clostridium, Kelbsiella, Desulfovibrio) or anaerobic Autotrophs (Chlorobium, Chromnatium, Rhodospirillum, Meihanobacterium etc).

- *Symbiotic (Associative)* : The organisms involved are Rhizobium, Bratfyrhizobium in legumes (aerobic) : Azospirillum (grasses), Actinomycetes frantic (with Casuarinas, Alder).

- *Soil microbes as biocontrol agents* : Several eco friendly bio-formulations of microbial origin are used in agriculture for the effective management of plant diseases, insect pests, weeds etc. *e.g.* : Trichoderma sp and Gleocladium sp are used for biologicalcontrol of seed and soil borne diseases. Fungal general Entomophthora, Beauveria, Metarrhizium and protozoa Maltesiagrandis. Malameba locustiae etc. are used in the management of insect pests. Nuclear polyhydrosis virus (NPV) is used for the control of Heliothis / American boll worm. Bacteria like Bacillusthuringiensis, Pseudomonas are used in cotton against Angular leaf spot and boll worms.

- *Degradation of pesticides in soil by micro-organisms* : Soil receives different toxic chemicals in various forms and causes adverse effects on beneficial soil micro-flora/micro-fauna, plants, animals and human beings. Various microbes present in soil act as the scavengers of these harmful chemicals in soil. The pesticides/chemicals reaching the soil are acted upon by several physical, chemical and biological forces exerted by microbes in the soil and they are degraded into non-toxic substances and there by minimise the damage caused by the pesticides to the ecosystem. For example, bacterial genera like Pseudomonas, Clostridium, Bacillus, Thiobacillus, Achromobacter etc. and fungal genera like Trichoderma, Penicillium, Aspergillus, Rhizopus, and Fusarium are playing important role in the degradation of the toxic chemicals / pesticides in soil.

- *Biodegradation of hydrocarbons* : Natural hydrocarbons insoil like waxes, paraffin's, oils etc. are degraded by fungi, bacteria and actinomycetes. *E.g.* ethane (C_2H_6) a paraffin hydrocarbon is metabolised and degraded by Mycobacteria, Nocardia, Streptomyces Pseudomonas, Flavobacterium and several fungi.

CHARACTERISATION OF SOIL MICROBIAL

Soil is a complex and dynamic biological system with up to 90 per cent of the processes in soil mediated by microbes. The microbial population in soil is diverse and contains up to 6000 bacterial genomes per gram of soil. Furthermore, it has recently been reported that soil microbial presence far exceeds currently accepted values and large contributions of microbial peptide/protein are found in the humic substance fraction of soils. Investigating microbial diversity is important as it is believed that differences in microbial community composition may also influence the chemical composition of organic materials in soil. Moreover, the structure of SOM is significantly impacted by the carbon input source, since the microbial and the plant derived biomass residues are believed to differ significantly in their molecular structures. In these contexts, the study of soil microbiology ecology is vital to our understanding of natural processes such as C and N cycling and their associated impacts on agricultural productivity and climate change.

Molecular markers, such as the 16S rRNA gene, have been extensively applied to detect, identify and measure microbial diversity from environmental samples. In most cases, the 16S rRNA is amplified from total DNA extracted from a sample by employing polymerase chain reaction (PCR). Universal primer pair 27f and 1525r, that allow amplification of nearly complete 16S rRNA genes from the majority of known bacteria have been used to study the types of bacteria present (composition), the number of types (richness), and the frequency of distribution or the relative abundance of types (structure) in a diverse range of habitats. In this chapter, the near complete 16S rRNA gene sequences of 40 isolates were determined and analysed for two soils used in subsequent studies.

Materials and Method of Soil Sampling

Soil samples were collected from two Irish field sites (Church and Big Bull Park) at the Teagasc, Oakpark Crops Research Centre. The soils, a light clay-loam (Big Bull Park) and heavy clay-loam (Church) were under similar management practices and agricultural regimes, and have been subjected to intensive tillage for over 20 years. The agricultural management practices and physico-chemical properties of these fields. Sampling was carried out according to a modified version of the protocol described by Joseph *et. al.* (2003). Composite samples (each composite sample composed of three samples) were collected at eight locations along transect lines following a 'Z' pattern. A 25-mm-diameter clean metal core was used to sample 100-mm long soil cores from the A horizon, which were transferred to sterile polyethylene bags and sealed at the collection sites. Soil cores were transported at the ambient temperature and processed within 24 h of collection. The upper 30 mm of each core was discarded, and large pieces of roots and stones were removed from the remainder, which was sieved through a stainless steel sieve with a 2-mm aperture (IMPACT Laboratory Test Sieve, UK). Sieved samples were pooled, homogenised and stored at 4°C at its field moisture content for further analysis.

Microbial Cultivation

Microbial cultivation was carried out according to a modified version of the protocol. Approximately 1 g of either soil was added to 100-ml aliquots of sterile distilled water in 250-ml conical flasks and dispersed by stirring with Teflon-coated magnetic bars (8 mm in diameter, 50 mm long) on a magnetic stirrer for 15 min at 400 rpm.

One-millilitre aliquots of soil suspension were added to 9-ml portions of dilute nutrient broth (DNB), containing gL-1 : Lab-Lemco' Powder 1.0; Yeast Extract 2.0; Peptone 5.0; and Sodium Chloride 5.0, at a concentration of 0.08 g per litre of distilled water (Oxoid Ltd., Hampshire, England).

Diluted soil suspensions were mixed by vortexing at approximately 150 rpm for 10 s, and used to prepare serial dilutions containing 10^{-2} to 10^{-4} g of soil suspension. One hundred-microlitre (100 µl) aliquots of each dilution series was plated on duplicate LB agar plates containing 0.5 per cent dripstone,

0.25 per cent yeast extract, 0.1 per cent D-glucose, 0.25 per cent NaCl and 1.5 per cent agar. Serially inoculated LB plates were incubated at RT for 48 h and all isolated colonies were selected from the 10^{-4} dilution of each soil type and used to inoculate 3.0 ml of LB broth. Cultures were incubated at RT for 48 h. All samples were done in duplicates.

Total DNA Separation and PCR Amplification

Total bacterial DNA was isolated from pure cultures according to the manufacturer's instructions (UltraClean™ DNA Extraction Kit, Mo Bio Laboratories). The region of the 16S rRNA between nucleotide position 27 and 1525 (Escherichia coli 16S rRNA gene sequence numbering), corresponding to almost the entire 16S rRNA gene, was targeted for PCR amplification from total genomic DNA. The amplification was primed with universal 16S rDNA primers 27f 5'-AGAGTTT-GATCMTGGCTCAG-3' and 1525r 5'-AAGGAGGTGWTCCARCC-3'. PCR reaction was prepared in a fifty microlitre (50 µl) reaction volume containing 2 ng of genomic DNA template, 0.2 mM concentrations of dNTPs, 1x GoTaq® Flexi Buffer1, 0.2 µM of each primer, 1.5 mM $MgCl_2$, 2.5 U of GoTaq® DNA Polymerase (Promega, Madison USA).

Amplification profiles consisted of an initial denaturation of one cycle of 95°C for 2 min, followed by 30 cycles of denaturation at 95°C for 1 min, annealing at 60°C for 1 min, and elongation at 72° for 2 min, 1 cycle of final extension at 72°C for 10 min (Biometra® TGradient, AnaChem, UK). PCR amplification products were electrophoresed in 1.0 per cent (wt/vol) agarose gels in 1x TAE buffer, stained with ethidium bromide (0.5µg/ml), destained in 1 mM MgSO4 and photographed using a gel documentation system (Pharmacia Biotech). A 1Kb DNA Ladder was used as the standard marker (Promega, Madison USA).

Nucleotide Sequence and Phylogenetic Analysis

The resulting PCR amplified products were purified by gel extraction (QIA-GEN®, Valencia, CA) and sequenced by commercial providers using primer pair 27f and 1525r (Macrogen, Korea). Partial 16S rRNA gene sequences were edited using Edit Seq (DNASTAR, Madison, WI USA), and contiguous 16S rRNA gene sequences assembled using SeqMan II (DNASTAR). The resulting sequences were submitted to the SEQUENCE MATCH programme of the Ribosomal Database Project (RDP [Maidak et. al., 1997]) and to the advanced BLAST search programme of the National Centre for Bio-technology Information (NCBI [Altschul et. al., 1997]) and the closest database relatives of all sequences retrieved for comparison, using the FASTA algorithm (Pearson and Lipman, 1988). Isolates exhibiting less than 95 per cent similarity to an existing Gen-Bank sequence were checked for chimeric artefacts by the CHIMERA_CHECK programme at the Ribosomal Database Project II (RDP-II [Maidak et. al., 2001]), using the default settings. The 16S rRNA gene sequences which were determined have been deposited into the Gen-bank under the accession numbers.

The low similarity of some isolate sequences to known organisms' sequences made classification into phyla based solely on these searches difficult: therefore, all isolate sequences were subjected to phylogenetic analysis with representative sequences of known phyla. Alignments of isolate sequences and reference organisms were created using the MegAlign module of DNASTAR, and only sequence data corresponding to E. coli bases 63 to 633 were considered for the phylogenetic trees (Furlong et. al., 2002). Trees were constructed using PAUP* Version 4.0 (Swofford, 1998). Phylogeny was by a distance method where a phylogenetic tree was constructed by the Neighbour Joining (NJ) method. A bootstrap analysis yielding bootstrap percentages was performed for each data set using 1000 bootstrap replicates and a 70 per cent confidence level to evaluate the statistical significance of branching. The resulting phylogenetic tree was visualised using TreeView (Page, 1996).

Results and Discussion of PCR Amplification

In order to assess the richness and relative abundance of bacteria considered in this study, totals of 21 and 19 bacterial 16S rDNA isolates of approximately 1.5 kb long, were amplified and sequenced from pure cultures of a heavy and light clay loam agricultural soil, respectively. The approximately 1.5 kb amplified fragment represented almost the entire 16S rRNA gene sequence.

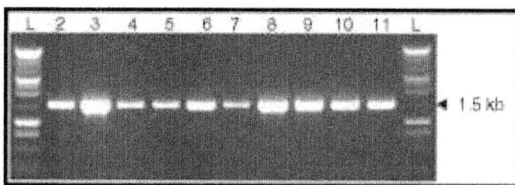

Fig. Polymerase Chain Reaction (PCR) Product of Teagasc Heavy Clay Loam Soil Bacterial 16S rRNA Genes Generated from Universal 16S Primer Pair 27f and 1525r. Lane L, 1 kb Ladder (Promega); Lanes 2-11, Teagasc Heavy Clay Loam Soil Bacterial 16S rRNA Gene.

Nucleotide Sequence and Phylogenetic Analysis

Isolate sequences from the heavy clay loam soil were analysed and compared with their closest database relatives. Of the 21 sequences analysed, 12 (57.1 per cent) had similarity values greater than 95 per cent compared with the available sequences from the database, while 3 (14.3 per cent) showed similarity values between 90 per cent and 95 per cent, and 6 (28.6 per cent) revealed nucleotide similarities ranging between 83 per cent and 87 per cent. Of the 19 light clay loam sequences analysed, 9 (47.4 per cent) had similarity values higher than 95 per cent when compared with the available sequences from the Gen-Bank, 6 (31.6 per cent) had similarity values between 90 per cent and 95 per cent, while 4 (21 per cent) revealed a similarity index of 86 per cent to 87 per cent. Further sequence analysis indicated that the isolates from both soils were affiliated with two major phylogenetic (candidate) groups of the eubacterial domain, which included members of Proteobacteria comprising the β and γ sub-divisions and Firmicutes phyla.

The Isolates of the heavy clay loam soil were dominated by members of the phylum Proteobacteria (81 per cent) with the γ- and β- sub-divisions accounting for approximately 57 and 24 per cent, respectively. The Firmicutes were less abundant, accounting for 21 per cent of the total heavy clay loam isolates. Sequences belonging to the γ-Proteobacteria (47.4 per cent) and β-Proteobacteria (26.3 per cent) comprised approximately 74 per cent of the total isolates from the light clay loam soil. The Firmicutes (26 per cent) represented the remaining light clay loam isolates. All isolates of the γ-Proteobacteria from the heavy and light clay loam soils (12 and 9 isolates, respectively) were associated with the family Pseudomonadaceae and, particularly, to the genus Pseudomonas, with sequence similarities of 85-99 per cent and 87-99 per cent, respectively. Isolates of the β-Proteobacteria of both soils (five isolates each) were grouped within the family Oxalobacteraceae of the order Burkholderiales and represented by the genus Janthinobacterium, with sequence similarities of 84-99 per cent and 86-99 per cent, respectively. The Firmicutes isolates of the heavy clay loam soil (four isolates) were all of the family Bacillaceae and associated with the Bacillus spp., with a sequence similarity of 83-99 per cent. The Firmicutes isolates of the light clay loam soil (5 isolates) were grouped within the class Bacilli and the families : Bacillaceae (4 isolates) and Staphylococcaceae (1 isolate), and were represented by the Bacillus spp. and Staphylococcus spp., respectively, with a sequence similarity of 87-98 per cent. All of these taxa are typical of soil microbial environment and the rhizosphere in particular. Isolate sequences examined with CHIMERA_CHECK were divided into two fragments. The results revealed that there was no significant difference between the highest binary association coefficient value of a full length sequence and its associated fragments, which indicated that the sequences were not chimeric.

In this study, 570 nucleotides of the almost complete 16S rRNA sequences of 21 heavy clay loam and 19 light clay loam soil isolates were used for phylogenetic analysis. Phylogenetic trees were constructed on the basis of these sequences and reference sequences obtained from the database. Both trees had three main clusters, with > 70 per cent bootstrap support. The first cluster of the heavy clay loam soil isolates contained 12 (57 per cent) isolates and grouped with known organisms of the γ-Proteobacteria sub-division. The second and third clusters contained five (24 per cent) and four (19 per cent) isolates and grouped with known organisms of the β-Proteobacteria sub-division and the Firmicutes phyla, respectively. The first and second clusters of the light clay loam soil isolates contained five (~26 per cent) and four (~21 per cent) isolates and grouped with known organisms of the Firmicutes phyla and the β-Proteobacteria sub-division, respectively. The third cluster of the light clay loam soil isolates contained 10 (~53 per cent) isolates that grouped with known organisms of the γ-Proteobacteria sub-division. These results were consistent with the initial assignment of both sets of isolates into candidate divisions based sequence matches from the RDP-II and nucleotide sequence similarities with known sequences from the database.

This work presents a molecular study of the diversity of micro-organisms cultured from two agricultural soils. The overall phylogenetic distribution of the bacterial isolates obtained from both soils indicated similar patterns of abundance and diversity in their community structures. The γ-Proteobacteria isolates from both soils was the dominant group, accounting for up to one half of the total number of isolates. Isolates of the β-Proteobacteria division were slightly less abundant in the heavy clay loam soil, while isolates of the Firmicutes division from both soils were of similar abundance. Except for a single isolate for the light clay loam soil associated with the Staphylococcus spp., all isolates of both soils were associated with the Pseudomonas species, Janthinobacterium species and Bacillus species.

Some members of the Bacillus species are also producers of secondary metabolites, previously discussed, however, they were less abundant when compared with isolates of the Pseudomonas species from both soils. The discrepancy in abundance may be due in part to the fact that Bacillus species often sporulate under adverse environmental conditions (Rajalakshmi and Shethna, 1980) and may require special pre-propagation treatments to encourage the germination of these spores. Such treatments could be undertaken to address these problems in future work.

While the community composition of both soils appeared typical of the microbial profile of the rhizosphere, a greater bacterial diversity might be expected from a nutrient-enriched root zone facilitated by root exudates. The identification of members of the α-Proteobacteria has been reported in many studies; however, this group of bacteria was not identified in the current study. The absence of this group may be due to the fact that it is mainly composed of slow growing bacterial species.

In addition, many of the members of the α-Proteobacteria respond to changes in environmental conditions by entering a viable by non-culturable state. Moreover, the α-Proteobacteria may require specific physicochemical conditions not found in simple growth media (Barbieri et. al., 2007). One suggestion for future work is to try and optimise the media and growth conditions to maximise the species richness of the culturable soil microbial fraction. Alternatively, microbes (culturable and non-culturable) may be extracted from soils; however, while this method favours microbial diversity studies, it must be understood that it is equally difficult to separate microbes from soil particles. Therefore, great care must be exercised to ensure that only microbial biomass is being considered for degradation. It is unlikely that cultivation-based diversity studies will reflect the true microbial community structure present in situ because of inherent qualitative and quantitative biases. However, the bacterial isolates obtained can be considered as, and provide relative measures of the natural bacterial diversity of both soil communities. It is also interesting to note that the diversity results presented in this chapter show some degree of similarity (an abundance of Proteobacteria and Firmicutes) to culture-independent studies of the bacterial community tightly associated to the gut wall of earthworms and within soil samples taken from pairs of two adjacent fields (arable and pasture) located at Johnstown Castle Estate, Wexford, Co. Wexford; Lyons Estate, Celbridge, Co. Kildare; and Teagasc, Oakpark Crops Research Centre, Carlow. Moreover, suggested that the microbial fingerprint of cultivable biomass is similar to that of microbes extracted from soils, and although only a small fraction of the total population can be cultured, the cultivable fraction is representative (at the bio-chemical input level) of the microbes that cannot be cultured.

The bio-chemical contribution of the culturable microbial fraction of the light clay-loam Oakpark soil to SOM is considered in greater details in subsequent chapters. In addition, the 16S rDNA sequences obtained from the bacterial isolates extended the taxonomic database of bacteria associated with agricultural soil, and more specifically the soils used in this work, for comparative systematic studies. In a recent study using culture-independent approaches, Thakuria et. al. (2008) demonstrated significant variations in the microbial diversity of Irish soils. Considering the taxonomic and metabolic diversity of soil microbes and the fact that microbes are inextricably linked to the chemical structure, location, and rates of decomposition of SOM, we would like to suggest that the focus of future work should be directed at characterising the spatial and temporal scales of influence of soil microbial biomass on SOM structure and composition in Irish soils and indeed soils across Europe.

SOIL MICRO-ORGANISM AND INHABITING

Actinomycetes

These are the organisms with characteristics common to both bacteria and fungi but yet possessing distinctive features to delimit them into a distinct category. In the strict taxonomicsense, actinomycetes are clubbed with bacteria the same class of Schizomycetes and confined to the order Actinomycetales. They are unicellular like bacteria, but produce a mycelium which is non-septate (coenocytic) and more slender, tike true bacteria they do not have distinct cell-wall and their cell wall is without chit in and cellulose (commonly found in the cell wall of fungi). On culture media unlike slimy distinct colonies of true bacteria which grow quickly, actinomycetes colonies grow slowly, show powdery consistency and stick firmly to agar surface. They produce hyphae and conidia / sporangia like fungi. Certain actinomycetes whose hyphae undergo segmentation resemble bacteria, both morphologically and physiologically. Actinomycetes are numerous and widely distributed in soil and are next to bacteria in abundance. They are widely distributed in the soil, compost etc. Plate count estimates give values ranging from 104 to 108 per gram of soil. They are sensitive to acidity/low pH (optimum pH range 6.5 to 8.0) and water logged soil conditions. The population of actinomycetes increases with depth of soil even up to horizon 'C' of a soil profiler They are heterotrophic, aerobic and mesophilic (25-30 °C) organisms and some species are commonly present in compost and manures are the rmophilic growing at 55-65 °C temperature (*e.g.* Thermoatinomycetes, Streptomyces). Actinomycetes belonging to the order of Actinomycetales are grouped under four families *viz.* Mycobacteriaceae, Actinomycetaceae, Streptomycetaceae and Actinoplanaceae. Actinomycetous genera which are agriculturally and industrially important are present in only two families of Actinomycetaceae and Strepotmycetaceae. In the order of abundance in soils, the common genera of actinomycetes are Streptomyces (nearly 70 per cent), Nocardia and Micro-monospora although Actinomycetes, Actinoplanes, Micro-monospora and Streptosporangium are also generally encountered.

Bacteria

Amongst the different micro-organisms inhabiting in the soil, bacteria are the most abundant and predominant organisms. These are primitive, prokaryotic, microscopic and unicellular micro-organisms without chlorophyll.

Morphologically, soil bacteria are divided into three groups *viz* :
- Bacilli
- Cocci (round/spherical), (rod-shaped)
- Spirilla/Spirllum (cells with long wavy chains).

Bacilli are most numerous followed by Cocci and Spirilla in soil. The most common method used for isolation of soil bacteria is the "dilution plate count" method which allows the enumeration of only viable/living cells in the soil. The size of soil bacteria varies from 0.5 to 1.0 micron in diameter and 1.0 to 10.0 microns in length. They are motile with locomotory organs flagella. Bacterial population is one-half of the total microbial biomass in the soil ranging from 1, 00000 to several hundred millions per gram of soil, depending upon the physical, chemical and biological conditions of the soil. Winogradsky, on the basis of ecological characteristics classified soil micro-organisms in general and bacteria in particular into two broad categories *i.e.* Autochnotus (Indigenous species) and the Zymogenous (fermentative). Autochnotus bacterial population is uniform and constant in soil, since their nutrition is derived from native soil organic matter (*e.g.* Arthrobacter and Nocardia whereas Zymogenous bacterial population in soil is low, as they require an external source of energy, *e.g.* Pseudomonas and Bacillus.

The population of Zymogenous bacteria increases gradually when a specific substrate is added to the soil. To this category belong the cellulose decomposers, nitrogen utilising bacteria and ammonifiers. As per the system proposed in the Bergey's Manual of Systematic Bacteriology, most of the bacteria which are predominantly encountered in soil are taxonomically included in the three orders, Pseudomonadales, Eubacteriales and Actinomycetales of the class Schizomycetes. The most common soil bacteria belong to the genera Pseudomonas, Arthrobacter, Clostridium, Achromobacter, Sarcina, Enterobacter etc. The another group of bacteria common in soils is the Myxobacteria belonging to the genera Micrococcus, Chondrococcus, Archangium, Polyangium, Cyptophaga. Bacteria are also classified on the basis of physiological activity or mode of nutrition, especially the manner in which they obtain their carbon, nitrogen, energy and other nutrient requirements.

They are broadly divided into two groups *i.e.* a) Autotrophs and b) Heterotrophs :

- Autotrophic bacteria are capable synthesising their food from simple inorganic nutrients, while heterotrophic bacteria depend on preformed food for nutrition. All autotrophic bacteria utilise CO_2 (from atmosphere) as carbon source and derive energy either from sunlight (photoautotrophs, *e.g.* Chromatrum. Chlorobium. Rhadopseudomonas or from the oxidation of simple inorganic substances present in soil (chemoautotrophs *e.g.* Nitrobacter, Nitrosomonas, Thiaobacillus).
- Majority of soil bacteria are heterotrophic in nature and derive their carbon and energy from complex organic substances/organic matter, decaying roots and plant residues. They obtain their nitrogen from nitrates and ammonia compounds (proteins) present in soil and other nutrients from soil or from the decomposing organic matter. Certain bacteria also require amino acids, B- Vitamins, and other growth promoting substances also.

Functions / Role of Bacteria

Bacteria bring about a number of changes and bio-chemical transformations in the soil and thereby directly or indirectly help in the nutrition of higher plants growing in the soil. The important transformations and processes in which soil bacteria play vital role are : decomposition of cellulose and other carbohydrates, ammonification (proteins ammonia), nitrification (ammonia-nitrites-nitrates), denitrification (release of free elemental nitrogen), biological fixation of atmospheric nitrogen (symbiotic and non-symbiotic) oxidation and reduction of sulphur and iron compounds. All these processes play a significant role in plant nutrition.

Process/Reaction	Bacterial Genera
Cellulose decomposition (celluloytic bacteria) most cellulose decomposers are mesophilic	a. Aerobic: Angiococcus, Cytophaga, Polyangium, Sporocytophyga, Bacillus, Achromobacter, Cellulomonas b. Anaerobic: Clostridium Methanosarcina, Methanococcus
Ammonification (Ammonifiers)	Bacillus, Pseudomonas
Nitrification (Nitrifying bacteria)	Nitrosomonas, Nilrobacter Nitrosococcus
Denitrification (Denitrifies)	Achromobacter, Pseudomonas, Bacillus, Micrococcus
Nitrogen fixing bacteria	a. Symbiotic- Rhizobium, Bradyrrhizobium b. Non-symbiotic: aerobic – Azotobacter Beijerinckia (acidic soils), anaerobic-Clostridium

Bacteria capable of degrading various plant residues in soil are :

Cellulose	Hemicelluloses	Lignin	Pectin	Proteins
Pseudomonas	Bacillus	Pseudomonas	Erwinia	Clostridium
Cytophaya	Vibrio	Micrococcus		Proteus
Spirillum	Pseudomonas	Flavobacteriumm		Pseudomonas
Actinomycetes	Erwinia	Xanthomonas		Bacillus
Cellulomonas		Streptomyces		

Algae

Algae are present in most of the soils where moisture and sunlight are available. Their number in soil usually ranges from 100 to 10,000 per gram of soil. They are photo-autotrophic, aerobic organisms and obtain CO_2 from atmosphere and energy from sunlight and synthesise their own food. They are unicellular, filamentous or colonial.

Soil algae are divided in to four main classes or phyla as follows :

- Cyanophyta (Blue-green algae)
- Chlorophyta (Grass-green algae)
- Xanthophyta (Yellow-green algae)
- Bacillariophyta (diatoms or golden-brown algae).

Out of these four classes/phyla, blue-green algae and grass-green algae are more abundant in soil. The green-grass algae and diatoms are dominant in the soils of temperate region while blue-green algae predominate in tropical soils. Green-algae prefer acid soils while blue green algae are commonly found in neutral and alkaline soils. The most common genera of green algae found in soil are Chlorella, Chlamydomonas, Chlorococcum, Protosiphon etc., and that of diatoms are Navicula, Pinnularia, Synedra, Frangilaria. Blue green algae are unicellular, photoautotrophic prokaryotes containing Phycocyanin pigment in addition to chlorophyll. They do not posses flagella and do not reproduce sexually. They are common in neutral to alkaline soils. The dominant genera of BGA in soil are : Chrococcus, Phormidium, Anabaena, Aphanocapra, Oscillatoria etc. Some BGA posses specialised cells known as "Heterocyst" which is the sites of nitrogen fixation. BGA fixes nitrogen (non-symbiotically) in puddle paddy/water logged paddy fields (20-30 kg/ha/season). There are certain BGA which possess the character of symbiotic nitrogen fixation in association with other organisms like fungi, mosses, liverworts and aquatic ferns Azolla, *e.g.* Anabaena-Azolla association fix nitrogen symbiotically in rice fields.

Functions/Role of Algae or BGA

- Plays important role in the maintenance of soil fertility especially in tropical soils.
- Add organic matter to soil when die and thus increase the amount of organic carbon in soil.
- Most of soil algae (especially BGA) act as cementing agent in binding soil particles and thereby reduce/prevent soil erosion.
- Mucilage secreted by the BGA is hygroscopic in nature and thus helps in increasing water retention capacity of soil for longer time/period.
- Soil algae through the process of photo-synthesis liberate large quantity of oxygen in the soil environment and thus facilitate the aeration in submerged soils or oxygenate the soil environment.
- They help in checking the loss of nitrates through leaching and drainage especially in un-cropped soils.
- They help in weathering of rocks and building up of soil structure.

Fungi

Fungi in soil are present as mycelial bits, rhizomorph or as different spores. Their number varies from a few thousand to a few million per gram of soil. Soil fungi possess filamentous mycelium composed of individual hyphae. The fungal hyphae maybe as eptate/coenocytic (Mastigomycotina and Zygomycotina) or septate (Ascomycotina, Basidiomycotina and Deuteromycotina).

As observed by C.K. Jackson, most commonly encountered genera of fungi in soil are : Alternaria, Aspergillus, Cladosporium, Cephalosporium, Botrytis, Chaetomium, Fusarium, Mucor, Penicillium, Verticillium, Trichoderma, Rhizopus, Gliocladium, Monilia, Pythium, etc. Most of these fungal genera belong to the sub-division Deuteromycotina / Fungi imperfeacta which lacks sexual mode of reproduction.

As these soil fungi are aerobic and heterotrophic, they require abundant supply of oxygen and organic matter in soil. Fungi are dominant in acid soils, because acidic environment is not conducive/suitable for the existence of either bacteria oractinomycetes. The optimum PH range for fungi lies-between 4.5 to 6.5. They are also present in neutral and alkaline soils and some can even tolerate pH beyond 9.0.

Functions/Role of Fungi :

- Fungi plays significant role in soils and plant nutrition.
- They plays important role in the degradation / decomposition of cellulose, hemi cellulose, starch, pectin, lignin in the organic matter added to the soil.
- Lignin which is resistant to decomposition by bacteria is mainly decomposed by fungi.
- They also serve as food for bacteria.
- Certain fungi belonging to sub-division Zygomycotina and Deuteromycotina are predaceous in nature and attack on protozoa and nematodes in soil and thus, maintain biologica lequilibrium in soil.
- They also plays important role in soil aggregation and in the formation of humus.
- Some soil fungi are parasitic and causes number of plant diseases such as wilts, root rots, damping-off and seedling blightseg. Pythium, Phyiophlhora, Fusarium, Verticillium etc.
- Number of soil fungi forms mycorrhizal association with the roots of higher plants (symbiotic association of a fungus with the roots of a higher plant) and helps in mobilisation of soil phosphorus and nitrogen *e.g.* Glomus, Gigaspora, Aculospora, (Endomycorrhiza) and Amanita, Boletus, Entoloma, Lactarius (Ectomycorrhiza).

DEFINITION OF SOIL MICROBIOLOGY AND SOIL IN VIEW OF MICROBIOLOGY

Definition : It is branch of science/microbiology which deals with study of soil micro-organisms and their activities in the soil.

Soil : It is the outer, loose material of earth's surface which is distinctly different from the underlying bedrock and the region which support plant life. Agriculturally, soil is the region which supports the plant life by providing mechanical support and nutrients required for growth. From the microbiologist view point, soil is one of the most dynamic sites of biological interactions in the nature. It is the region where most of the physical, biological and bio-chemical reactions related to decomposition of organic weathering of parent rock take place.

Components of Soil

Soil is an admixture of five major components *viz.* organic mater, mineral matter, soil-air, soil water and soil micro-organisms/living organisms. The amount/proposition of these components varies with locality and climate.

- *Mineral / Inorganic Matter* : It is derived from parent rocks/bed rocks through decomposition, disintegration and weathering process. Different types of inorganic compounds containing various minerals are present in soil. Amongst them the dominant minerals are Silicon, Aluminum and iron and others like Carbon, Calcium Potassium, Manganese, Sodium, Sulphur, Phosphorus etc. are in trace amount. The proportion of mineral matter in soil is slightly less than half of the total volume of the soil.
- *Organic matter/components* : Derived from organic residues of plants and animals added in the soil. Organic matter serves not only as a source of food for micro-organisms but also supplies energy for the vital processes of metabolism which are characteristics of all living organisms. Organic matter in the soil is the potential source of N, P and S for plant growth. Microbial decomposition of organic matter releases the unavailable nutrients in available from. The proportion of organic matter in the soil ranges from 3-6 per cent of the total volume of soil.
- *Soil Water* : The amount of water present in soil varies considerably. Soil water comes from rain, snow, dew or irrigation. Soil water serves as a solvent and carrier of nutrients for the plant growth. The micro-organisms inhabiting in the soil also require water for their metabolic activities. Soil water thus, indirectly affects plant growth through its effects on soil and micro-organisms. Percentage of soil-water is 25 per cent total volume of soil.
- *Soil air (Soil gases)* : A part of the soil volume which is not occupied by soil particles *i.e.* pore spaces are filled partly with soil water and partly with soil air. These two components (water and air) together only accounts for approximately half the soil's volume. Compared with atmospheric air, soil is lower in oxygen and higher in carbon dioxide, because CO_2 is continuous recycled by the micro-organisms during the process of decomposition of organic matter. Soil air comes from external atmosphere and contains nitrogen, oxygen CO_2 and water vapour (CO_2 > oxygen). CO_2 in soil air (0.3-1.0 per cent) is more than atmospheric air (0.03 per cent). Soil aeration plays important role in plant growth, microbial population, and microbial activities in the soil.
- *Soil micro-organisms* : Soil is an excellent culture media for the growth and development of various micro-organisms. Soil is not an inert static material but a medium pulsating with life. Soil is now believed to be dynamic or living system.

Soil contains several distinct groups of micro-organisms and amongst them bacteria, fungi, actinomycetes, algae, protozoa and viruses are the most important. But bacteria are more numerous than any other kinds of micro-organisms. Micro-organisms form a very small fraction of the soil mass and occupy a volume of less than one per cent. In the upper layer of soil (top soil up to 10-30 cm depth *i.e.* Horizon A), the microbial population is very high which decreases with depth of soil. Each organisms or a group of organisms are responsible for a specific change

/ transformation in the soil. The final effect of various activities of micro-organisms in the soil is to make the soil fit for the growth and development of higher plants.

Living organisms present in the soil are grouped into two categories as follows :

- Soil flora (micro flora) *e.g.* Bacteria, fungi, Actinomycetes, Algae and
- Soil fauna (micro fauna) animal like *e.g.* Protozoa, Nematodes, earthworms, moles, ants, rodents.

Relative proportion / percentage of various soil micro-organisms are : Bacteria-aerobic (70 per cent), anaerobic (13 per cent), Actinomycetes (13 per cent), Fungi / molds (03 per cent) and others (Algae Protozoa viruses) 0.2-0.8 per cent. Soil organisms play key role in the nutrient transformations.

Types of Micro-organisms in Soil

Living organisms both plants and animals, constitute an important component of soil. The pioneering investigations of a number of early micro-biologists showed for the first time that the soil was not an inert static material but a medium pulsating with life. The soil is now believed to be a dynamic or rather a living system, containing a dynamic population of organisms/micro-organisms. Cultivated soil has relatively more population of micro-organisms than the fallow land, and the soils rich in organic matter contain much more population than sandy and eroded soils. Microbes in the soil are important to us in maintaining soil fertility / productivity, cycling of nutrient elements in the biosphere and sources of industrial products such as enzymes, antibiotics, vitamins, hormones, organic acids etc. At the same time certain soil microbes are the causal agents of human and plant diseases.

The soil organisms are broadly classified in to two groups *viz.* soil flora and soil fauna, the detailed classification of which is as follows :

Soil Organisms

Soil Flora

- *Micro-flora* : 1. Bacteria 2. Fungi, Molds, Yeast, Mushroom 3. Actinomycetes, Stretomyces 4. Algae *e.g.* BGA, Yellow Green Algae, Golden Brown Algae. 1. Bacteria is again classified in I) Heterotrophic *e.g.* symbiotic and non-symbiotic N_2 fixers, Ammonifier, Cellulose Decomposers, Denitrifiers II) Autrotrophic *e.g.* Nitrosomonas, Nitrobacter, Sulphur oxidisers, etc.
- Macro-flora : Roots of higher plants

Soil Fauna

- *Micro-fauna* : Protozoa, Nematodes
- *Macrofauna* : Earthworms. moles, ants and others.

As soil inhabit several diverse groups of micro-organisms, but the most important amongst them are : bacteria, actinomycetes, fungi, algae and protozoa.

SOIL MICRO-ORGANISMS IN BIODEGRADATION OF PESTICIDES AND HERBICIDES

Pesticides are the chemical substances that kill pests and herbicides are the chemicals that kill weeds. In the context of soil, pests are fungi, bacteria insects, worms, and nematodes etc., that cause damage to field crops. Thus, in broad sense pesticides are insecticides, fungicides, bactericides, herbicides and nematicides that are used to control or inhibit plant diseases and insect pests. Although widescale application of pesticides and herbicides is an essential part of augmenting crop yields; excessive use of these chemicals leads to the microbial imbalance, environmental pollution and health hazards. An ideal pesticide should have the ability to destroy target pest quickly and should be able to degrade non-toxic substances as quickly as possible.

The ultimate "sink" of the pesticides applied in agriculture and public health care is soil. Soil being the storehouse of multitudes of microbes, in quantity and quality, receives the chemicals in various forms and acts as a scavenger of harmful substances. The efficiency and the competence to handle the chemicals vary with the soil and its physical, chemical and biological characteristics.

Effects of Pesticides

Pesticides reaching the soil in significant quantities have direct effect on soil micro-biological aspects, which in turn influence plant growth. Some of the most important effects caused by pesticides are :

- Alterations hi ecological balance of the soil micro-flora,
- Continued application of large quantities of pesticides may cause ever lasting changes in the soil micro-flora,
- Adverse effect on soil fertility and crop productivity,
- Inhibition of N_2 fixing soil micro-organisms such as Rhizobium, Azotobacter, Azospirillum etc. and cellulolytic and phosphate solubilising micro-organisms,
- Suppression of nitrifying bacteria, Nitrosomonas and Nitrobacter by soil fumigants ethylene bromide, Telone, and vapam have also been reported,
- Alterations in nitrogen balance of the soil,
- Interference with ammonification in soil,
- Adverse effect on mycorrhizal symbioses in plants and nodulation in legumes, and
- Alterations in the rhizosphere micro-flora, both quantitatively and qualitatively.

Persistence of Pesticides in Soil

How long an insecticide, fungicide, or herbicide persists in soil is of great importance in relation to pest management and environmental pollution. Persistence of pesticides in soil for longer period is undesirable because of the reasons : a) accumulation of the chemicals in soil to highly toxic levels, b) may be assimilated by the plants and get accumulated in edible plant products, c) accumulation in the edible portions of the root crops, d) to be get eroded with soil particles and may enter into the water streams, and finally leading to the soil, water and air pollutions. The effective persistence of pesticides in soil varies from a week to several years depending upon structure and properties of the constituents in the pesticide and availability of moisture in soil. For instance, the highly toxic phosphates do not persist for more than three months while chlorinated hydrocarbon insecticides (*e.g.* DOT, aldrin, chlordane etc.) are known to persist at least for 4-5 years and some times more than 15 years.

From the agricultural point of view, longer persistence of pesticides leading to accumulation of residues in soil may result into the increased absorption of such toxic chemicals by plants to the level at which the consumption of plant products may prove deleterious / hazardous to human beings as well as livestock's. There is a chronic problem of agricultural chemicals, having entered in food chain at highly inadmissible levels in India, Pakistan, Bangladesh and several other developing countries in the world. For example, intensive use of DDT to control insect pests and mercurial fungicides to control diseases in agriculture had been known to persist for longer period and thereby got accumulated in the food chain leading to food contamination and health hazards. Therefore, DDT and mercurial fungicides has been, banned to use in agriculture as well as in public health department.

Biodegradation of Pesticides in Soil

Pesticides reaching to the soil are acted upon by several physical, chemical, and biological forces. However, physical and chemical forces are acting upon/ degrading the pesticides to some extent, micro-organism's plays major role in the degradation of pesticides. Many soil micro-organisms have the ability to act upon pesticides and convert them into simpler non-toxic compounds. This process of degradation of pesticides and conversion into non-toxic compounds by micro-organisms is known as "biodegradation". Not all pesticides reaching to the soil are biodegradable and such chemicals that show complete resistance to biodegradation are called "recalcitrant".

The chemical reactions leading to biodegradation of pesticides fall into several broad categories which are discussed in brief in the following paragraphs.

- *Detoxification* : Conversion of the pesticide molecule to a non-toxic compound. Detoxification is not synonymous with degradation. Since a single chance in the side chain of a complex molecule may render the chemical non-toxic.

- *Degradation* : The breaking down / transformation of a complex substrate into simpler products leading finally to mineralisation. Degradation is often considered to be synonymous with mineralisation, *e.g.* Thirum (fungicide) is degraded by a strain of Pseudomonas and the degradation products are dimethlamine, proteins, sulpholipaids, etc.
- *Conjugation (complex formation or addition reaction)* : In which an organism make the substrate more complex or combines the pesticide with cell metabolites. Conjugation or the formation of addition product is accomplished by those organisms catalysing the reaction of addition of an amino acid, organic acid or methyl crown to the substrate, for *e.g.*, in the microbial metabolism of sodium dimethly dithiocarbamate, the organism combines the fungicide with an amino acid molecule normally present in the cell and thereby inactivate the pesticides/ chemical.
- *Activation* : It is the conversion of non-toxic substrate into a toxic molecule, for *e.g.* Herbicide, 4-butyric acid (2, 4-D B) and the insecticide Phorate are transformed and activated micro-biologically in soil to give metabolites that are toxic to weeds and insects.
- *Changing the spectrum of toxicity* : Some fungicides/pesticides are designed to control one particular group of organisms / pests, but they are metabolised to yield products inhibitory to entirely dissimilar groups of organisms, for *e.g.* the fungicide PCNB fungicide is converted in soil to chlorinated benzoic acids that kill plants.

Biodegradation of pesticides / herbicides is greatly influenced by the soil factors like moisture, temperature, pH and organic matter content, in addition to microbial population and pesticide solubility. Optimum temperature, moisture and organic matter in soil provide congenial environment for the break down or retention of any pesticide added in the soil. Most of the organic pesticides degrade within a short period (3-6 months) under tropical conditions. Metabolic activities of bacteria, fungi and actinomycetes have the significant role in the degradation of pesticides.

Criteria for Bioremediation / Biodegradation

For successful biodegradation of pesticide in soil, following aspects must be taken into consideration.

- Organisms must have necessary catabolic activity required for degradation of contaminant at fast rate to bring down the concentration of contaminant,
- The target contaminant must be bio-availability,
- Soil conditions must be congenial for microbial / plant growth and enzymatic activity and
- Cost of bioremediation must be less than other technologies of removal of contaminants.

Principle of microbial infallibility, for every naturally occurring organic compound there is a microbe / enzyme system capable its degradation.

Strategies for Bioremediation

For the successful biodegradation / bioremediation of a given contaminant following strategies are needed.

- *Passive/ intrinsic Bioremediation* : It is the natural bioremediation of contaminant by tile indigenous micro-organisms and the rate of degradation is very slow.
- *Biostimulation* : Practice of addition of nitrogen and phosphorus to stimulate indigenous micro-organisms in soil.
- *Bioventing* : Process/way of Bio-stimulation by which gases stimulants like oxygen and methane are added or forced into soil to stimulate microbial activity.
- *Bioaugmentation* : It is the inoculation/introduction of micro-organisms in the contaminated site/soil to facilitate biodegradation.
- *Composting* : Piles of contaminated soils are constructed and treated with aerobic thermophilic micro-organisms to degrade contaminants. Periodic physical mixing and moistening of piles are done to promote microbial activity.
- *Phytoremediation* : Can be achieved directly by planting plants which hyperaccumulate heavy metals or indirectly by plants stimulating micro-organisms in the rhizosphere.
- *Bioremediation* : Process of detoxification of toxic/unwanted chemicals / contaminants in the soil and other environment by using micro-organisms.
- *Mineralisation* : Complete conversion of an organic contaminant to its inorganic constituent by a species or group of micro-organisms.

Chapter 4

ORGANIC CONSTITUENTS OF PLANTS

When the water naturally existing in plants is expelled by exposure to the air or a gentle heat, the residual dry matter is found to be composed of a considerable number of different substances, which have been divided into two great classes, called the organic and the inorganic, or mineral constituents of plants. The former are readily combustible, and on the application of heat, catch fire, and are entirely consumed, leaving the inorganic matters in the form of a white residuum or ash.

All plants contain both classes of substances; and though their relative proportions vary within very wide limits, the former always greatly exceed the latter, which in many cases form only a very minute proportion of the whole weight of the plant.

Owing to the great preponderance of the organic or combustible matters, it was at one time believed that the inorganic substances formed no part of the true structure of plants, and consisted only of a small portion of the mineral matters of the soil, which had been absorbed along with their organic food; but this opinion, which probably was never universally entertained, is now entirely abandoned, and it is no longer doubted that both classes of substances are equally essential to their existence.

Although they form so large a proportion of the plant, its organic constituents are composed of no more than four elements, *viz.* : —

- Carbon.
- Hydrogen.
- Nitrogen.
- Oxygen.

The inorganic constituents are much more numerous, not less than thirteen substances, which appear to be essential, having been observed.

These are :

- Potash.

- Soda.
- Lime.
- Magnesia.
- Peroxide of Iron.
- Silicic Acid.
- Phosphoric Acid.
- Sulphuric Acid.
- Chlorine.

And more rarely :

- Manganese.
- Iodine.
- Bromine.
- Fluorine.

Several other substances, among which may be mentioned alumina and copper, have also been enumerated; but there is every reason to believe that they are not essential, and the cases in which they have been found are quite exceptional.

It is to be especially noticed that none of these substances occur in plants in the free or uncombined state, but always in the form of compounds of greater or less complexity, and extremely varied both in their properties and composition.

It would be out of place, in a work like the present, to enter into complete details of the properties of the elements of which plants are composed, which belongs strictly to pure chemistry, but it is necessary to premise a few observations regarding the organic elements, and their more important compounds.

CARBON

When a piece of wood is heated in a close vessel, it is charred, and converted into charcoal. This charcoal is the most familiar form of carbon, but it is not absolutely pure, as it necessarily contains the ash of the wood from which it was made.

In its purest form it occurs in the diamond, which is believed to be produced by the decomposition of vegetable matters, and it is there crystallised and remarkably transparent; but when produced by artificial processes, carbon is always black, more or less porous, and soils the fingers.

It is insoluble in water, burns readily, and is converted into carbonic acid. Carbon is the largest constituent of plants, and forms, in round numbers, about 50 per cent of their weight when dry.

CARBONIC ACID

This, the most important compound of carbon and oxygen, is best obtained by pouring a strong acid upon chalk or limestone, when it escapes with effervescence.

It is a colourless gas, extinguishing flame, incapable of supporting respiration, much heavier than atmospheric air, and slightly soluble in water, which takes up its own volume of the gas.

It is produced abundantly when vegetable matters are burnt, as also during respiration, fermentation, and many other processes. It is likewise formed daring the decay of animal and vegetable matters, and is consequently evolved from dung and compost heaps.

HYDROGEN

It occurs in nature only in combination. Its principal compound is water, from which it is separated by the simultaneous action of an acid, such as sulphuric acid and a metal, in the form of a transparent gas, lighter than any other substance.

It is very combustible, burns with a pale blue flame, and is converted into water. It is found in all plants, although in comparatively small quantity, for, when dry, they rarely contain more than four or five per cent. Its most important compound is water, of which it forms one-ninth, the other eight-ninths consisting of oxygen.

NITROGEN

It exists abundantly in the atmosphere, of which it forms nearly four-fifths, or, more exactly, 79 per cent. It is there mixed, but not combined with oxygen; and when the latter gas is removed, by introducing into a bottle of air some substance for which the former has an affinity, the nitrogen is left in a state of purity.

It is a transparent gas, which is incombustible and extinguishes flame. It is a singularly inert substance, and is incapable of directly entering into union with any other element except oxygen, and with that it combines with the greatest difficulty, and only by the action of the electric spark—a peculiarity which has very important bearings on many points we shall afterwards have to discuss. Nitrogen is found in plants to the extent of from 1 to 4 per cent.

NITRIC ACID

This, the most important compound of nitrogen and oxygen, can be produced by sending a current of electric sparks through a mixture of its constituents, but in this way it can be obtained only in extremely small quantity.

It is much more abundantly produced when organic matters are decomposed with free access of air, in which case the greater proportion of their nitrogen combines with the atmospheric oxygen. This process, which is known by the name of nitrification, is greatly promoted by the presence of lime or some other substance, with which the nitric acid may combine in proportion as it is formed. It takes place, to a great extent, in the soil in India and other hot climates; and our chief supplies of saltpetre, or nitrate of potash, are derived from the soil in these countries, where it has been formed in this manner. The same change occurs, though to a much smaller extent, in the soil in temperate climates.

AMMONIA

Ammonia is a compound of nitrogen and hydrogen, but it cannot be formed by the direct union of these gases. It is a product of the decomposition of organic substances containing nitrogen, and is produced when they are distilled at a high temperature, or allowed to putrefy out of contact of the air.

In its pure state it is a transparent and colourless gas, having a peculiar pungent smell, and highly soluble in water. It is an alkali resembling potash and soda, and, like these substances, unites with the acids and forms salts, of which the sulphate and muriate are the most familiar. In these salts it is fixed, and does not escape from them unless they be mixed with lime, or some other substance possessing a more powerful affinity for the acid with which it is united.

OXYGEN

It is one of the most widely distributed of all the elements, and, owing to its powerful affinities, is the most important agent in almost all natural changes. It is found in the air, of which it forms 21 per cent, and in combination with hydrogen, and almost all the other chemical elements. In the pure state it possesses very remarkable properties. All substances burn in it with greater brilliancy than they do in atmospheric air, and its affinity for most of the elements is extremely powerful.

When diluted with nitrogen, it supports the respiration of animals; but in the pure state it proves fatal after the lapse of an hour or two. It is found in plants, in quantities varying from 30 to 36 per cent. It is worthy of observation, that of the four organic elements, carbon only is fixed, and the other three are gases; and likewise, when any two of them unite, their compound is either a gaseous or a volatile substance.

The charring of organic substances, which is one of their most characteristic properties, and constantly made use of by chemists as a distinctive reaction, is due to this peculiarity; for when they are heated, a simpler arrangement of their particles takes place, the hydrogen, nitrogen, and oxygen unite among themselves, and carry off a small quantity of carbon, while the remainder is left behind in the form of charcoal, and is only consumed when access of the external air is permitted.

Now, in order that a plant may grow, its four organic constituents must be absorbed by it, and that this absorption may take place, it is essential that they be presented to it in suitable forms. A seed may be planted in pure carbon, and supplied with unlimited quantities of hydrogen, nitrogen, oxygen, and inorganic substances, and it will not germinate; and a plant, when placed in similar circumstances, shows no disposition to increase, but rapidly languishes and dies.

The obvious inference from these facts is, that these substances cannot be absorbed when in the *elementary* state, but that it is only after they have entered into certain forms of combination that they acquire the property of being readily taken up, and assimilated by the organs of the plant.

It was at one time believed that many different compounds of these elements might be absorbed and elaborated, but later and more accurate experiments have reduced the number to four—namely, carbonic acid, water, ammonia, and nitric acid. The first supplies carbon, the second hydrogen, the two last nitrogen, while all of them, with the exception of ammonia, may supply the plant with oxygen as well as with that element of which it is the particular source.

There are only two sources from which these substances can be obtained by the plant, *viz.* the atmosphere and the soil, and it is necessary that we should here consider the mode in which they may be obtained from each.

THE ATMOSPHERE AS A SOURCE OF THE ORGANIC CONSTITUENTS OF PLANTS

Atmospheric air consists of a mixture of nitrogen and oxygen gases, watery vapour, carbonic acid, ammonia, and nitric acid. The two first are the largest constituents, and the others, though equally essential, are present in small, and some of them in extremely minute quantity. When deprived of moisture and its minor constituents, 100 volumes of air are found to contain 21 of oxygen and 79 of nitrogen.

Although these gases are not chemically combined in the air, but only mechanically mixed, their proportion is exceedingly uniform, for analyses completely corresponding with these numbers have been made by Humboldt, Gay-Lussac, and Dumas at Paris, by Saussure at Geneva, and by Lewy at Copenhagen; and similar results have also been obtained from air collected by Gay-Lussac during his ascent in a balloon at the height of 21,430 feet, and by Humboldt on the mountain of Antisano in South America at a height of 16,640 feet.

In short, under all circumstances, and in all places, the relation subsisting between the oxygen and nitrogen is constant; and though, no doubt, many local circumstances exist which may tend to modify their proportions, these are so slow and partial in their operations, and so counter balanced by others acting in an opposite direction, as to retain a uniform proportion between the main constituents of the atmosphere, and to prevent the undue accumulation of one or other of them at any one point.

No such uniformity exists in the proportion of the minor constituents. The variation in the quantity of watery vapour is a familiar fact, the difference between a dry and moist atmosphere being known to the most careless observer, and the proportions of the other constituents are also liable to considerable variations.

Carbonic Acid

The proportion of carbonic acid in the air has been investigated by Saussure. From his experiments, made at the village of Chambeisy, near Geneva, it appears that the quantity is not constant, but varies from 3·15 to 5·75 volumes in 10,000; the mean being 4·15. These variations are dependent on different circumstances. It was found that the carbonic acid was always more abundant during the night than during the day — the mean quantity in the former case being 4·32, in the latter 3·38.

The largest quantity found during the night was 5·74, during the day 5·4. Heavy and continued rain diminishes the quantity of carbonic acid, by dissolving and carrying it down into the soil. Saussure found that in the month of July 1827, during the time when nine millimetres of rain fell, the average quantity of carbonic acid amounted to 5·18 volumes in 10,000; while in September 1829, when 254 millimetres fell, it was only 3·57.

A moist state of the soil, which is favourable to the absorption of carbonic acid, also diminishes the quantity contained in the air, while, on the other hand, continued frosts, by retaining the atmosphere and soil in a dry state, have an opposite effect. High winds increase the carbonic acid to a small extent.

It was also found to be greater over the cultivated lands than over the lake of Geneva; at the tops of mountains than at the level of the sea; in towns than in the country. The differences observed in all these cases, though small, are quite distinct, and have been confirmed by subsequent experimenters.

Ammonia

The presence of ammonia in the atmosphere appears to have been first observed by Saussure, who found that when the sulphate of alumina is exposed to the air, it is gradually converted into the double sulphate of alumina and ammonia.

Liebig more recently showed that ammonia can always be detected in rain and snow water, and it could not be doubted that it had been absorbed from the atmosphere. Experiments have since been made by different observers with the view of determining the quantity of atmospheric ammonia, which gives the quantity found in a million parts of air.

For this reason, more recent experimenters have endeavoured to arrive at conclusions bearing more immediately upon agricultural questions, by determining the quantity of ammonia brought down by the rain. The first observations on this subject were made by Barral in 1851, and they have been repeated during the years 1855 and 1856 by Mr. Way.

In 1853, Boussingault also made numerous experiments on the quantity of ammonia in the rain falling at different places, as well as in dew and the moisture of fogs. It thus appears that in Paris the quantity of ammonia in rain-water is just six times as great as it is in the country, a result, no doubt, due to the ammonia evolved during the combustion of fuel, and to animal exhalations, and to the same cause, the large quantity contained in the moisture of fogs in Paris may also be attributed.

Barral and Way have made determinations of the quantity of ammonia carried down by the rain in each month of the year, the former using for this purpose the water collected in the rain-gauges of the Paris Observatory, and representing, therefore, a town atmosphere; the latter, that from a large rain-gauge at Rothamsted, at a distance from any town.

According to Barral the ammonia annually deposited on an acre of land amounts to 12·28 lbs., a quantity considerably exceeding that obtained by Way, whose experiments being made at a distance from towns, must be considered as representing more accurately the normal condition of the air. His results for the years 1855 and 1856 are given below, along with the quantities of nitric acid found at the same time.

Nitric Acid

The presence of nitric acid in the air appears to have been first observed by Priestley at the end of the last century, but Liebig, in 1825, showed that it was always to be found after thunder-storms, although he failed to detect it at other times.

In 1851 Barral proved that it is invariably present in rain-water, and stated the quantity annually carried down to an acre of land at no less than 41·29 lbs. But at the time his experiments were made, the methods of determining very minute quantities of nitric acid were exceedingly defective, and Way, by the adoption of an improved process, has shown that the quantity is very much smaller than Barral supposed, and really falls short of three pounds.

No attempts have been made to determine the proportion of nitric acid in air, but its quantity is undoubtedly excessively minute, and materially smaller than that of ammonia. At least this conclusion seems to be a fair inference from Way's researches, as well as the recent experiments of Boussingault on the proportion of nitric acid contained in rain, dew, and fog, made in a manner exactly similar to those on the ammonia, already quoted.

Although it thus appears that Barral's results have been only partially confirmed, enough has been ascertained to show that the quantity of ammonia and nitric acid in the air is sufficient to produce a material influence in the growth of plants. The large amount of these substances contained in the dew is also particularly worthy of notice, and may serve to some extent to explain its remarkably invigourating effect on vegetation.

Carburetted Hydrogen

Gay-Lussac, Humboldt, and Boussingault have shown, that when the whole of the moisture and carbonic acid have been removed from the air, it still contains a small quantity of carbon and hydrogen; and Saussure has rendered it probable that they exist in a state of combination as carburetted hydrogen gas.

No definite proof of this position has, however, as yet been adduced, and the function of the compound is entirely unknown. It is possible that the presence of carbon and hydrogen may be due to a small quantity of organic matter; but, whatever be its source, its amount is certainly extremely small.

Sulphuretted Hydrogen and Phosphuretted Hydrogen

The proportion of these substances is almost infinitesimal; but they are pretty general constituents of the atmosphere, and are apparently derived from the decomposition of animal and vegetable matters.

The preceding statements lead to the important conclusion, that the atmosphere is capable of affording an abundant supply of all the organic elements of plants, because it not only contains nitrogen and oxygen in the free state, but also in those forms of combination in which they are most readily absorbed, as well as a large quantity of carbonic acid, from which their carbon may be derived.

At first sight it may indeed appear that the quantity of the latter compound, and still more that of ammonia, is so trifling as to be of little practical importance. But a very simple calculation serves to show that, though relatively small, they are absolutely large, for the carbonic acid contained in the whole atmosphere amounts in round numbers to 2,400,000,000,000 tons, and the ammonia, assuming it not to exceed one part in fifty millions, must weigh 74,000,000 tons, quantities amply sufficient to afford an abundant supply of these elements to the whole vegetation of our globe.

The Soil as a Source of the Organic Constituents of Plants

When a portion of soil is subjected to heat, it is found that it, like the plant, consists of a combustible and an incombustible part; but while in the plant the incombustible part or ash is small, and the combustible large, these proportions are reversed in the soil, which consists chiefly of inorganic or mineral matters, mixed with a quantity of combustible or organic substances, rarely exceeding 8 or 10 per cent, and often falling considerably short of this quantity.

The organic matter exists in the form of a substance called humus, which must be considered here as a source of the organic constituents of plants, independently of the general composition of the soil, which will be afterwards discussed. The term *humus* is generic, and applied by chemists to a rather numerous group of substances, very closely allied in their properties, several of which are generally present in all fertile soils.

They have been submitted to examination by various chemists, but by none more accurately than by Mulder and Herman, to whom, indeed, we owe almost all the precise information we possess on the subject. The organic matters of the soil may be divided into three great classes; the first containing those substances which are soluble in water; the second, those extracted by means of caustic potash; and the third, those insoluble in all menstrua.

When a soil is boiled with a solution of caustic potash, a deep brown fluid is obtained, from which acids precipitate a dark brown flocculent substance, consisting of a mixture of at least three different acids, to which the names of humic, ulmic, and geic acids have been applied.

The fluid from which they have been precipitated contains two substances, crenic and apocrenic acids, while the soil still retains what has been called insoluble humus. The acids above named do not differ greatly in chemical characters, but they have been subdivided into the humic, geic, and crenic groups, which present some differences in properties and composition.

They are compounds of carbon, hydrogen, and oxygen, and are characterised by so powerful an affinity for ammonia that they are with difficulty obtained free from that substance, and generally exist in the soil in combination with it.

They are all products of the decomposition of vegetable matters in the soil, and are formed during their decay by a succession of changes, which may be easily traced by observing the course of events when a piece of wood or any other vegetable substance is exposed for a length of time to air and moisture. It is then found gradually to disintegrate with the evolution of carbonic acid, acquiring first a brown and finally a black colour.

At one particular stage of the process it is converted into one or other of two substances, called humin and ulmin, both insoluble in alkalies, and apparently identical with the insoluble humus of the soil; but when the decomposition is more advanced the products become soluble in alkalies, and then contain humic, ulmic, and geic acids, and finally, by a still further progress, crenic and apocrenic acids are formed as the result of an oxidation occurring at certain periods of the decay.

The roots and other vegetable debris remaining in the soil undergo a similar series of changes, and form the humus, which is found only in the surface soil, that is to say, in the portion which is now or has at some previous period been occupied by plants, and the quantity of humus contained in any soil is mainly dependent on the activity of vegetation on it.

Numerous analyses of humus compounds extracted from the soil have been made, and have served to establish a number of minor differences in the composition even of those to which the same name has been applied, due manifestly to the fact that their production is the result of a gradual decomposition, which renders it impossible to extract from the soil one pure substance, but only a variable mixture of several, so similar to one another in properties, that their separation is very difficult, if not impossible.

For this reason great discrepancies exist in the statements made regarding them by different observers, but this is a matter of comparatively small importance, as their exact composition has no very direct bearing on agricultural questions, and it will suffice to give the names and chemical formulæ of those which have been analysed and described,—

Ulmic acid from long Frisian turf C_{40} Humic acid from hard turf $H_{18}O_{16}$

H Humic acid from arable soil C_{40}H Humic acid from a pasture field $C_{40}15O_{15}16O_{16}$ Geic acid C_{40} Apocrenic acid C_{48} Crenic acid C_{24}

$C_{40}H_{14}O_{14}H_{15}O_{17}H_{12}O_{24}$ $H_{12}O_{16}$

It is only necessary to observe further, that these formulæ indicate a close connection with woody fibre, and the continuous diminution of the hydrogen and increase of oxygen shows that they must have been produced by a gradually advancing decay.

The earlier chemists and vegetable physiologists attributed to the humus of the soil a much more important function than it is now believed to possess. It was formerly considered to be the exclusive, or at least the chief source of the organic constituents of plants, and by absorption through the roots to yield to them the greater part of their nutriment.

But though this view has still some supporters, among whom Mulder is the most distinguished, it is now generally admitted that humus is not a direct source of the organic constituents of plants, and is not absorbed as such by their roots, although it is so indirectly, in as far as the decomposition which it is constantly undergoing in the soil yields carbonic acid, which can be absorbed.

The older opinion is refuted by many well-ascertained facts. As regards the exclusive origin of the carbon of plants from humus, it is easy to see that this at least cannot be true, for humus, as already stated, is itself derived solely from the decomposition of vegetable and animal matters; and if the plants on the earth's surface were to be supported by it alone, the whole of their substance would have to return to the soil in the same form, in order to supply the generation which succeeds them.

But this is very far from being the case, for the respiration of animals, the combustion of fuel, and many other processes, are annually converting a large quantity of these matters into carbonic acid; and if there were no other source of carbon but the humus of the soil, the amount of vegetable life would gradually diminish, and at length become entirely extinct.

Schleiden, who has discussed this subject very fully, has made an approximative calculation of the total quantity of humus on the earth's surface, and of the carbon annually converted into carbonic acid by the respiration of man and animals, the combustion of wood for fuel, and other minor processes; and he draws the conclusion that, if there were no other source of carbon except humus, the quantity of that substance existing in the soil would only support vegetation for a period of sixty years. The particular phenomena of vegetation also afford abundant evidence that humus cannot be the only source of carbon.

Thus Boussingault has shown that on the average of years, the crops cultivated on an acre of land remove from it about one ton more organic matter than they receive in the manure applied to them, although there is no corresponding diminution in the quantity of humus contained in the soil. An instance which leads still more unequivocally to the same conclusion is given by Humboldt.

He states that an acre of land, planted with bananas, yields annually about 152,000 pounds weight of fruit, containing about 32,000 pounds, or almost exactly 14 tons of carbon; and as this production goes on during a period of twenty years, there must be withdrawn in that time no less than 280 tons of carbon.

But the soil on an acre of land weighs, in round numbers, 1000 tons, and supposing it to contain 4 per cent of humus, the total weight of carbon in it would amount to little more than 20 tons. It is obvious from these and many other analogous facts that humus cannot be the only or even a considerable source of the carbon of plants, although it is still contended by some chemists that it may be absorbed to a small extent.

But even this is at variance with many well-known facts. For if humus were absorbed, it might be expected that vegetation would be most luxuriant on soils containing abundance of that substance, especially if it existed in a soluble and readily absorbable form; but so far from this being the case, nothing is more certain than that peat, in which these conditions are fulfilled, is positively injurious to most plants.

On the other hand, our daily experience affords innumerable examples of plants growing luxuriantly in soils and places where no humus exists. The sands of the sea-shore, and the most barren rocks, have their vegetation, and the red-hot ashes which are thrown out by active volcanoes are no sooner cool than a crop of plants springs up on them. The conclusions to be drawn from these considerations have been further confirmed by the direct experiments of different observers.

Boussingault sowed peas, weighing 15.60 grains, in a soil composed of a mixture of sand and clay, which had been heated red-hot, and consequently contained no humus, and after 99 days' growth, during which they had been watered with distilled water, he found the crop to weigh 68·72 grains, so that there had been a four-fold increase. Similar experiments have been made by Prince Salm Horstmar, on oats and rape sown in a soil deprived of organic matter by ignition, in which they grew readily, and arrived at complete maturity.

One oat straw attained a height of three feet, and bore 78 grains; another bore 47; and a third 28—in all 153. These when dried at 212° weighed 46.302 grains, and the straw 45.6 grains. The most satisfactory experiments, however, are those of Weigman and Polstorf, these observers having found that it was possible to obtain a two-hundred-fold produce of barley in an entirely artificial soil, provided care was taken to give it the physical characters of a fertile soil.

They prepared a mixture of six parts of sand, two of chalk, one of white bole, and one of wood charcoal; to which was added a small quantity of felspar, previously fused with marble and some soluble salts, so as to imitate as closely as possible the inorganic parts of a soil, and in it they planted twelve barley pickles.

The plants grew luxuriantly, reaching a height of three feet, and each bearing nine ears, containing 22 pickles. The grain of the twelve plants weighed 2040 grains. These experiments show that plants can grow and produce seed when the most scrupulous care is taken to deprive them of every trace of humus. But Saussure has gone further, and shown that even when present, humus is not absorbed. He allowed plants of the common bean and the Polygonum Persicaria to grow in solutions of humate of potash, and found a very trifling diminution in the quantity of humic acid present; but the value of his experiments is invalidated by his having omitted to ascertain whether the diminution of humic acid which he observed was really due to absorption by the plant.

This omission has been supplied by Weigman and Polstorf. They grew plants of mint (Mentha undulata) and of Polygonum Persicaria in solutions of humate of potash, and placed beside the glass containing the plant, another perfectly similar, and containing only the solution of humate of potash.

The solution, which contained in every 100 grains, 0·148 grains of solid matter, consisting of humate of potash, etc. was found to become gradually paler, and at the end of a month, during which time the plants had increased by 6–1/2 inches, the quantity of solid matter in 100 grains had diminished to 0·132.

But the solution contained in the other glass, and in which no plant had grown, had diminished to 0·136, so that the absorption could not have amounted to more than 0·004 grains for every 100 grains of solution employed.

This quantity is so small as to be within the limits of error of experiment, and we are consequently entitled to draw the conclusion that humus, even under the most favourable circumstances, is not absorbed by plants.

But though not directly capable of affording nutriment to plants, it must not, on that account, be supposed that humus is altogether devoid of importance, for it is constantly undergoing decomposition in the soil, and thus becomes a source of carbonic acid which can be absorbed, and, as we shall afterwards more particularly see, it exercises very important functions in bringing the other constituents of the soil into readily available forms of combination.

It has been already observed that carbon, hydrogen, nitrogen, and oxygen, cannot be absorbed by plants when uncombined, but only in the forms of water, carbonic acid, ammonia, and nitric acid. It is scarcely necessary to detail the grounds on which this conclusion has been arrived at in regard to carbon and hydrogen, for practically it is of little importance whether they can be absorbed or not, as the former is rarely, the latter never, found uncombined in nature.

Neither can there be any doubt that water and carbonic acid are the only substances from which these elements can be obtained. Every-day experience convinces us that water is essential to vegetation; and Saussure, and other observers, have shown that plants will not grow if they are deprived of carbonic acid, and that they actually absorb that substance abundantly from the atmosphere. The evidence for the non-absorption of oxygen lies chiefly in the fact that plants obtain, in the form of water and carbonic acid, a larger quantity of that element than they require, and in place of absorbing, are constantly exhaling it.

The form in which nitrogen may be absorbed has given rise to much difference of opinion. In the year 1779, Priestley commenced the examination of this subject, and drew from his experiments the conclusion, that plants absorb the nitrogen of the air.

Saussure shortly afterwards examined the same subject, and having found, that when grown in a confined space of air, and watered with pure water, the nitrogen of the plants underwent no increase, he inferred that they derived their entire supplies of that element from ammonia, or the soluble nitrogenous constituents of the soil or manure.

Boussingault has since re-examined this question, and by a most elaborate series of experiments, in which the utmost care was taken to avoid every source of fallacy, he was led to the conclusion, that when haricots, oats, lupins, and cresses were grown in calcined pumice-stone, mixed with the ash of plants, and supplied with air deprived of ammonia and nitric acid, their nitrogen underwent no increase.

It has been objected to these experiments, that the plants being confined in a limited bulk of air, were placed in an unnatural condition, and Ville has recently repeated them with a current of air passing through the apparatus, and found a slight increase in the nitrogen, due, as he thinks, to direct absorption.

It is much more probable, however, that it depends on small quantities of ammonia or nitric acid which had not been completely removed from the air by the means employed for that purpose, for nothing is more difficult than the complete abstraction of these substances, and as the gain of nitrogen was only 0·8 grains, while 60,000 gallons of air, and 13 of water, were employed in the experiment, which lasted for a considerable time, it is reasonable to suppose that a sufficient quantity may have remained to produce this trifling increase.

While these experiments show that plants maintain only a languid existence when grown in air deprived of ammonia and nitric acid, and hence, that the direct absorption of nitrogen, if it occur at all, must do so to a very small extent, the addition of a very minute quantity of the former substance immediately produces an active vegetation and rapid increase in size of the plants.

Among the most striking proofs of this are the experiments of Wolff, made by growing barley and vetches in a soil calcined so as to destroy organic matters, and then mixed with small quantities of different compounds of ammonia.

These experiments not only prove that ammonia can be absorbed, but they also indirectly confirm the statement already made, that humus is not necessary; for in some instances the produce was higher than that obtained from the uncalcined soil with the same manures, although it contained four per cent of humus.

On such experiments Liebig rests his opinion that ammonia is the exclusive source of the nitrogen of plants, and although he has recently admitted that it may be replaced by nitric acid, it is obvious that he considers this a rare and exceptional occurrence.

The evidence, however, for the absorption of nitric acid appears to rest on as good grounds as that of ammonia, for experience has shown that nitrate of soda acts powerfully as a manure, and its effect must be due to the nitric acid, and not to the soda, for the other compounds of that alkali have no such effect. Wolff has illustrated this point by a series of experiments on the sunflower, of which we shall quote one.

He took two seeds of that plant, and sowed them on the 10th May, in a soil composed of calcined sand, mixed with a small quantity of the ash of plants, and added at intervals during the progress of the experiment, a quantity of nitrate of potash, amounting in all to 17.13 grains.

The plants were watered with distilled water, containing carbonic acid in solution, and the pot in which they grew was protected from rain and dew by a glass cover. On the 19th August one of the plants had attained a height of above 28 inches, and had nine fine leaves and a flower-bud; the other was about 20 inches high, and had ten leaves. On the 22d August, one of the plants having been accidentally injured, the experiment was terminated.

The plants, which contained 103·16 grains of dry matter, were then carefully analysed, and the quantity of nitrogen contained in the soil after the experiment and in the seed was determined.

Hence, the nitrogen contained in the plants must, in this instance, have been obtained entirely from the nitrate of potash, for the quantity contained in it and in the seeds is exactly equal to that in the plants and the soil, the difference of 0·03 grains being so small that it may be safely attributed to the errors inseparable from such experiments.

For the sake of comparison, an exactly similar experiment was made on two seeds grown without nitrate of potash, and in this instance, after an equally long period of growth, the largest plant had only attained a height of 7.5 inches, and had three small pale and imperfectly developed leaves.

They contained only 0.033 grains of nitrogen, while the seeds contained 0.032—indicating that, under these circumstances, there was no increase in the quantity of that element. But, independently of these experimental results, it may be inferred from general considerations, that nitric acid must be one of the sources from which plants derive their nitrogen.

It has been already stated, that the humus contained in the soil consists of the remains of decayed plants, and there is every reason to suppose that the primeval soil contained no organic matters, and that the first generation of plants must have derived the whole of their nitrogen from, the atmosphere.

If, therefore, it be assumed that ammonia is the only source of the nitrogen of plants, it would follow, that as that substance cannot be produced by the direct union of its elements, the quantity of ammonia in the air could only remain undiminished in the event of the whole of the nitrogen of decaying plants returning into that form.

But this is certainly not the case, for every time a vegetable substance is burned, part of its nitrogen is liberated in the free state, and in certain conditions of putrefaction, nitric acid is produced. Now, if ammonia be the only form in which nitrogen is absorbed, there must be a gradual diminution of the quantity contained in the air; and further, there must either be some continuous source of supply by which its quantity is maintained, or there must be some other substance capable of affording nitrogen in a form fitted for the maintenance of plant life.

As regards the first alternative, it must be stated that we know of no source other than the decomposition of plants from which ammonia can be derived, and we are therefore compelled to adopt the second alternative, and to admit that there must be some other source of nitrogen, and it cannot be doubted, from what has been already stated, that it is from nitric acid only that it can be obtained.

It must be admitted, then, that carbonic acid, ammonia, nitric acid, and water, are the great organic foods of plants. But while they have afforded to them an inexhaustible supply of the last, the quantity of the other three available for food are limited, and insufficient to sustain their life for a prolonged period. It has been shown by Chevandrier, that an acre of land under beech wood accumulates annually about 1650 lb. of carbon.

Now, the column of air resting upon an acre of land contains only about 15,500 lb. of carbon, and the soil may be estimated to contain 1 per cent., or 22,400 lb. per acre, and the whole of this carbon would therefore be removed, both from the air and the soil, in the course of little more than 23 years.

But it is a familiar fact, that plants continue to grow with undiminished luxuriance year after year in the same soil, and they do so because neither their carbon nor their nitrogen are permanently absorbed; they are there only for a period, and when the plant has finished its functions, and dies, they sooner or later return into their original state.

Either the plant decays, in which case its carbon and nitrogen pass more or less rapidly into their original state, or it becomes the food of animals, and by the processes of respiration and secretion, the same change is indirectly effected.

In this way a sort of balance is sustained; the carbon, which at one moment is absorbed by the plant, passes in the next into the tissues of the animal, only to be again expired in that state in which it is fitted to commence again its round of changes. But while there is thus a continuous circulation of these constituents through both plants and animals, there are various changes which tend to liberate in the free state a certain quantity both of the carbon and nitrogen of plants, and these being thus removed from the sphere of organic life, there would be a gradual diminution in the amount of vegetation at the earth's surface, unless this loss were counterbalanced by some corresponding source of gain.

In regard to carbonic acid the most important source is volcanic action, but the loss of nitrogen, which is far more important and considerable, is restored by the direct combination of its elements. The formation of nitric acid during thunder storms has been long familiar; but it would appear from the recent experiments of Clöez, which, should they be confirmed by farther enquiry, will be of much importance, that this compound is also produced without electrical action when air is passed over certain porous substances, saturated with alkaline and earthy compounds.

Fragments of calcined brick and pumice stone were saturated with solution of carbonate of potash, with carbonates of lime and magnesia and other mixtures, and a current of air freed from nitric acid and ammonia passed over them for a long period, at the end of which notable quantities of nitric acid were detected.

SOURCE OF THE INORGANIC CONSTITUENTS OF PLANTS

The inorganic constituents of plants being all fixed substances, it is sufficiently obvious that they can only be obtained from the soil, which, as we shall afterwards see, contains all of them in greater or less abundance, and has always been admitted to be the only substance capable of supplying them.

The older chemists and physiologists, however, attributed no importance to these substances, and from the small quantities in which they are found in plants, imagined that they were there merely accidental impurities absorbed from the soil along with the humus, which was at that time considered to be their organic food.

This opinion, sufficiently disproved by the constant occurrence of the same substances in nearly the same proportions, in the ash of each individual plant, has been further refuted by the experiments of Prince Salm Horstmar, who has established their importance to vegetation, by experiments upon oats grown on artificial soils, in each of which one inorganic constituent was omitted.

He found that, without silica, the grain vegetated, but remained small, pale in colour, and so weak as to be incapable of supporting itself; without lime, it died when it had produced its second leaf; without potash and soda, it grew only to the height of three inches; without magnesia, it was weak and incapable of supporting itself; without phosphoric acid, weak but upright; and without sulphuric acid, though normal in form, the plant was feeble, and produced no fruit.

Manner in which the Constituents of Plants are Absorbed

Having treated of the sources of the elements of plants, it is necessary to direct attention to the mode in which they enter their system.

Water

The absorption of water by plants takes place in great abundance, and is connected with many of the most important phenomena of vegetation. It is principally absorbed by the roots, and passes into the tissues of the plant, where a part of it is decomposed, and goes to the formation of certain of its organic compounds; while by far the larger quantity, in place of remaining in it, is again exhaled by the leaves.

The extent to which this takes place is very large. Hales found that a sunflower exhaled in twelve hours about 1 lb. 5 oz. of water, but this quantity was liable to considerable variation, being greater in dry, and less in wet weather, and much diminished during the night.

Saussure made similar experiments, and observed that the quantity of water exhaled by a sunflower amounted to about 220 lb. in four months. The exhalation of plants has recently been examined with great accuracy by Lawes.

His experiments were made by planting single plants of wheat, barley, beans, peas, and clover, in large glass jars capable of holding about 42 lb. of soil, and covered with glass plates, furnished with a hole in the centre for the passage of the stem of the plant. Water was supplied to the soil at certain intervals, and the jars were carefully weighed.

It further appears, that the exhalation is not uniform, but increases during the active growth of the plant, and diminishes again when that period is passed. Similar experiments were made with the same plants in soils to which certain manures had been added, and with results generally similar. Calculating from these experiments, we are led to the apparently anomalous conclusion that the quantity of water exhaled by the plants growing on an acre of land greatly exceeds the annual fall of rain; although it is obvious that of all the rain which falls, only a small proportion can be absorbed by the plants growing on the soil, for a large quantity is carried off by the rivers, and never reaches their roots.

It has been calculated, for instance, that the Thames carries off in this way at least one-third of the annual rain that falls in the district watered by it, and the Rhine nearly four-fifths. Of course, this large exhalation must depend on the repeated absorption of the same quantity of water, which, after being exhaled, is again deposited on the soil in the form of dew, and passes repeatedly through the plant.

This constant percolation of water is of immense importance to the plant, as it forms the channel through which some of its other constituents are carried to it.

Carbonic Acid

While the larger part of the water which a plant requires is absorbed by its roots, the reverse is the case with carbonic acid. A certain proportion no doubt is carried up through the roots by the water, which always contains a quantity of that gas in solution, but by far the larger proportion is directly absorbed from the air by the leaves.

A simple experiment of Boussingault's illustrates this absorption very strikingly. He took a large glass globe having three apertures, through one of which he introduced the branch of a vine, with twenty leaves on it.

With one of the side apertures a tube was connected, by means of which the air could be drawn slowly through the globe, and into an apparatus in which its carbonic acid was accurately determined. He found, in this way, that while the air which entered the globe contained 0·0004 of carbonic acid, that which escaped contained only 0·0001, so that three-fourths of the carbonic acid had been absorbed.

Ammonia and Nitric Acid

Little is known regarding the mode in which these substances enter the plant. It is usually supposed that they are entirely absorbed by the roots, and no doubt the greater proportion is taken up in this way, but it is very probable that they may also be absorbed by the leaves, at least the addition of ammonia to the air in which plants are grown, materially accelerates vegetation.

It is probable, however, that the rain carries down the ammonia to the roots, and there is no doubt that that derived from the decomposition of the nitrogenous matters in the soil is so absorbed.

Inorganic Constituents

The inorganic constituents of course are entirely absorbed by the roots; and it is as a solvent for them that the large quantity of water continually passing through the plants is so important. They exist in the soil in particular states of combination, in which they are scarcely soluble in water.

But their solubility is increased by the presence of carbonic acid contained in the water, and which causes it to dissolve, to some extent, substances otherwise insoluble. It is in this way that lime, which occurs in the soil principally as the insoluble carbonate, is dissolved and absorbed. And phosphate of lime is also taken up by water containing carbonic acid, or even common salt in solution.

The amount of solubility produced by these substances is extremely small; but it is sufficient for the purpose of supplying to the plant as much of its mineral constituents as are required, for the quantity of water which, as we have already seen, passes through a plant is very large when compared with the amount of inorganic matters absorbed.

It has been shown by Lawes and Gilbert, that about 2000 grains of water pass through a plant for every grain of mineral matter fixed in it, so that there is no difficulty in understanding how the absorption takes place.

It is worthy of notice, however, that the absorption of the elements of plants takes place even though they may not be in solution in the soil, the roots apparently possessing the power of directly acting on and dissolving insoluble matters; but a distinction must be drawn between this and the view entertained by Jethro Tull, who supposed that they might be absorbed in the solid state, provided they were reduced to a state of sufficient comminution.

It is now no longer doubted that, whatever action the roots may exert, the constituents of the plant must be in solution before they can pass into it—experiment having distinctly shown that the spongioles or apertures through which this absorption takes place are too minute to admit even the smallest solid particle.

THE CONSTITUENTS OF PLANTS

The substances absorbed by the plant, which are of simple composition, and contain only two elements, are elaborated within it, and converted into the many complicated compounds of which its mass is composed. Some of these, as, for example, the colouring matters of madder and indigo, the narcotic principle of the poppy, are confined to a single species, or small group of plants, while others are found in all plants, and form the main bulk of their tissues.

The latter are the only substances which claim notice in a treatise like the present. They have been divided into three great classes, of widely different properties, composition, and functions.

The Saccharine and Amylaceous Constituents

These substances are compounds of carbon, hydrogen, and oxygen, and all possess a certain degree of similarity in composition, the quantities of hydrogen and oxygen they contain being always in the proportion required to form water, so that they may be considered as compounds of carbon and water; not that it can be asserted that they actually do contain water, as such, for of that there is no evidence, but only that its elements are present in the proportion to form it.

Cellulose

This substance forms the fundamental part of all plants. It is the principal constituent of woody fibre, and is found in a state of purity in the fibre of cotton and flax, and in the pith of plants; but in wood it is generally contaminated with another substance, which has received the name incrusting matter, because it is deposited in and around the cells of which the plant is in part composed.

Cellulose is insoluble in all menstrua, but, when boiled for a long time with sulphuric acid, is converted into a substance called dextrine. Cellulose consists of —

From pith of Elder-tree. Spongioles of roots. Carbon 43·37 43·00 Hydrogen 6·04 6·18 Oxygen 50·59 50·82 — — — — — — 100·00 100·00

It is represented chemically by the formula, $C\{24\}H\{21\}O\{21\}$, which shows it to be a compound of 24 atoms of carbon with 21 of hydrogen and 21 of oxygen.

Incrusting Matter

Large quantities of this substance enter into the composition of all plants. Of its chemical nature little is known, as it cannot be obtained separate from cellulose, but it is analogous to that substance in its composition, and probably contains hydrogen and oxygen in the proportion to form water.

Starch

Starch is one of the most abundant constituents of plants, and is found in most seeds, as those of the cereals and the leguminous plants; in the tubers of the potatoe, the bulbs of tulips. It is obtained by placing a quantity of wheat flour in a bag, and kneading it under a gentle stream of water. When the water is allowed to stand, it deposits the starch as a fine white powder, which, when examined by the microscope, is found to be composed of minute grains, formed of concentric layers deposited on one another.

These grains vary considerably in size and structure in different plants; but in the same plant they are generally so much alike as to admit of their recognition by a practised observer. They were formerly believed to be composed of an external coating of a substance insoluble in water, and containing in their interior a soluble kernel; but this opinion has been refuted, and distinct evidence been brought to show that the exterior and interior of the globules are identical in chemical properties.

Organic Constituents of Plants 79

Starch is insoluble in cold water, but by boiling, it dissolves, forming a thick paste. By long continued boiling with water containing a small quantity of acid, it is completely dissolved and converted into dextrine, and eventually into sugar. The same change is produced by the action of fermenting substances, such as the extract of malt; when heated in the dry state to a temperature of about 390 Fahr., it becomes soluble in cold water. It is distinguished by giving a brilliant blue compound with iodine.

Starch Contains

Carbon 44·47 Hydrogen 6·28 Oxygen 49·25 — — — 100·00 and its composition is represented by the formula $C_{12}H_{10}O_{10}$, so that it differs but little from cellulose in composition, although its chemical functions in the plant are extremely different. It is connected with some of the most important changes which occur in the growing plants, and by a series of remarkable transformations is converted into sugar and other important compounds.

Lichen Starch is found in most species of lichens, and is distinguished from common starch by producing a green colour with iodine. Its composition is the same as that of ordinary starch.

Inuline

The species of starch to which this name is given is characterised by its dissolving in boiling water, and giving a white pulverulent deposit in cooling. It is found in the tuber of the dahlia, in the dandelion, and some other plants. Its composition is identical with that of cellulose, and its formula is

$$C_{24}H_{21}O_{21}.$$

Gum is excreted from various plants as a thick fluid, which dries up into transparent masses. Its composition is identical with that of starch. It dissolves readily in cold water, and is converted into sugar by long continued boiling with acids. Its properties are best marked in gum arabic, which is obtained from various species of acacia; that from other plants differs to some extent, although its chemical composition is the same.

Dextrine

When starch is exposed to a heat of about 400°, or when treated with sulphuric acid, or with a substance extracted from malt called *diastase*, it is converted into dextrine. It may also be obtained from cellulose by a similar treatment.

The dextrine so obtained has the same composition as the starch from which it is produced, but its properties more nearly resemble those of gum. It plays a very important part in the process of germination, and may be converted into sugar on the one hand, and apparently also into starch on the other.

Sugar

Under this name are included four or five distinct substances, of which the most important are, cane sugar, grape sugar, and the uncrystallisable sugar found in many plants.

Cane Sugar

This variety of sugar, as its name implies, is found most abundantly in the sugar cane, but it occurs also in the maple, beet-root, and various species of palms, from all of which it is extracted on the large scale. It is extremely soluble in water, and can be obtained in large transparent prismatic crystals, as in common sugar-candy. It swells up, and is converted into a brown substance called caramel, when heated, and by contact with fermenting substances, yields alcohol and carbonic acid.

It contains— Carbon 42·22 Hydrogen 6·60 Oxygen 51·18 – – – 100·00 and its chemical formula is $C_{12}H_{11}O_{11}$.

Grape Sugar

Grape Sugar is met with in the grape, and most other fruits, as well as in honey. It is produced artificially when starch is boiled for a long time with sulphuric acid, or treated with a large quantity of diastase. It is less soluble in water than cane sugar, and crystallises in small round grains. Its composition, when dried at 284°, is— Carbon 40·00 Hydrogen 6·66 Oxygen 53·34 – – – 100·00 and its formula is $C_{12}H_{12}O_{12}$; but when crystallised it contains two equivalents of water, and is then represented by the formula

$$C_{12}H_{12}O_{12} + 2H_2O.$$

The uncrystallisable sugar of plants is closely allied to grape sugar, and, so far as at present known, has the same composition, although, from the difficulty of obtaining it quite free from crystallised sugar, this is still uncertain. *Mucilage* is the name applied to the substance existing in linseed, and in many other seeds, and which communicates to them the property of swelling up and becoming gelatinous when treated with water.

It is found in a state of considerable purity in gum tragacanth and some other gums. Its composition is not known with absolute certainty, but it is either C_{24}, or $C_{12}H_{19}O_{19}H_{10}O_{10}$; and in the latter case it must be identical with starch and gum. It will be observed that all the substances belonging to this class are very closely related in chemical composition, some of them, as starch and gum, though easily distinguished by their properties, being identical in constitution, while others only differ in the quantity of water, or of its elements which they contain.

The relation between these substances being so close, it is not difficult to understand how one may be converted into another by the addition or subtraction of water. Thus, cellulose has only to absorb an equivalent of water to become grape sugar, or to lose an equivalent in order to be converted into starch, and we shall afterwards see that such changes do actually occur in the plant during the process of germination.

Pectine and Pectic Acid

These substances are met with in many fruits and roots, as, for instance, in the apple, the carrot, and the turnip. They differ from the starch group in containing more oxygen than is required to form water along with their hydrogen; but their exact composition is still uncertain, and they undergo numerous changes during the ripening of the fruit.

OILY OR FATTY MATTERS

The oily constituents of plants form a rather extensive group of substances all closely allied, but distinguished by minor differences in properties and constitution. Some of them are very widely distributed throughout the vegetable kingdom, but others are almost peculiar to individual plants. They are all compounds of carbon, hydrogen, and oxygen, and are at once distinguished from the preceding class, by containing much less oxygen than is required to form water with their hydrogen.

The principal constituents of the fatty matters and oils of plants are three substances, called stearine, margarine, and oleine, the two former solids, the latter a fluid; and they rarely, if ever, occur alone, but are mixed together in variable proportions, and the fluidity of the oils is due principally to the quantity of the last which they contain.

If olive oil be exposed to cold, it is seen to become partially solid; and if it be then pressed, a fluid flows out, and a crystalline substance remains; the former is oleine, though not absolutely pure, and the latter margarine. The perfect separation of these substances involves a variety of troublesome chemical processes; and when it has been effected, it is found that each of them is a compound of a peculiar acid, with another substance having a sweet taste, and which has received the name of glycerine, or the sweet principle of oil.

Glycerine, as it exists in the fats, appears to be a compound of $C_3 H O_2$, and its properties are the same from whatever source it is obtained. The acids separated from it are known by the names of margaric, stearic, and oleic acids.

Margaric Acid

Margaric Acid is best obtained pure by boiling olive oil with an alkali until it is saponified, and decomposing the soap with an acid, expressing the margaric acid, which separates,cxccxcx and crystallising it from alcohol. It is a white crystalline fusible solid, insoluble in water, but soluble in alcohol and in solutions of the alkalies. Its composition is —

Carbon 75·56 Hydrogen 12·59 Oxygen 11·85 — — — 100·00 and its formula $C_{34}H_{34}O_4$.

Stearic Acid

Although this acid exists in many plants, it is most conveniently extracted from lard. It is a crystalline solid less fusible than margaric acid, but closely resembling it in its other properties. Its formula is $C_{36}H_{36}O_4$.

Oleic Acid

Under this name two different substances appear to be included. It has been applied generally to the fluid acids of all oils, while it would appear that the drying and non-drying oils actually contain substances of different composition. The acid extracted from olive oil appears to have the formula, *while that from linseed oil is* C_{46}, but this is still doubtful.

$$C_{36}H_{34}O_4 \quad H_{38}O_6$$

Other fatty acids have been detected in palm oil, cocoa-nut oil, which so closely resemble margaric and stearic acids as to be easily confounded with them. Though presenting many points of interest, it is unnecessary to describe them in detail here. *Wax* is a substance closely allied to the oils. It consists of two substances, cerine and myricine, which are separated from one another by boiling alcohol, in which the former is more soluble.

They are extremely complex in composition, the former consisting principally of an acid similar to the fatty acids, called cerotic acid, and containing $C_{54}H_{54}O_4$. The latter has the formula $C_{92}H_{92}O_4$. The wax found in the leaves of the lilac and other plants appears to consist of myricine, while that extracted from the sugar-cane is said to be different, and to have the formula $C_{48}H_{50}O_2$ wax. It is probable that other plants contain different sorts of wax, but their investigation is still so incomplete, that nothing definite can be said regarding them.

Wax and fats appear to be produced in the plant from starch and sugar; at least it is unquestionable that the bee is capable of producing the former from sugar, and we shall afterwards see that a similar change is most probably produced in the plant. The fatty matters contained in animals are identical with those of plants.

Nitrogenous or Albuminous Constituents of Plants and Animals

The nitrogenous constituents of plants and animals are so closely allied, both in properties and composition, that they may be most advantageously considered together.

Organic Constituents of Plants

Albumen

Vegetable albumen is found dissolved in the juices of most plants, and is abundant in that of the potato, the turnip, and wheat. In these juices it exists in a soluble state, but when its solution is heated to about 150°, it coagulates into a flocky insoluble substance. It is also thrown down by acids and alcohol.

Coagulated albumen is soluble in alkalies and in nitric acid. Animal albumen exists in the white of eggs, the serum of blood, and the juice of flesh; and from all these sources is scarcely distinguishable in its properties from vegetable albumen.

It is a substance of very complicated composition, and chemists are not agreed as to the formula by which its constitution is to be expressed, a difficulty which occurs also with most of the other nitrogenous compounds.

Closely allied to vegetable albumen is the substance known by the name of glutin, which is obtained by boiling the gluten of wheat with alcohol. It appears to be a sort of coagulated albumen, with which its composition completely agrees.

Vegetable Fibrine

If a quantity of wheat flour be tied up in a piece of cloth, and kneaded for some time under water, the starch it contains is gradually washed out, and there remains a quantity of a glutinous substance called gluten. When this is boiled with alcohol, the glutinabove referred to is extracted, and vegetable fibrine is left. It dissolves in dilute potash, and on the addition of acetic acid is deposited in a pure state. Treated with hydrochloric acid, diluted with ten times its weight of water, it swells up into a jelly-like mass.

When boiled or preserved for a long time under water, it cannot be distinguished from coagulated albumen.

Caseine

Vegetable caseine exists abundantly in most plants, especially in the seeds, and remains in the juice after albumen has been precipitated by heat, from which it may be separated in flocks by the addition of an acid. It has been obtained for chemical examination, principally from peas and beans, and from the almond and oats. When prepared from the pea it has been called legumine, from almonds emulsine, and from oats avenine; but they are all three identical in their properties, although formerly believed to be different, and distinguished by these names.

Vegetable caseine is best obtained by treating peas or beans with hot water, and straining the fluid. On standing, the starch held in suspension is deposited, and the caseine is retained in solution in the alkaline fluid; by the addition of an acid it is precipitated as a thick curd. Caseine is insoluble in water, but dissolves readily in alkalies; its solution is not coagulated by heat, but, on evaporation, becomes covered with a thin pellicle, which is renewed as often as it is removed.

Animal Caseine

Animal Caseine is the principal constituent of milk, and is obtained by the cautious addition of an acid to skimmed milk, by which it is precipitated as a thick white curd. It is also obtained by the use of rennet, and the process of curding milk is simply the coagulation of its caseine. It is soluble in alkalies, and precipitated from its solution by acids, and in all other respects agrees with vegetable caseine.

The composition of animal caseine has been well ascertained, but considerable doubt still exists as to that of vegetable caseine, owing to the difficulty of obtaining it absolutely pure. Other results differ considerably from these, and some observers have even obtained as much as eighteen per cent of nitrogen and fifty-three of carbon. The composition of animal caseine differs from this principally in the amount of carbon.

The most cursory examination of these analytical numbers is sufficient to show that a very close relation subsists between the different substances just described. Indeed, with the exception of vegetable caseine, they may be said all to present the same composition; and, as already mentioned, there are analyses of it which would class it completely with the others. While, however, the quantities of carbon, hydrogen, nitrogen, and oxygen are the same, differences exist in the sulphur and phosphorus they contain, and which, though very small in quantity, are indubitably essential to them. Much importance has been attributed to these constituents by various chemists, and especially by Mulder, who has endeavoured to make out that all the albuminous substances are compounds of a substance to which he has given the name of proteine, with different quantities of sulphur and phosphorus.

The composition of proteine, according to his newest experiments, is — Carbon 54·0 Hydrogen 7·1 Nitrogen 16·0 Oxygen 21·4 Sulphur 1·5 — — 100·0 and is exactly the same from whatever albuminous compound it is obtained. Although the importance of proteine is probably not so great as Mulder supposed, it affords an important illustration of the close similarity of the different substances from which it is obtained, the more especially as there is every reason to believe that the different albuminous compounds are capable of changing into one another, just as starch and sugar are mutually convertible; and the possibility of this change throws much light on many of the phenomena of nutrition in plants and animals.

Indeed, it would seem probable that these compounds are formed from their elements by plants only, and are merely assimilated by animals to produce the nitrogenous constituents they contain. Diastase is the name applied to a substance existing in malt, and obtained by macerating that substance with cold water, and adding a quantity of alcohol to the fluid, when the diastase is immediately precipitated in white flocks. It is produced during the malting process, and is not found in the unmalted barley. Its chemical composition is unknown, but it is nitrogenous, and is believed to be produced by the decomposition of gluten. If a very small quantity of diastase be mixed with starch suspended in hot water, the starch is found gradually to dissolve, and to pass first into the state of dextrine, then into that of sugar.

The change thus effected takes place also in a precisely similar manner in the plant, diastase being produced during the process of germination of all seeds and tubers, for the purpose of effecting this change, and to fulfil other functions less understood, but no doubt equally important. Diastase is found in the seeds only during the period when the starch they contain is passing into sugar; as soon as that change has taken place, its function is ended, and it disappears.

CHEMICAL CHANGES DURING PLANT GROWTH

The simple compounds which the plant absorbs from the atmosphere and soil are elaborated within its system, and converted into the various complex substances of which its tissues are composed, by a series of changes, the details of which are still in some respects imperfectly known, although their general nature is sufficiently well understood. They may be best rendered intelligible by reference, in the first instance, to the changes occurring during germination, when the young plant is nourished by a supply of food stored up in the seed, in sufficient quantity to maintain its existence until the organs by which it is afterwards to draw its nutriment from the air and soil are sufficiently developed to serve that purpose.

Changes Occurring During Germination

When a seed is placed in the soil under favourable circumstances, it becomes the seat of an important and remarkable series of chemical changes, which result in the production of the young plant. Experiment and observation have shown that heat, moisture, and air, are necessary to the production of these changes, and though probably not absolutely essential, the absence of light is favourable in the early stages. The temperature required for germination varies greatly in different seeds, some germinating readily at a few degrees above the freesing point, and others requiring a tolerably high temperature.

The rapidity with which it takes place appears to increase with the temperature; but this is true only within very narrow limits, for beyond a certain point heat is injurious, and when it exceeds 120° or 130° Fahrenheit, entirely prevents the process. The presence of oxygen is also essential, for it has been shown that if seeds are placed in a soil exposed to an atmosphere deprived of that element, or if they be buried so deep that the air does not reach them, they may lie without change for an unlimited period; but so soon as they are exposed to the air, germination immediately commences.

Illustrations of this fact are frequently observed where earth from a considerable depth has been thrown up to the surface, when it often becomes covered with plants not usually seen in the neighbourhood, which have sprung from buried seeds.

When all the necessary conditions for germination are fulfilled, the seed absorbs moisture, swells up, and sends out a shoot which rises to the surface, and a radicle which descends—the one destined to develop the leaves, the other the roots, by which the plant is afterwards to derive its nutriment from the air and the soil. But until these organs are properly developed, the plant is dependent on the matters contained in the seed itself. These substances are mostly insoluble, but are brought into solution by the atmospheric oxygen acting upon the gluten, and converting it into a soluble substance called diastase, which in its turn reacts upon the starch, converting it first into dextrine, and then into cellulose, and the latter is finally deposited in the form of organised cells, and produces the first little shoot of the plant.

At the first moment of germination, the oxygen absorbed appears simply to oxidise the constituents of the seed, but this condition exists only for a very limited period, and is soon followed by the evolution of carbonic acid, water being at the same time formed from the organic constituents of the seed, which gradually diminishes in weight. The amount of this diminution is different with different plants, but always considerable. Boussingault found that the loss of dry substance in the pea amounted in 26 days to 52 per cent, and in wheat to 57 per cent in 51 days. Against this, of course, is to be put the weight of the young plant produced; but this is never sufficient to counterbalance the diminished weight of the seed, for Saussure found that a horse bean and the plant produced from it weighed, after 16 days, less by 29 per cent than the seed before germination.

The same phenomenon is observed in the process of malting, which is in fact the artificial germination of barley, the malt produced always weighing considerably less than the grain from which it was obtained. It was believed by Saussure, and the older investigators, that the carbonic acid evolved was entirely produced from starch and sugar; and as these substances may be viewed as compounds of carbon and water, the change was very simply explained by supposing that the carbon was oxidised and converted into carbonic acid and its water eliminated.

But this hypothesis is incapable of explaining all the phenomena observed; for woody fibre, which is one of the chief constituents of the young plant, contains more carbon than the starch and sugar from which it must have been produced, and we are, therefore, forced to admit that the action must be more complicated.

There is every reason to believe that the nitrogenous constituents of the seed are most abundantly oxidised, for they are remarkably prone to change; but the action of the air is not confined to them, and it appears most probable that all the substances take part in the decomposition, and the process of germination may, in some respects, be compared to decay or putrefaction, which, like it, is attended by the absorption of oxygen and evolution of carbonic acid; but while in the latter case the residual substances remain in a useless state, in the former they at once become part of a new organism.

Changes Occurring During the After-growth of the Plant

When the plant has developed its roots and leaves, and exhausted the store of materials laid up for it in the seed, it begins to derive its subsistence from the surrounding air, and to absorb carbonic acid, water, ammonia, and nitric acid, and to decompose and convert them into the different constituents of its tissues.

Organic Constituents of Plants

These changes take place slowly at first, and more rapidly as the organs fitted for the elaboration of its food are developed. The roots and the leaves are equally active in performing this duty, the former absorbing the mineral matters along with the carbonic acid, ammonia, nitric acid, and moisture in the soil, or the manure added to it; the latter gathering the gaseous substances existing in the air. Each of these undergoes a series of changes claiming our consideration.

Decomposition of Carbonic Acid

Carbonic acid, which appears to be absorbed with equal readiness by the roots, leaves, and stems, undergoes immediate decomposition, its carbon being retained, and its oxygen, in whole or in part, evolved into the air. This decomposition occurs only under the action of the sun's rays, and has been found to be proportionate to the amount of light to which the plant is exposed.

It takes place only in the green parts of plants, for though the roots absorb carbonic acid, they cannot decompose it, or evolve oxygen; and the coloured parts, the flowers, fruits, etc., have an entirely opposite effect, absorbing oxygen and giving off carbonic acid.

The absorption of carbonic acid and escape of oxygen has been proved by numerous direct experiments by Saussure and others, in which both atmospheric air and artificial mixtures containing an increased quantity of carbonic acid have been employed.

Saussure allowed seven plants of periwinkle (Vinca minor) to vegetate in an atmosphere containing 7·5 per cent of carbonic acid for six days, during each of which the apparatus was exposed for six hours to the sun's rays.

In this experiment the whole of the carbonic acid, amounting to 431 volumes, was absorbed, but only 292 volumes of oxygen were given off. Had the carbonic acid been entirely decomposed, and all its oxygen eliminated, its volume would have been equal to that of the acid, or 431, so that in this instance 139 volumes of the oxygen of the carbonic acid have been retained to form part of the tissues of the plant.

On the other hand, the nitrogen is found to be increased after the experiment. It might be supposed that the nitrogen evolved had been derived from the decomposition of the nitrogenous constituents of the plant, but this cannot be the true explanation, because in this particular case it greatly exceeded the whole nitrogen contained in the plants experimented on. Its source is not well understood, but Boussingault supposes it to have existed in the interstices of the plant, and to have escaped during the course of the experiment. Saussure found that the oak, the horse-chesnut, and other plants, absorb oxygen and give off carbonic acid in less volumes than the oxygen, while the house-leek and the cactus absorb oxygen without evolving carbonic acid.

The absorption and decomposition of carbonic acid takes place only during the day, and matters are entirely reversed during the night, when oxygen is absorbed and carbonic acid eliminated from all parts of the plants. Although the action occurring during the night is the reverse of that which takes place during the day, it is in no degree to be attributed to a re-oxidation of the carbon which had been deposited in the tissues of the plant.

It appears, on the contrary, to be a purely mechanical, and not a chemical process. During the night the sap continues to circulate through the vessels of the plant, and moisture, carrying with it carbonic acid in solution, is absorbed by the roots; but when it reaches the leaves, where the sun's light would have caused its decomposition during the day, it is again exhaled unchanged. The oxygen absorbed during the night must, however, take part in some chemical processes, for if it were merely mechanical, the absorption would not be confined to that gas alone, but would be participated in by the other constituents of the air. Moreover, the amount of absorption varies greatly in different plants—being scarcely appreciable in some, and very abundant in others.

Plants containing volatile oils, which are readily converted into resins by the action of oxygen, or those containing tannin or other readily oxidisable substances, take up the largest quantity. This is remarkably illustrated by an experiment in which the leaves of the Agave americana, after twenty-four hours' exposure in the dark, were found to have absorbed only 0·3 of their volume of oxygen, while those of the fir, in which volatile oil is abundant, had taken up twice, and those of the oak, containing tannin, eighteen times as much oxygen.

In the flowers, both by day and night, there is a constant absorption of oxygen, and evolution of carbonic acid. In fact, an active oxidation is going on, attended by the evolution of heat, which, in the Arum maculatum and some other plants, is so great as to raise the temperature of the flower 10° or 12° above that of the surrounding air.

Decomposition of Water in the Plant

In addition to the function which water performs in the plant, as the solvent of the different substances which form its nutriment, and hence as the medium through which they pass into its organs, it serves also as a direct food, undergoing decomposition, and yielding hydrogen to the organic substances. Its constituents, along with those of the carbonic acid absorbed, undergo a variety of transformations, and form the principal part of the non-nitrogenous constituents.

It has been already observed that starch, sugar, and the other allied substances, may be considered as compounds of carbon with water; and they might be supposed to owe their origin to the carbonic acid losing the whole of its oxygen, and direct combination then ensuing between the residual carbon and a certain proportion of water; but this would imply that the latter substance undergoes no decomposition, and though undoubtedly the simplest view of the case, it is by no means the most probable.

It is much more likely that the carbonic acid is only partially decomposed, half its oxygen being separated, and replaced by hydrogen, produced by the decomposition of a certain quantity of water into its elements. It must not be supposed that we are in a condition to assert that sugar is really produced in the manner here shown, the illustration being given merely for the purpose of pointing out how it may be supposed to occur, and on a similar principle it is possible to explain the formation of most other vegetable compounds; and this subject has been very fully discussed by the late Dr. Gregory, in his "Handbook of Organic Chemistry."

That water must be decomposed, is evident from the fact, established by analysis, that the hydrogen of the plant generally exceeds the quantity required to form water with its oxygen, so that this excess at least must be produced by the decomposition of water. The hydrogen of the volatile oils, many of which contain no oxygen, and that of the fats, which contain only a small quantity, must manifestly be obtained in a similar manner.

Decomposition of Ammonia

The nitrogenous or albuminous compounds of vegetables must necessarily obtain their nitrogen from the decomposition either of ammonia or nitric acid, experiment having distinctly shown that they are incapable of absorbing it in the free state from the atmosphere. It has been clearly ascertained that the albuminous substances do not contain ammonia, and it is hence apparent that a complete decomposition of that substance must take place in the plant. No doubt carbonic acid and water take part with it in these changes, which must be of a very complex character, and in the present state of our knowledge it seems hopeless to attempt any explanation of them.

Decomposition of Nitric Acid

Chemists are not entirely at one as to whether nitric acid is directly absorbed by the plant, or is first converted into ammonia. But there are certain facts connected with the chemistry of the soil, to be afterwards referred to, which seem to us to leave no doubt that it may be directly absorbed; and in that case it must be decomposed, its oxygen being eliminated, and the nitrogen taking part with carbon and hydrogen in the formation of the organic compounds.

It must be clearly understood that while such changes as those described manifestly must take place, the explanations of them which have been attempted by various chemists are not to be accepted as determinately established *facts*; they are at present no more than hypothetical views which have been expressed chiefly with the intention of presenting some definite idea to the mind, and are unsupported by absolute proof; they are only inferences drawn from the general bearings of known facts, and not facts themselves.

Although, therefore, they are to be received with caution, they have advantages in so far as they present the matter to us in a somewhat more tangible form than the vague general statements which are all that could otherwise be made.

THE INORGANIC CONSTITUENTS OF PLANTS

When treating of the general constituents of plants, it has been already stated that the older chemists and vegetable physiologists, misled by the small quantity of ash found in them, entertained the opinion that mineral matters were purely fortuitous components of vegetables, and were present merely because they had been dissolved and absorbed along with the humus, which was then supposed to enter the roots in solution, and to form the chief food of the plant.

This supposition, which could only be sustained at a time when analysis was imperfect, has been long since disproved and abandoned, and it has been distinctly shown by repeated experiment that not only are these inorganic substances necessary to the plant, but that every one of them, however small its quantity, must be present if it is to grow luxuriantly and arrive at a healthy maturity.

The experiments of Prince Salm Horstmar, before alluded to, have established beyond a doubt, that while a seed may germinate, and even grow, to a certain extent, in absence of one or more of the constituents of its ash, it remains sickly and stunted, and is incapable of producing either flower or seed.

Of late years the analysis of the ash of different plants has formed the subject of a large number of laborious investigations, by which our knowledge of this subject has been greatly extended. From these it appears that the quantity of ash contained in each plant or part of a plant is tolerably uniform, differing only within comparatively narrow limits, and that there is a special proportion belonging to each individual organ of the plant.

Fruits

Plum 0·40 Cherry 0·43 Strawberry 0·41 Pear 0·41 Apple 0·27 Chesnut 0·99 Cucumber 0·63 Vegetable Marrow 5·10 on examining this table it may be observed that, notwithstanding the very great variety in the proportion of ash in different plants, some general relations may be traced.

A certain similarity may be observed between those belonging to the same natural family, the seeds of all the cereal grains, for instance, containing in round numbers two per cent of inorganic matters. Leguminous seeds (peas and beans) contain about three per cent, while in rape-seed, linseed, and the other oily seeds, it reaches four per cent. In the stems and straws less uniformity exists, but with the exception of a few extreme cases, the quantity of ash in general approaches pretty closely to five per cent.

Still more diversified results are obtained from the entire plants; but this diversity is probably much more apparent than real, and must be, in part at least, dependent on the proportion existing between the stem and leaves, for the leaves are peculiarly rich in ash, and a leafy plant must necessarily yield a higher total percentage of ash, although, if stems and leaves were separately examined, they might not show so conspicuous a difference. The leaves surpass all other parts of plants, in the proportion of inorganic constituents they contain, the table showing that in some instances, as in the maple and Jerusalem artichoke, they exceed one-fourth of the whole weight of the dry matter. In other leaves, and more especially in those of the coniferæ, the proportion is much smaller.

Taking the average of all the analyses hitherto made, it appears that leaves contain about thirteen per cent of ash, but the variations on either side are so large that little value is to be attached to it except as an indication of the general abundance of mineral matters. In roots and tubers the variations are less, and all, except the potato and the turnip, contain about seven per cent of ash.

The smallest proportion of mineral matter is found in wood. In one case only does the proportion reach five per cent, while the average scarcely exceeds one, and in the fir the quantity amounts to no more than one six-hundredth of the dry matter. In the bark the quantity is much larger, and may be stated at seven per cent.

The differences in the quantity of ash contained in different parts of plants are obviously intended to serve a useful purpose, and it is interesting to observe that the wood which is destined to remain for a long period, sometimes for several centuries, a part of the plant, contains the smallest proportion, and it is not improbable that what it does contain is really due, not to the actual woody matter itself, but to the sap which permeates its vessels.

By this arrangement but a small proportion of these important mineral matters, which the soil supplies in very limited quantity, is locked up within the plant, and those which are absorbed, after circulating through it, and fulfilling their allotted functions, are accumulated in the leaves, and annually returned to the soil.

The different proportions of mineral matters contained in the individual organs of plants is most strikingly illustrated when parallel experiments are made on the same species; but the number of instances in which a sufficiently extensive series of analyses has been made to show this, is comparatively limited, and is confined to the oat, the orange-tree, and the horse chesnut— each of which has formed the subject of a very elaborate investigation.

The specimens of oats on which these analyses were made were from different districts of country, grown on soils of different quality, and were, further, of different varieties; and yet they show, on the whole, a remarkable similarity in the proportion of ash in each part, and indicate that there is a normal quantity belonging to it.

Such a series of analyses also affords the most convincing proof that the inorganic matters cannot be fortuitous, and merely absorbed from the soil along with their organic food, as the old chemists supposed, because, in that case, they ought to be uniformly distributed throughout the entire plant, and not accumulated in particular proportions in each individual organ.

Not only does the proportion of ash vary in the different parts of a plant, but even in the same part it is greatly influenced by its period of growth. The laws which regulate these variations are very imperfectly known, but in general it is observed that during the period of active growth the quantity of ash is largest.

Thus, it has been found that in early spring the wood of the young shoots of the horse-chesnut contains 9·9 per cent of ash. In autumn this has diminished to 3·4, and the last year's twigs contain only 1·1 per cent, while in the old wood the quantity does not exceed 0·5. Saussure has also observed that the quantity of ash diminishes in certain plants when the seed has ripened.

The increase is here principally confined to the leaves and chaff, while the stalks, which owe their strength to a considerable extent to the inorganic matters they contain, are equally supplied at all periods of their growth. In the grain only is there a diminution, but this is apparent and not real, and is due to the fact that the determination of the quantity of ash, as made on the grain with its husk, and the former, which contains only a small quantity of mineral matters, increases much more rapidly in weight than the latter, when it approaches the period of ripening, and it is accordingly during the last three weeks of its growth that this diminution becomes apparent.

Although those differences are very large, especially in the straw, and must be attributed to the soil, it has hitherto been found impossible to ascertain the nature of the relation subsisting between it and the crops it yields; indeed, it must obviously be dependent on very complicated questions, which cannot at present be solved, for it may be observed that the increase in the grain does not occur simultaneously with that in the straw, and in several cases a large proportion of ash in the former is associated with an unusually small amount in the latter.

A priori, it might be expected that those soils which are especially rich in the more important constituents of the ash should yield a produce containing more than the average quantity, but this is very far from being an invariable occurrence, and not unfrequently the very reverse is the case.

In the majority of instances we fail to establish any connection between the nature of the soil and the plants it yields, chiefly because we are still very deficient in analyses of those grown on uncultivated soils; and on cultivated land it is impossible to draw conclusions, because the nature of the manure exerts an influence quite as great, if not greater, than that of the soil itself.

It is particularly to be observed that some of the constituents of the ash are not invariably present, and two at least—namely, alumina and manganese—are found so rarely as to justify the inference that they are not indispensable. Of the other substances, iodine is restricted exclusively to sea-plants, but to them it appears to be essential.

Oxide of iron, which occurs only in small quantities, has sometimes been considered fortuitous, but it is almost invariably present, and the experiments of Prince Salm Horstmar leave no doubt that it is essential to the plant.

Its function is unknown, but it is an important constituent of the blood of herbivorous animals, and may be present in the plant, less for its own benefit than for that of the animal of which it is destined to become the food.

Soda appears to be a comparatively unimportant constituent of the ash, of which it generally forms but a small proportion, although the instances of its entire absence are rare. In the cruciferous plants (turnip, rape, etc.) it is found abundantly, and to them it appears indispensable, but in most other plants it admits of replacement by potash. It seems probable that where the soil is rich in the latter substance, plants will select that alkali in preference to soda; but as they must have a certain quantity of alkali, the latter may supply the place of the former where it is deficient.

The soda having almost entirely disappeared in the cultivated plant, while a corresponding increase had taken place in the quantity of potash. Potash is one of the most important elements of the ash of all plants, rarely forming less than 20, and sometimes more than 50 per cent of its weight. The latter proportion occurs chiefly in the roots and tubers, but it is also abundant in all seeds and in the grasses.

The straw, and particularly the chaff of the cereals, and the leaves of most plants, contain it in smaller quantity, although exceptions to this are not unfrequent, one of the most curious being the case of poppy-seed, which contains only about 12 per cent, while the leaves yield upwards of 37 per cent. The proportion of lime varies within very wide limits, being sometimes as low as 1, and in other plants reaching 40 per cent of their ash.

The former proportion occurs in the grains of the cerealia, and the latter in the leaves of some plants, and more especially in the Jerusalem artichoke. The turnip and some of the leguminous plants also contain it abundantly. Magnesia is generally found in small quantity. It is largest in the grains, amounting in them to about 12 or 13 per cent of the ash, but in other plants it varies from 2 to 4 per cent. Although small in quantity, it is an important substance, and apparently cannot be dispensed with; at least there is no instance known of its entire absence.

Chlorine

Chlorine is by no means an invariable constituent of the ash, although it is generally present, and sometimes in considerable quantity. It is most abundant when the proportion of soda is large, and exists in the ash principally in combination with that base as common salt.

The relation between these two elements may be traced more or less distinctly throughout the whole table of analyses, and conspicuously in that of mangold-wurzel, where the common salt amounts to almost exactly one-half of the whole mineral matter.

The analyses of the cultivated and uncultivated asparagus also show that a diminution in the soda is accompanied by a reduction in the proportion of chlorine.

Sulphuric Acid

Sulphuric Acid is an essential constituent of the ash. But it is to be observed that it is in some instances entirely, and in all partially, a product of the combustion to which the plant has been submitted in order to obtain the ash.

It is partly derived from the sulphur contained in the albuminous compounds, which is oxidised and converted into sulphuric acid during the process of burning the organic matter, and remains in the ash. The quantity of sulphuric acid found in the ash is, however, no criterion of that existing in the plant, for a considerable quantity of it escapes during burning.

The extent to which this occurs in particular instances is well illustrated by reference to the case of white mustard, which yields an ash containing only 2·19 of sulphuric acid, equivalent to 0·9 of sulphur; and if calculated on the seed itself, this will amount to no more than 0·039 per cent, while experiments made in another manner prove it to contain about thirty times as much, or more than 1 per cent.

Phosphoric Acid

Phosphoric acid, which may be looked upon as the most important mineral constituent of plants, is found to be present in very variable proportions.

The straws, stems, and leaves contain it in comparatively small quantity, but in the seeds of all plants it is very abundant. In these of the cereals it constitutes nearly half of their whole mineral components, and it rarely falls below 30 per cent.

Carbonic Acid

Carbonic acid occurs in very variable quantities in the ash. It is of comparatively little importance in itself, and is really produced by the oxidation of part of the carbonaceous matters of the plant; but it has a special interest, in so far as it shows that part of the bases contained in the plant must in its natural state have been in union with organic acids, or combined in some way with the organic constituents of the plant.

Silica

Silica is an invariable constituent of the ash, but in most plants occurs but in small quantity. The cereals and grasses form an exception to this rule, for in them it is an abundant and important element. It is not, however, uniformly distributed through them, but is accumulated to a large extent in the stem, to the strength and rigidity of which it greatly contributes.

The hard shining layer which coats the exterior of straw, and which is still more remarkably seen on the surface of the bamboo, consists chiefly of silica; and in the latter plant this element is sometimes so largely accumulated, that concretions resembling opal, and composed entirely of it, are found loose within its joints.

The necessity for a large supply of silica in the stems of other plants does not exist, and in them it rarely exceeds 5 or 6 per cent, but in some leaves it is more abundant. A knowledge of the composition of the ash of plants is of considerable importance in a practical point of view, and enables us in many instances to explain why some plants will not grow upon particular soils on which others flourish.

Thus, for instance, a plant which contains a large quantity of lime, such as the bean or turnip, will not grow in a soil in which that element is deficient, although wheat or barley, which require but little lime, may yield excellent crops.

Again, if the soil be deficient in phosphoric acid, those plants only will grow luxuriantly which require but a small quantity of that element, and hence it follows that on such a soil plants cultivated for the sake of their stems, roots, or leaves, in which the quantity of phosphoric acid is small, may yield a good return; while others, cultivated for the sake of their seed, in which the great proportion of that constituent of the ash is accumulated, may yield a very small crop.

It is obvious also that even where a soil contains a proper quantity of all its ingredients, the repeated cultivation of a plant which removes a large quantity of any individual element, may, in the course of time, so far reduce the amount of that substance as to render the soil incapable of any longer producing that plant, although, if it be replaced by another which requires but little of the element thus removed, it may again produce an abundant crop.

On this principle also, attempts have been made to explain the rotation of crops, which has been supposed to depend on the cultivation in successive years of plants which abstract from the soil preponderating quantities of different mineral matters. But though this has unquestionably a certain influence, we shall afterwards see reason to doubt whether it affords a sufficient explanation of all the observed phenomena.

It may be observed, that some plants are especially rich in alkalies, while in others lime or silica preponderate, and it would therefore be the object of the farmer to employ, in succession, crops containing these elements in different proportions.

The special application of these facts must be reserved till we come to treat of the rotation of crops. It is manifest that, as the crops removed from the soil all contain a greater or less amount of inorganic matters, they must be continually undergoing diminution, and at length be completely exhausted unless their quantity is maintained from some external source. In many cases the supply of these substances is so large that ages may elapse before this becomes apparent, but where the quantity is small, a system of reckless cropping may reduce a soil to a state of absolute sterility.

A remarkable illustration of this fact is found in the virgin soils of America, from which the early settlers reaped almost unheard-of crops, but, by injudicious cultivation, they were soon exhausted and abandoned, new tracts being brought in and cultivated only to be in their turn abandoned. The knowledge of the composition of the ash of plants assists us in ascertaining how this exhaustion may be avoided, and indicates the mode in which such soils may be preserved in a fertile state.

Chapter 5

SOIL TAXONOMY

INTRODUCTION

According to the United States Department of Agriculture's Soil Taxomony, the soils of the world fall into 12 major categories, called soil orders. Soils within one soil order share similar characteristics. With 11 of the 12 soil orders, has more soil orders than any other state in the United States. Within the 1,159 mi² land area alone, there are 7 different soil orders. In comparison, the entire state of Maine, which encompasses 35,387 mi², has only four soil oders. Knowledge of soil behaviour is very important in nutrient management because this knowledge enables you to predict how your management strategies will affect your soil.

Soil taxonomy is the practice of describing, categorising, and naming soils. Like the taxonomy of living organisms, soil taxonomy is designed to make it easier for people to communicate information about different kinds of soils, how they are used, their properties, and where they are found. The United States Department of Agriculture (USDA) has developed a complex soil taxonomy system which is widely used, and the organisation publishes keys which can be used to identify soils, as well as mediating when disputes about the taxonomy of various soils arise.

Under the USDA's soil taxonomy system, soils are organised into orders, suborders, great groups, sub-groups, families, and series, with orders being the broadest category, while series are the smallest. Some examples of the 12 orders in the system include : Gelisols, Oxisols, Vertisols, Aridisols, and Inceptisols. When soils are taxonomised, their composition is a key feature, but scientists also evaluate their location, and factors like the climate where the soil is found. Features like permafrost can be important to soil taxonomy, for example, as can extreme dryness or humidity. Soil composition is based on a number of factors, including minerals in the area, decayed organic material, underlying geology, and so forth, and these influences can be quite diverse, with thousands of soil types recognised under the USDA system.

A number of functions are served by soil taxonomy. The ability to use taxonomic nomenclature is critical to people when they want to communicate with each other about soils, as rather than using a term like "loose, loamy soil," they can select the appropriate series using a key, precisely communicating the details of the soil type in a name. This is useful in the preparation of environmental reports and a wide variety of other documents, allowing someone anywhere in the world to immediately understand the soil conditions in a given area when they are described taxonomically.

Soil taxonomy is also important because it creates a framework which people can use to understand soils. The hierarchical organisation can be used to examine the relationships between different types of soils, for example, and researchers can use this information to explore geology, agricultural techniques, and a wide variety of other topics. Soil scientists usesoil taxonomy extensively in their work, to do everything from describing the soil in someone's back yard and discussing the implications for gardening to exploring the loss of top-soil as a result of heavy winds, desertification, or flooding.

ARID SOILS (ARIDISOL)

Typical Characteristics

- *Distribution* : Arid soils are one of the most prevalent soil orders of the world.
- *Climate* : Arid soils are most characterised by their water deficiencies. Most arid soils contain sufficient amounts of water to support plant growth for no more than 90 consecutive days.
- *Mineralogy* : Arid soils typically contain high levels of calcium carbonates, gypsum, as well as sodium.
- During soil formation, soluble salts at translocated to a subsurface horizon. However, since there is insufficient rainfall to leach salts from the soil profile, soluble salts accumulate.
- Aridisols often have a surface horizon that is pale in colour and low in organic matter.
- *Fertility* : Due to limited moisture and accumulated soluble salts, these soils are generally not suited for major crop production. However, if properly managed and irrigated, these soil may become productive. There are notable soils that are classified as arid soils, but are notable for their unique fertility.

Arid Soils

- Water saturated, salty arid soils (Aquisalid)
- Kealia Series

- Minimal typical arid soil (Haplocambid)
- Keahua Series.

Unique Characteristics of Arid Soils

Of the two types of arid soils, the agriculturally significant soil is the Keahua Series.

- *Location* : The Keahua Series is lies within the central isthmus in the uplands, at elevations ranging from 600 to 1,500 ft. It extends 13,000 acres.
- *Texture* : Silty clay loam.
- *Temperature and Rainfall* : Its average annual temperature is 73 degrees F, and the average annual rainfall is between 15 to 25 inches.
- *Slope* : Gentle to moderate.
- *Parent material* : Keahua Series formed from basic igneous rock with some volcanic ash.
- *Depth* : Surface layer is generally 10 inches. Sub-soil extends 50 inches in depth.
- *Physical Traits* : The soil is well drained and characterised by medium run-off and moderate permeability.
- *Colour of Surface Horizon* : Dark reddish brown.
- *Fertility* : If irrigated, Keahua soils contain plant nutrients and are highly suitable for crop production.
- *Acidity* : Slightly acidic.
- *Natural Vegetation* : Grasses, ilima, kiawe, and lantana.
- *Crops* : Sugarcane, pineapple, pasture.

As demonstrated in the following data table, although the Keahua Series is high in base cations, its organic carbon levels are not high enough to meet the criteria of a Mollisol, or highly fertile grassland soils.

Depth (cm)	Ca	Mg	Na	K	Org. C	Sum Cations	Base Sat	pH water
0-31	15.5	5.4	0.9	1.2	3.98*	52.4	44	6.1
31-46	11.2	5.2	0.9	0.3	2.05	41.4	43	6.6
46-58	4.6	3.5	0.8	0.1	1.01	27.3	33	6.1
58-88	2.6	2.7	1.0	0.1	0.87	25.6	25	5.6
88-160	1.1	2.6	1.5	0.1	0.91	32.0	17	5.8

SOIL TAXONOMY — CLASSIFYING SOILS AND THE SOIL ORDERS

Soil scientists use a hierarchical classification system called Soil Taxonomy to differentiate soils based on a variety of properties exhibited within a soil.

Nomenclature used in this classification system may seem cumbersome to the novice user, but it communicates a significant amount of information about any particular soil being studied. Within Soil Taxonomy.

There are six levels of classification :

- Order (the most general level);
- Sub-order;
- Great group; 4 Subgroup; 5 Family; and 6 Soil series (the most specific level). In all cases, names of soil orders end with the syllable "sol", which is derived from the Latin name solum or soil. The syllable(s) at the beginning of a soil order name convey information to a scientist about a general soil characteristic. Take the soil order Mollisol as an example. Mollisols are soils generally formed under prairie grasses. They contain high organic matter content making them soft in nature and mollis is the Latin word for soft. Thus, when someone mentions that a soil is a Mollisol, a soil scientist has an immediate picture in mind of the vegetation that is or was on this soil during formation (prairie grasses) and what the soil will generally look like and feel like (thick, dark surface horizon that is high in organic matter content and rather soft in nature). The material presented below will guide you through the soil orders found in Missouri and those that are found elsewhere. It will also provide you some general information typically associated with each soil order.

Soil Orders Found in Missouri

Entisols

Entisols are soils that are young or recently formed. They exhibit little soil development as evidenced by the lack of soil horizons found within the soil profile. Entisols are typically found in floodplains where alluvium is deposited during flooding events. They may also be found forming in areas where bedrock has more recently been exposed. Entisols found in floodplains can be highly productive for agriculture and close proximity to large rivers reduces costs associated with food and fibre transport.

Inceptisols

Inceptisols are more weathered than Entisols and exhibit greater development of soil horizons as noted by formation of a B horizon. Thus, they are often considered to be slightly older than Entisols and may form from Entisols with time and weathering. This soil order ranges from low to high in natural fertility. Inceptisols may be planted with crops, used as rangelands, or managed for timber.

Alfisols

Alfisols exhibit a greater degree of soil development than Inceptisols and Entisols as noted by evidence that silicate clay has moved from the surface downward into subsoil horizons through a process called illuviation. Subsoil horizons in Alfisols show some evidence of clay illuviation in the form of clay skins (coatings of clay on the face of soil structural units) or increases in per cent clay content relative to soil horizons located immediately above the soil horizon where clay has accumulated. Alfisols are generally slightly acidic in nature but contain a higher proportion of exchangeable base cations (calcium - Ca, potassium - K, magnesium - Mg, and sodium -Na) than Ultisols. Alfisols are typically developed under deciduous forests, although the current vegetation may not be deciduous trees (*e.g.*, the trees may have been cleared for row crop agriculture or pasture). Due to the high base status of these soils, Alfisols are fertile soils that can support good crop or tree growth although some liming may be required to increase soil pH to more optimal levels.

Ultisols

Ultisols are more highly weathered and older soils than Alfisols. They are also more acidic than Alfisols, more red or orange in colour due to a higher concentration of iron (Fe) and aluminum (Al) oxide minerals, and they have a lower proportion of base cations (<35 per cent of the soil exchange capacity is saturated with Ca, K, Mg, and Na) due to leaching of these elements out of the soil with time. Evidence of clay illuviation into subsoil horizons may be expressed more greatly in Ultisols than Alfisols. Forests are generally considered to be the primary vegetation growing on Ultisols during development. Large areas of Ultisols can be found in the South-eastern United States. Ultisols may be used to support forested ecosystems. These soils may also be used for row-crop agriculture or pasture. However, use of these soils for agronomic purposes requires liming and fertilizer addition due to the acidic nature of these soils and their low nutrient status.

Mollisols

Mollisols are formed under native prairie vegetation and in some poorly drained areas in forests where organic matter may accumulate. These soils are high in organic matter content and base cations, particularly Ca, which is attributed to the dense root structure produced by native grassland plants. Structural units of soil (called soil peds) are soft and easily crushed between the forefinger and thumb even when dry. Due to the high productivity of Mollisols, most of the native grasslands in the United States have been converted to agriculture. The use of Mollisols for agronomic purposes is so great that few native grasslands still exist in the U.S. Tucker Prairie east of Columbia, MO on I-70 is an example of untouched native grassland. Efforts are underway in Missouri and elsewhere to re-establish native grasslands, and the Missouri Department of Conservation's Prairie Fork Conservation Area is a fine example of prairie restoration efforts ongoing in Missouri.

Vertisols

Vertisols are clayey soils with exhibiting high shrink swell potential. Vertisols develop from calcium and magnesium rich parent materials such as limestone or basalt. Soils of this order are found in sub-humid and semi-arid climates where dry periods last several months. During dry periods, clays within the soil shrink resulting in the formation of deep, wide cracks. Water enters the cracks during precipitation events causing the clays to swell and the cracks to seal shut. Due to high shrink-well properties of Vertisols, construction of roads and buildings is extremely difficult and high maintenance may be needed to ensure the safety associated with these structures. Vertisols are rather fertile and the major land uses associated with this soil order are wetlands, crops and rangeland.

Histosols

Histosols are organic soils formed in wetland environments (swamps, marshes, and bogs) from vegetative materials that accumulate but decompose rather slowly. Some refer to locales where Histosols are found as peat lands or peat bogs. It should be noted, however, that not all wetland soils are Histosols, but nearly all Histosols occur in wetland ecosystems. Lack of oxygen (anoxic conditions) in saturated environments retards microbial decomposition of the vegetative materials derived from sedges, grasses, mosses, cattails, trees, etc. Subsequently, organic materials accumulate with time resulting in formation of Histosols that may have several feet of organic material overlying mineral soil or bedrock.

Formation of Histosols can occur at nearly all latitudes on Earth, provided that there is sufficient moisture for plant growth and the development of anoxic conditions. However, Histosols are more predominant in colder climatic regions due to reduction in microbial decomposition of organic materials with decreasing temperature. Additionally glaciation of a region enhances Histosol formation by disrupting or truncating drainage ways formed before an area was glaciated.

This prevents some areas from draining rapidly or moderately and allows water to accumulate in low lying areas. Relevant examples of Histosols formed due to glacial activity are found around the Great Lakes Region of the U.S and in Canada. Glaciation is not a requirement for Histosol formation as evidenced by the presence of this soil order in Florida and Louisiana. In many parts of the U.S. and the world, Histosols have been drained for a variety of reasons including agricultural production, mining of landscaping and potting material, and even for use as a fuel source. From an agricultural perspective, Histosols range from moderate to high in soil fertility but they must be drained and carefully managed to ensure continued productivity. The major land uses associated with these soils are wetlands and crop production.

Soil Orders Found Outside Missouri

Andisols

Andisols are soils formed from volcanic tephra (ash and cinders) or they have properties associated with volcanic ash. As one would expect, the geographic extent of Andisols on Earth is limited to regions having experienced geologically recent volcanic activity. Similar to Entisols and Inceptisols, Andisols are considered to be relatively young soils that may ultimately weather to other soil orders as the five soil forming factors act upon the soil.

After deposition of tephra ejected from the volcano, volcanic ash begins to weather rather rapidly to form poorly crystalline aluminum silicate minerals such as allophane and imogolite. High contents of poorly crystalline iron oxides are an additional mineralogical feature associated with soils weathered from tephra. The presence of significant quantities of the aforementioned minerals in addition to (1) the presence of volcanic glass and (2) high phosphate (PO_4^{3-}) retention by the soil are, cumulatively, referred to as andic soil properties. Laboratory tests used to investigate the degree to which a soil expresses andic properties are used as a factor for classifying a soil as an Andisol.

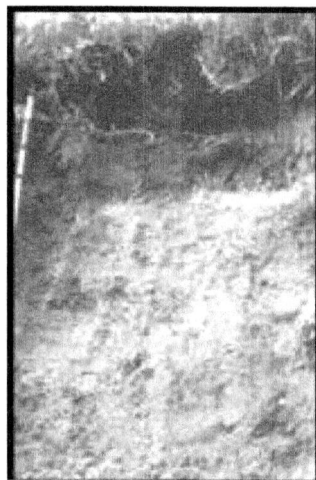

Andisols also exhibit some other interesting soil properties. They are often high in organic matter content due to the formation of stable complexes between organic matter and the poorly crystalline minerals present in Andisols. Andisols have a low load bearing capacity, thus, proper engineering of structures built on these soils is critical. Andisols create several problems in the Pacific Northwest with timber harvest activities. For decades, stringent management practices have been enforced to minimise compaction of the soil.

Andisols usually have rather high natural soil fertility. However, fertility is dependent upon the types and chemical composition of ejected tephra. One of the major issues associated with the fertility of Andisols is the fixation or high retention of phosphate by the poorly crystalline soil minerals present in these soils. Proper soil nutrient management practices are required to ensure adequate plant growth and yield response for crops grown on Andisols. The major land uses for Andisols include forests, crops, and rangelands.

Aridisols

Aridisols are soils found in deserts or arid environments. A common misnomer associated with Aridisols is that they are devoid or almost devoid of vegetation. That is not true; vegetation present on Aridisols is usually scattered and plants are specifically adapted to living in these harsh environments (*e.g.*, cacti, yucca, agave, mesquite trees, and creosote bush). Scarce vegetation does, however, result in low organic matter content in Aridisols. Temperature is not a requirement for Aridisol formation as evidenced by the formation of these soils in both hot and cool temperature regimes. For example, the Sonoran Desert in the Southwestern U.S. is desert that reaches very high air temperatures (>40 °C) and the Gobi Desert of Northern China and Southern Mongolia is a cold desert with temperatures reaching -30 °C.

Due to the lack of precipitation, soluble components in the soil are often not leached out of the soil profile. This results in an accumulation of these components in sub-soil. This can result in high soluble or exchangeable salt content or the formation of minerals such as calcium carbonate or gypsum. It is also possible for carbonates to accumulate to such a degree that the soil becomes cemented. This soil horizon is commonly called caliche or a petrocalcic horizon by soil scientists. Aridisols may also show evidence of clay illuviation in the soil profile.

Aridisols can be very productive for agronomic purposes, provided that they are properly irrigated to prevent buildup of soluble salts in the root zone. Use of Aridisols for rangeland is very common in the South-western U.S., however, animal carrying capacity is very low. Thus, large land areas are required for grasing a relatively low number of animals.

Gelisols

Gelisols are soils containing permafrost (frozen soil) within 1 meter of depth from the soil surface for several consecutive years. Cold and freesing temperatures slow the rate of soil formation, thus these soils are generally considered young. Accumulation of organic matter in Gelisols may be quite substantial due reduced microbial activity caused by low soil temperatures. Gelisols are know for stability management issues and patterned ground formations caused by freesing and thawing. Short growing seasons prevent substantial agricultural operations from occurring on these soils and most remain vegetated with native tundra species.

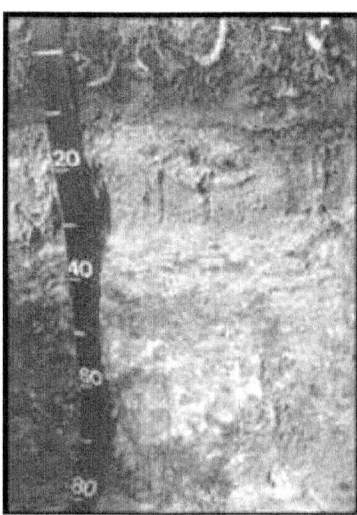

UNIVERSAL SOIL CLASSIFICATION SYSTEM

Most natural sciences struggle for a common classification system such as botany, anthropology and astronomy. Natural classification systems should be accepted and used globally. Soil science and soil classification are viewed as

National Systems yet none have received full international acceptance. Common reasons given for universal systems are pleas from a discipline to work together towards a common understanding and to provide a common language for communication.

Currently when we look at soil classification around the world we have what might be called an 'adobe tower of Babel'. The international politics that hampered global collaboration in soil science in the last Century has slowly mellowed towards a movement for more harmony and to develop an internationally acceptable nomenclature and methodology. This process was not merely to standardise terminology but required evaluation and changes in the whole process including methods of soil analysis and choice of criteria. The time is ripe for acceptance of standard soil terminology, concepts and rationale in elaborating the system and for linkages with current systems.

A group of international soil scientists while attending "Bridging the Centuries Conference in Gödöllo, Hungary, in 2009 agreed that we should submit the following declaration to the International Union of Soil Sciences : The "Bridging the Centuries 1909-2009" events were organised to celebrate the 100th anniversary of the 1st International Conference of Agrogeology and to overview the last 100 years of advances in soil sciences :

The purpose of the 1st conference was to discuss the different approaches to field and laboratory methods, soil descriptions soil classification and soil mapping. An important objective was to gain a common understanding of methods and language, and to develop common soil classification and mapping schemes.

Although much has been achieved in the subsequent 100 years, the participants of the 2009 Centenary Conference concluded that soil science community is still lacking commonly accepted and used field and laboratory standards in soil characterisation and classification, making communication and data exchange difficult within soil science and other disciplines.

Therefore the participants of the "Bridging the Centuries : 1909-2009" conference, declared that there is a need to develop a "Universal Soil Classification" (USC) system for the effective transfer of soil information.

It was recommended that a proposal for the development of such a system be addressed to the International Union of Soil Sciences (IUSS) Council at the 2010 World Soil Congress in Brisbane, Australia. The system would be developed under the scientific auspices of the IUSS in the form of a working group which would effectively be composed of representatives from countries from all continents, and representatives from key international and national agencies.

It was further recommend that the USC should be based on the experiences of existing broadly used national classification systems and to build on the experiences of the World Reference Base for Soil Resources (WRB), the correlation system of IUSS, as well as on accumulated soil information and state-of-the-art observation and data processing tools.

History

Many countries developed a national soil classification system, among others, Argentina, Australia, Brazil, Canada, China, France, Germany, Russia, Scotland, South Africa, and the United States and many of them have a long history.

A first international approach was the FAO Legend for the Soil Map of the World. This map had three levels, which was very apt for the scale of the map (1 : 5,000,000). Some countries have used the same legend for national mapping and encountered many difficulties. The scale of the potential use has to be reflected together with the detail of the categoric level that is selected. For classification purposes, the FAO Legend is meanwhile replaced by the World Reference Base for Soil Resources (WRB) that is maintained by a Working Group of the IUSS. WRB is a flexible system with a flat hierarchy and two categoric levels, the first level using a key, the second level using independent combinations of characteristics. The original purpose was to serve as a tool for correlation of national systems and help international communication. Some countries however adopted it also for mapping purposes. Lacking a common guideline however, the results were not satisfactory. In 2009, mapping guidelines have been established which allow using WRB for constructing small-scale map legends (1 : 250,000 and smaller) if the relevant soil data are available.

The Example of the United States

Standards are very important to have in all systems. There is need for a common and consistent way of describing, collecting and measuring soils. We should build upon existing systems. Moving away from an accepted system presents problems including psychological ones. The viable process that will enable change is to build upon existing systems and not to make dramatic changes that alarms users. For example, in the US we minimised the disruption to the Soil Survey Programme when we replaced the 1938 system of soil classification with Soil Taxonomy by, in so far as was possible, accommodating the soil series that were established at that time as the lowest level of the new classification system.

One of the reasons for the success and acceptance of Soil Taxonomy was that there was ownership and an institutional guardian, the USDA Soil Conservation Service, and users were invited to contribute to developing the system. There are examples of individuals in countries who have tried to propose systems of their own; apart from being academic contributions, these quickly faded away in pedological history. A second important reason becomes evident and that is "for making and interpreting soil surveys". Just making an inventory of the soils with lines is not enough. By strongly linking the inventory with data to a system of making interpretations is the real strength of soil surveys. In the US soil interpretations became the important tool of soil survey and soil classification was only the vehicle. In countries where soil classification was a theoretical academic exercise, the system fell into disuse very quickly. A third reason that Soil Taxonomy is successful is that because it uses the properties of the soil itself for defining taxa, and not theories of the soil's genesis, any competent soil scientist, whether a junior or senior member of the soil survey programme, can classify the soil accurately and consistently. These are important considerations when considering a global system.

Soil Taxonomy was developed to accommodate all soils, not just those known in the United States. There have been eleven International Committees formed over the years to improve Soil Taxonomy. Six committees were devoted to improving the system at the Order level (Andisols Aridisols Gelisols Oxisols, Spodosols, and Vertisols.). Five other international committees were established to improve Soil Taxonomy with regard to specific characteristics including low activity clays, the aquic moisture regime, family-level classes, moisture and temperature regimes, and anthropogenic soils. The intent of these committees has been to continue to strive to achieve the eighth stated attribute of the system which is "to provide for all known soils, wherever they may be."

Towards a Universal Soil Classification System

We should consider adopting the most modern systems that have been inherited with a conceptual diagnostic approach, with established terminology, and existing structural elements. We need to look for a starting point for a USC that is the most documented existing system. It should have the highest amount and most accurate data collected to support the science. We should share existing documents and documentation that represents the starting point for standards that will support our USC System. We also need an accepted approach by the IUSS in developing the new system, one that is fair and based on the best science of today one that is decided by a group of experts and political leaders and one that is accepted internationally. We should however be cognizant of new developments. We see no reason why the numerical concepts that have been developed over the last 30 years cannot be investigated and if found fit for purpose, incorporated into the new system. These include the concepts of pedotaxonomic distance, numerical polythetic allocation and where appropriate, continuous classes. Existing taxa could be used as the starting point for such approaches.

Of crucial importance for the success of any global soil classification system is that there must be a solid and long-term support from an institution or group of institutions assuring the necessary resources for the development, maintenance and implementation of the USC worldwide. One of the main reasons for the failure of previous attempts was the lack of such an operational support. Any future global system will have to have behind its establishment one, or a group of, solid and committed institutions willing to mobilise the needed resources (staff and financial means) to maintain such a system on a long-term.

The USC system must be dynamic and innovative. It must be continuously used and continuously tested and numerical approaches facilitate this. The classification should not be viewed as being just a name. The relevance and implications of the name and the kinds of accessory information incorporated in the name makes the system more powerful. The detail in information required would depend on the kind of use and the scale of observations. During the processes of developing the system, an agreement on the number of categories is necessary. The system may use a Key that enables the selection of taxa, and the classification may consist of categories where the user can navigate with the aid of the Key. Standards for Terminology and Definitions (common data dictionary) are needed. Components of any system and users must adhere to the agreed terms and definitions to use the system accurately and reproducibly. This includes common methods of characterisation of soil analysis and common methods for soil descriptions. Scale of mapping, 1 : 12,000 to 1 : 250,000 must be agreed upon also the way we make observations *e.g.*, field morphology *versus* micromorphology.

Soil Scientists from around the world have expressed the desire and need to develop a common USC System. But, even more important, we need a global soil classification system that will be adopted by the major National soil survey and mapping agencies currently actively engaged in operational soil survey activities. The future USC should not be solely an academic exercise, but should be developed together with the major agencies supporting soil survey and mapping in the world.

Finally, an important consideration is the practicality of the process. A core group will be responsible for binding decisions into the USC System but they must be supported by a number of specialised groups that provide inputs for the different components. An advantage of the proposed effort is that many classification systems exist, each have been tested and enhanced and deficiencies are known by the authors or users. Inputs by these experts can merge the different systems into a universally acceptable system. A focus on use and management from the system should be the ultimate goal. Linkages to existing national soil survey programmes are very important. It is important for us to achieve "Buy In" into a system that is active and established such as the US "National Co-operative Soil Survey" or the Mexican "Instituto Nacional de Estadística y Geografía", which are by far the most developed and extended operational soil survey programmes in the world.

With the experience and enthusiasm that exists today, a USC System is feasible and will have international acceptance. The generation that developed the current systems is leaving the scene through retirement. Today the opportunity arises for the current generation to collaborate and realise the dream of one Universal Soil Classification System.

APPLYING SOIL TAXONOMY

Definition and Purpose

The national system of soil classification identifies sets of soil properties and groups them in taxonomic classes. The system is dynamic and amended as needed. The purpose of soil classification is to order, name, organise, understand, remember, transfer, and use information about soils.

Policy and Responsibilities

- The Natural Resources Conservation Service (NRCS) maintains and provides leadership for amending Soil Taxonomy and for maintaining the soil series

classification files. All soil surveys within the National Co-operative Soil Survey must utilise Soil Taxonomy.
- The MLRA office is responsible for :

    Maintaining accurate and current descriptions of soil series,

    Approving changes to the type location of soil series,

    Soil series classification and the official soil series description files,

    Approving the names of soil series, and

    approving all official series descriptions.

    All users are responsible for reviewing and recommending the disposition of proposals to amend the soil classification system.
- The National Soil Survey Center is responsible for :

    Leadership for maintaining and amending Soil Taxonomy and for part 615 NSSH : Amendments to Soil Taxonomy,

    Maintaining the soil series classification and official soil series description file system, and

    Maintaining standards on the use of soil classification within soil survey.

National Soil Classification System

The national soil classification system has two parts :

- The first part is Soil Taxonomy : A Basic System of Soil Classification for Making and Interpreting Soil Surveys, second edition, Agriculture Handbook No. 436, referred to as Soil Taxonomy 2nd edition, latest revision. This part provides definitions and nomenclature for classifying soils. The National Co-operative Soil Survey (NCSS) adopted this system in January 1965. The amendments to the system are in the NSSH part 615 until placed into the revised edition of Keys to Soil Taxonomy and the Web version of Soil Taxonomy.

- The second part consists of the official soil series descriptions. The Soil Survey Division maintains the official soil series description file and the soil series classification file. These files list the classification of established, tentative, and inactive soil series of the United States, Puerto Rico, the Pacific Basin, and the U.S. Virgin Islands. The official soil series description file is the official reference to soil series descriptions. The soil series classification file is the official source for the classification of the soil series. Both the official soil series description file and the soil series classification file are accessible by computer.

Use of the National Soil Classification System in Soil Surveys

- Soil surveys use Soil Taxonomy to provide :

    A connotative naming system that enables those users familiar with the nomenclature to remember selected properties of soils,

A means for understanding the relationships among soils within a given area and in different areas,

A means of communicating concepts of soils and soil properties,

A means of projecting experience with soils from one area to another, and

Names that can be used as reference terms to identify soil map unit components.

- The names of soil taxa are reference terms for naming soil components of map unit in most soil surveys. Soil taxa are classes at any categorical level in the multi-categorical system of Soil Taxonomy. The name that is used is generally from a taxon of the lowest category that identifies the dominant kind or kinds of soil. Because soil taxa names can have several meanings, the names must be clearly understood. Even though names of one or more taxonomic classes identify map units, the map units are not the same as soil taxa. If the fixed limits of soil taxa are superimposed on the pattern of soils in nature, the limits of taxonomic classes rarely, if ever, coincide precisely with mappable areas. In addition to the named component or components, a map unit contains components of minor extent that are inclusions of other soils that may be similar or dissimilar to the named soil.
- Distinguish a map unit name from a soil taxon name by adding one or more phase terms to the soil taxon reference name. For example, Gamma is a soil taxon; Gamma silt loam, saline, 0 to 2 per cent slopes, is a map unit name.

Soil Taxonomy Committees, Work Groups, and Referees

- *Regional Soil Taxonomy Committees* : Each group of states within the experiment station region has a soil taxonomy committee (or other standards-related committee) as part of the Regional Cooperative Soil Survey Conference. The membership and operational procedures of the committee should be described in the regional conference by-laws. These committees work on standards-related issues that are identified as being important within the region, and also review proposed amendments that are referred to them from time to time by the National Leader for Soil Classification and Standards.
- *National Soil Taxonomy Committee* : The National NCSS Conference has a Standing Committee on Standards that includes some members from the regional committees as well as other members appointed by the Conference Steering Committee. The membership and operational procedures of the committee is described in the national NCSS conference by-laws. This committee works on standards-related issues that are identified by the Conference Steering Committee as being important, considers business items refered to it by the regional committees, and also reviews proposed amendments that are referred to them from time to time by the National Leader for Soil Classification and Standards
- *National ad hoc work groups* : The Director, Soil Survey Division, appoints work groups as needed. They review reports from regional soil taxonomy

committees and recommend additional study or implementation of proposed amendments. Membership includes representatives of State and Federal agencies and may include international representatives. The groups have :

> A chairperson, usually a member of the National Soil Survey Center staff, and
>
> Additional members, depending upon the nature of the recommended changes and the expertise needed.

- *International committees* : The Director, Soil Survey Division, establishes international committees if major national and international users of Soil Taxonomy identify a need for major additions or changes in the soil classification system. The Director appoints a chairperson. Membership is open to any user of Soil Taxonomy who chooses to participate and usually includes representatives of State and Federal agencies as well as international co-operators.
- *Referees* : The Director may request referees to prepare position papers on proposed amendments. The referee requests, as needed, a review by peers and assumes the responsibility for decisions regarding the proposal.

PROCEDURES FOR AMENDING SOIL TAXONOMY

Submitting Proposed Amendments

- Proposals may be made by anyone using Soil Taxonomy from within or outside the United States. Submit proposals that originate in the United States to the National Leader, Soil Classification and Standards or to the appropriate regional soil taxonomy committee chair.
- Submit proposals that originate outside the United States to the appropriate international committee or to the National Leader, Soil Classification and Standards, at the National Soil Survey Center, Federal Building, Room 152, 100 Centennial Mall North, Lincoln, Nebraska, 68508-3866.

Documenting Proposed Amendments

- *Above the family level* : The minimum supporting evidence for all proposed classes must include pedon descriptions, the impact on interpretations, an estimate of geographical extent, and certain laboratory data. The laboratory data must be on at least the critical parts of diagnostic horizons in the proposed new class if the limits between the proposed class and the other recognised classes cannot be adequately identified using field criteria alone.
- *New family criteria* : The minimum supporting evidence includes about 10 pedon descriptions or a description of a proposed soil series and the expected impact on interpretations for the intended use. Submit laboratory data on at least the critical parts of the proposed new class if the limits between the proposed class and the other recognised classes cannot be adequately identified using field criteria alone.

Evaluating Proposed Amendments

- The National Leader, Soil Classification and Standards, at the National Soil Survey Center circulates the proposed amendment to all co-operators for review. Review and comment is welcome from any interested co-operators. Those who are current members of the regional taxonomy committees have a special obligation to review and comment on proposals. Co-operators recommend
- Approval without change,
- Approval with change, or
- Rejection. Recommendations to change or reject the proposal are documented. The National Leader, Soil Classification and Standards, reviews the recommendations and either makes a decision to return the proposal to the originator with reasons for the rejection or includes the proposal in a part 615 NSSH issue. The Deputy Chief for Soil Survey and Resource Assessments signs the cover letter for the distribution of the issue and thus also gives final official approval for the changes.
- The National Leader, Soil Classification and Standards, evaluates all proposals from the international committees and other proposals that originate outside the United States, arranges for a review of these proposals by co-operators or work groups, and makes disposition of the proposals.

Distributing Amendments

- The publication of the amendments constitutes final approval. Amendments to Soil Taxonomy. Updates of Keys to Soil Taxonomy include these amendments. All soil scientists of the NCSS and to other soil scientists, both national and international receive copies of amendments.
- The originator receives proposed amendments that are rejected along with recommendations for disposition to.

CATEGORY OF THE SOIL SERIES

The soil series is the lowest category of the national soil classification system. The name of a soil series or the phase of a soil series is the most common reference term used in soil map unit names. The name of a soil series is also the most common reference term used as a soil map unit component. The purpose of the soil series category is closely allied to the interpretive uses of the system, though map unit components provide the interpretive applications within soil survey for most detailed purposes. Soil series are the most homogeneous classes in the system of taxonomy.

Establishing Norms and Class Limits for Soil Series

In developing or revising soil series concepts, systematic procedures are essential. They reduce the possibility of recognising more soil series than are necessary to organise and present existing knowledge about soil behaviour. The distinctions between one soil series and its competitors must be large enough to be consistently recognised and to be recorded clearly. Clearly differentiate each soil series from all other soil series. Simplify this differentiation by using the systematic procedure described in this section.

Assemble and study all available information on morphology, composition, position on the landscape, and geographic distribution of the soils being considered. Compare the available information with the concepts of existing soil series, and evaluate possible concepts for new soil series. Refine soil characteristics that define higher categories of soil taxonomy to differentiate one soil series from another. These characteristics reflect the kind and sequence of horizons that can be observed, or they associate with characteristics that are observable and that can be consistently measured. Only use those characteristics that are observed or measured within the soil series control section to differentiate soil series. A significant soil characteristic is one that has genetic implication, such as the nature or arrangement of horizons or the absence of horizons, or one that has an influence on use and management, such as per cent of gravel or reaction. Exercise judgement in the selection and weighing of soil characteristics used to set apart soil series.

Competing soil series are those that are in the same family as the soil series under study. Changing the concept of one soil series likely stimulates modification to the concepts of other soil series in the family. When proposing a new series, conceptualise a model of it. Develop a model with a specific norm and range in characteristics for the proposed soil series description. Some of the characteristics of the new series will overlap the characteristics of an existing series; however, the range for differentiating characteristics cannot overlap with that of an existing soil series in the same family. Limits of the range in soil characteristics for the proposed soil series may be as wide as those permitted in the family to which it belongs. Generally keep the range in differentiating soil characteristics of the soil series narrower than that for the span of the family. The permissible ranges should not be defined too narrow for precise and consistent identification. They must be practical. Select a pedon that is typical for the soil series concept. The typical pedon is a reference specimen that illustrates the central concept for the soil series. This pedon, along with other very similar pedons, forms the model for the soil series class. Thus, the selection of a typical pedon is a very important process and is done with great care. Base it on the arrayed data on morphology, composition, and geographic distribution. No pedon is likely to be central for all ranges, but the representative pedon should lie reasonably near the center of the ranges for most physical and chemical properties and for the geographic distribution. If the pedon selected to typify a soil series has one or more properties unusual for the soil series class, record the properties as part of the range of characteristics and note them in the "Remarks" section of the description.

After selecting the typical pedon, define the permissible ranges in soil characteristics. Use the arrayed information on morphology and composition of the soils, especially the profile descriptions, field notes, and laboratory analyses. Only part of the set of observed properties define the classification of any soil, but consider all properties when defining the soil series. Not all observable soil properties are necessarily definitive for a soil series class. The definitive properties that set a soil series apart from similar competing soil series are essential. Emphasise these properties in the statement of the range of characteristics. Also describe the ranges in significant properties that do not differentiate between the soil series being described and its competing soil series. Next, test the soil series concept. Check the norm and ranges in characteristics against the class limits for the family to which the soil series belongs. Do not cross the limits of the family with the ranges specified for the soil series. The distinctions in definitive characteristics between the norms for the proposed soil series and the norms for competing soil series must be clearly greater than what may be normal errors of observation or be based on laboratory data and geomorphic or geographic information. Do not overlap ranges in differentiating characteristics.

Differences in a single characteristic seldom set apart soil series. Use the distinctions in several characteristics to separate soil series. Some may have greater influence than others. Justify a new soil series if the differences in morphology and composition are clearly greater than what are normal errors of observation and affect use and management. It is hard to decide whether or not to propose a new soil series when two or more properties of the soils to be classified are outside but near the limits of an existing soil series. Propose new soil series if the soils differ in characteristics that have practical significance to use and management.

Normal Errors of Observation

As a general guide, a new soil series differs appreciably in either morphology or composition, or both, from already defined soil series. Differences in relevant characteristics must be larger than what may be normal errors of observation or estimates. The following paragraphs give examples of allowed normal errors of observation and tolerance. Soils within these tolerances do not need a new series, nor do they need to be named as taxadjuncts. Identification of soil colour in the field is subject to errors because of (i) changes in the quality of light and in soil moisture, (ii) differences in the visual acuity and skill of individuals, and (iii) limitations in the standards used to determine colour. Field observations of soil colour are at different times of the day and have differing soil moisture contents. These variables could result in differences as large as a full interval between chips in the Munsell colour system. The differences in identification of soil colour resulting from one person looking at the same specimen at different times and under different conditions or from a group of individuals looking at the same specimen together are an example of normal errors of observation. Optimum field conditions allow soil colour to be matched to within one-half interval between chips on the colour chart. The normal range of difference between careful observations is plus or minus a half interval between chips of the same hue or between chips of the same value and chroma on adjacent hues. Colour distinctions, if definitive, between the soils of two soil series, must be greater than this normal range.

Field estimates of textures are commonly within plus or minus one-half class of the actual texture, though errors by highly skilled individuals are smaller. To set apart soil series that are based in part on differences in texture, use distinctions that are greater than the probable error of field estimates or use laboratory data and geomorphic or geographic information. This rule applies to the entire soil series control section and any of its parts. Not all differences among soil series are obvious. The limit between fine-loamy and fine particle-size classes is a clay content of 35 per cent. The experienced mapper has little difficulty in distinguishing between 30 per cent and 40 per cent clay. Only the laboratory can consistently distinguish between 34 per cent and 36 per cent. If this is the only difference, the distinction is not important for most uses of the soil map. Name the delineation for either of the two soil series that have a common conceptual boundary at 35 per cent clay. Differences that are no greater than the normal errors of observation cause many needless decisions, even for an experienced mapper. If the estimate of the properties varies by these normal errors, the similar inclusions that result do not seriously affect the use of the map if the map units are defined to allow for the variation.

Proposing and Naming a Soil Series

Soil scientists in the National Cooperative Soil Survey write and complete descriptions of new soil series and their accompanying estimated properties. The soil series classification (SC) file contains a complete list of active and inactive soil series. The soil series classification file provides the official classification for all soil series in the official soil series description file (OSD). When naming a proposed series, give preference to the names of geographic places as a source of possible names.

Avoid the following kinds of names :
- Names consisting of very long words or of two words;
- Bizarre, discriminatory, comical, and vulgar words;
- Geological terms, such as the names of rocks, minerals, landforms, and the formations of a locality;
- Names of animals and birds;
- Given names of persons, unless the name is a known geographic location;
- Copyrighted names and registered trademarks; and
- Names essentially identical in pronunciation or similar in spelling to a name already in use.

Coin names if sufficient names of geographic places do not exist in a survey area or in the nearby area. Geographic place names must also avoid all restrictions listed above. Coined names must be consistent with American usage and free from the restrictions listed above.

After review of the proposed soil series description within the MLRA office region, the MLRA office approves the name and reserves the name by entering the name and classification into the soil series classification (SC) file. The soil series description is identified as tentative. The MLRA office enters the soil series description into the official soil series description (OSD) file, where it is available for adjoining MLRA offices and co-operators to review and comment. A notification and request for comments is sent to adjoining MLRA offices and to all other MLRA offices that have soil series in the same family as the proposed series.

The MLRA office evaluates any comments and prepares a revision of the soil series description. The revised description is transmitted to the official soil series description file. If the decision is made not to use the series, remove the tentative soil series from the soil series classification (SC) file. This will cause the tentative soil series description in the official soil series description (OSD) file to move to an inaccessible file.

The MLRA office resolves disagreements on concepts of soil series. They assemble and evaluate available evidence on the points in question, and, if necessary, request additional information about the soils under consideration from one or more MLRA regions. If the soil series is in dispute or if the questions about the soil series concept are of considerable importance, a joint field study may be necessary. After the differences have been resolved, the MLRA office updates the soil series description in the official soil series description file.

Revising Official Soil Series Descriptions

Soil scientists must revise soil series descriptions if one or more of the following conditions exists :

- Change in the concept of the soil series, including the range in characteristics;
- Change in the classification of the soil series; and
- Change in the type location of the soil series.

Any soil scientist in the NCSS can write revisions of soil series descriptions. Submit these descriptions to the MLRA office assigned to the type location for the series. Base the revision on pedon descriptions, laboratory data, and other available sources of information about the soils that represent the series.

If the soil series classification or type location is changed, the MLRA office reviews these changes within MLRA office region and with other MLRA office regions in which the soil series or competing series are known or expected to occur. After critical review, reviewing scientists return comments to the originating MLRA office. The MLRA office soil scientist evaluates the comments and makes the necessary changes in the revised description of the soil series. The MLRA office updates the classification of the soil series in the soil series classification file, if necessary, and then transmits the revised description to the official soil series description file.

Inactivating an established soil series : The MLRA office places established soil series on the inactive list when appropriate. Support the decision to inactive a soil series with documentation as to why the soil series should be made inactive and include a recommendation for the disposition of the soils that have been classified as the inactive series. Before placing a soil series on the inactive list, the MLRA office sends a memorandum of intentions and supporting reasons to affected MLRA offices. The MLRA office notifies other disciplines and co-operators who may use the series name in databases and publications. Allow forty-five days for filing objections to the recommendation. If the MLRA office determines that the soil series should be made inactive, they notify the affected regions. The memorandum includes the reclassification to the appropriate soil series or to a taxon of a higher category of all pedons in the inactive series that have been sampled and analysed by the NRCS, co-operating universities, highway departments, or other laboratories. List inactive soil series in the soil series classification file.

Reactivating an inactive soil series name : Do not reuse the name of a soil series that is placed on the inactive list unless the series concept is the same as in the previous description. If an MLRA office wants to reactivate a soil series name, they follow the procedure that is used to propose a soil. Make a notation under "Remarks" that the soil series name is being reactivated.

Dropping a tentative soil series : Drop a tentative soil series from the soil series classification list if it duplicates an already recognised series. If multiple MLRA offices use the soil series, the MLRA office with the series type location requests concurrence from user MLRA offices to drop the series. Upon concurrence, the MLRA office notifies the users that the series is dropped. The notification includes a statement of reasons for dropping the series. Note the name of the dropped series in the correlation document of the soil survey area that has the type location. If only the originating MLRA office is using a soil series listed as tentative, drop the series by listing it as dropped in the correlation document of the survey area that has the type location. Remove the name and record from the soil series classification file, this causes the description in the official soil series description file to move to an inaccessible file. Do not list a tentative soil series as inactive.

Transferring Responsibility for a Soil Series and Changing the Type Location

Approval for transfer of the responsibility for a soil series and change of type location is as follows :

- The MLRA office approves changes within the MLRA office region.
- Mutual consent of the MLRA offices allows transfers between MLRA office regions.
- All transfers of a soil series responsibility and change of type location require a series description using the new type location. The MLRA office enters the new description into the database.
- Establishing a soil series. A soil series is established when it is used in the correlation of a survey area and the correlation document is approved and signed by the MLRA office. The correlation document contains a list

of the soil series that are established by that correlation. If a soil series is established by a correlation, the responsible MLRA office changes the status of the series in the official soil series description file and soil series classification file from "tentative" to "established" and also changes the heading "SERIES PROPOSED" to "SERIES ESTABLISHED". The MLRA office also enters in the official soil series file the year that the soil series is established and the name of the survey area in which it is established. The MLRA office transmits the updated description to the official soil series description file and the soil series classification file. If a tentative soil series is not used and established in correlation document the survey area in which it was proposed and no other potential use is pending, remove the soil series from the soil series classification file.

Official Soil Series Descriptions

"Official soil series description" is a term applied to the description approved by the MLRA offices, which defines a specific series in the United States. The description follows a prescribed format. Revise an official soil series description if more information about the soils in the series is available or if the classification of the series changes because of revisions to the national system of soil classification. All soil scientists working in the NCSS must be familiar with the requirements for adequate soil series descriptions.

The official soil series descriptions are descriptions of the taxa in the series category of the national system of soil classification. They mainly serve as specifications for identifying and classifying soils. Field soil scientists must have access to all the existing official soil series descriptions that are applicable to their soil survey areas and other official soil series descriptions that include soils in adjacent or similar survey areas commonly. Scientists in other disciplines, such as agronomists, horticulturists, engineers, planners, and extension specialists, also use the descriptions to learn about the properties of soils in a particular area.

The format for descriptions and the order in which the major items appear are as follows :

- Location line,
- Status of soil series (tentative or established),
- Initials of authors,
- Name of soil series,
- Introductory paragraph,
- Taxonomic class,
- Typical pedon,
- Type location,
- Range in characteristics,
- Competing series,

- Geographic setting,
- Geographically associated soils,
- Drainage and saturated hydraulic conductivity,
- Use and vegetation,
- Distribution and extent,
- MLRA office responsible,
- Series proposed or series established,
- Remarks on diagnostic horizons and features recognised in the pedon, and
- Additional data.

Every official soil series description includes all but the "additional data" item, which is used only as needed. Each description must be complete and as brief as possible without omitting any essential information. It must clearly differentiate between the series being described and all other series. It states the present concept of a soil series rather than past concepts or its evolution.

The description must record the soil properties that :
- Define the soil series,
- Distinguish it from other soil series,
- Serve as the basis for the placement of that soil series in the soil family, and
- Provide a record of the soil properties needed to prepare soil interpretations.

In the competing series paragraph, give differentiae used to separate other soils in the same family in terms of soil properties, diagnostic horizons, or features. Use the standard terminology that is defined in the Soil Survey Manual as appropriate. If applicable, use terms defined in Soil Taxonomy. The rule for the use of standard terms applies to all parts of soil series descriptions but is especially important for descriptions of individual horizons. Some soil descriptions need to use some terms that are not defined in the Soil Survey Manual or Soil Taxonomy. Use such terms in their ordinary, standard, dictionary sense.

Chapter 6
SOIL CLASSIFICATION

INTRODUCTION

Soils can be grouped into categories based on their present properties. The most general soil category is called order. All world soils are place into 10 orders.

- *Entisols* : Those soils that have natal, if any, profile development are known as entisols. Soils in desert belong to this classification. The productivity of these soils varies with their location and properties. With controlled water supply and proper fertilization, these soils have good productivity and good for vegetables, groundnut, citrus, wheat, paddy, etc.

- *Inceptisols* : These soils have better profile development than entisols but are less developed. The horizons are formed mostly from alteration of the parent materials with accumulation of clay. The productivity is limited due to poor drainage. Found in humid regions.

- *Histosols* : These are organic soils (pleats and mucks) consisting of variable depths of accumulated plant remains in bogs, marshes and swamps that have developed under water saturated environment. Highly rich in organic matter *i.e.* Org. C ranges from 12 to 18 per cent in soils with low to more than 50 per cent clay content.

- *Aridisols* : Soils found in arid or dry areas with light in colour, poor inorganic matter and are not subjected to leaching, used for cultivation with irrigation. Process a horizon of $CaCO_3$ (lime), Calcium sulphate (Gypsum) or more soluble salts. These are desert soils.

- *Mellisols* : Mostly these are grasslands having thick surface horizon of dark colour, dominated by divalent cations. Process normal granular or crub structure, do not harden on drying and with moderate to have fertilization soil are productive.

- *Vertisols* : These have a high content of clays that swell when wetted (more than 30 per cent). During the dry season, these soils on tract and give rise to deep cracks which disappear in the wet season or after irrigation. Found in sub-humid or semi-arid (Temperate to tropical) climates where temp. are moderate to high. Good for crop production with fine texture which are plastic and sticky when wet and hard when dry. Difficult to manage due to very little time for their proper preparation by tilling good for the production of cotton, millet, sorghum, wheat, paddy, etc.

- *Alfisols* : Develop in humid and sub humid climates (500 mm to 1300 mm rainfall) with gray to brown surface horizons. Soils are slightly too moderately acid and quite productive with good texture. Soils are frequently under forest vegetation.

- *Spodosols* : Soils belong to forests with low content of bases, having coarse texture (sandy). Found in humid climates where temperatures are low. The subsurface horizons have accumulation of org. matter and sesquioxide.

- *Ultisols* : These are strongly acid, normally forest soils with low content of bases extensively weathered soils of tropical and sub-tropical climates, respond to good mgt. practices, have clay of 1 : 1 type and give good crop production with adequate fertilization.

- *Oxisols* : These are most developing in tropical and subtropical climates. The subsurface horizons are high in clay and acid. The soils are productive with supplements of 'P' micro-nutrients.

Soils, like other naturally occurring things, come in great variety and exhibit great ranges of properties. Using measurable and observable properties, such as the kind and arrangement of soil horizons, soils can be characterised and named. The soil series is the lowest category within soil taxonomy (classification system). All soils within a single series have uniform differentiating characteristics and arrangement of horizons. This does not mean that all soils within a series are identical; it does mean that they have the same horizonation, but the horizons may be of different thickness, colour, structure, etc. within prescribed limits. Some 15,000 soil series have been described and named in the United States. Most series names are taken from the name of a town, city, county, river or other constructed or natural feature near the location where the soil is first described and named.

All of the soils within a series will have developed in the same kind of parent material with comparable drainage characteristics and will be of similar age. The effects of climate and biological activity will have been very similar. Consequently, the soils within a series exhibit like properties and respond in like fashion to usage or manipulation. Higher levels of classification are the family, sub-group, great group, sub-order and order. All these categories are given generic names which convey as much information as possible about the soil series to be classified within the group. Eleven soil orders form the highest level of classification. Soils classified within each order show only small differences in the kinds and relative strengths of processes that tend to develop soil horizons.

Considerable effort by soil scientists has been expended, and continues to be expended, in surveying soil resources. These surveys, when done in a detailed manner, require a soil scientist to traverse the landscape at frequent intervals, stopping periodically to auger or dig into the soil. The soil surveyor plots the occurrence of soils on a map which is subsequently formalised and eventually published in a soil survey report which includes not only the soil maps for a given area, usually a county, but all types of information about the county as well as descriptions of the properties of the soils in the area, their present and potential uses, and potential problems associated with the utilisation of the soils for both agricultural and engineering purposes. Such reports are very expensive in terms of labour and money, but they contain information that is of great value to those who utilise soils for any purpose. Unfortunately, many who could use the information in soil survey reports to great advantage do not know that the reports are available. With increased emphasis on planning, which includes land use, on county, state and regional bases, more utilisation of soil survey information is being made and even more detailed and more modern soil surveys are being urgently requested in many areas. Much of this information is being digitised for electronic transmission.

SOIL DESCRIPTION AND CLASSIFICATION

It is necessary to adopt a formal system of soil description and classification in order to describe the various materials found in ground investigation. Such a system must be comprehensive (covering all but the rarest of deposits), meaningful in an engineering context (so that engineers will be able to understand and interpret) and yet relatively concise. It is important to distinguish between description and classification : Description of soil is a statement describing the physical nature and state of the soil. It can be a description of a sample, or a soil *in situ*. It is arrived at using visual examination, simple tests, observation of site conditions, geological history, etc. Soil classification is the separation of soil into classes or groups each having similar characteristics and potentially similar behaviour. A classification for engineering purposes should be based mainly on mechanical properties, *e.g.* permeability, stiffness, strength. The class to which a soil belongs can be used in its description.

Basic Characteristics of Soils

Soils consist of grains (mineral grains, rock fragments, etc.) with water and air in the voids between grains. The water and air contents are readily changed by changes in conditions and location : soils can be perfectly dry (have no water content) or be fully saturated (have no air content) or be partly saturated (with both air and water present). Although the size and shape of the solid (granular) content rarely changes at a given point, they can vary considerably from point to point.

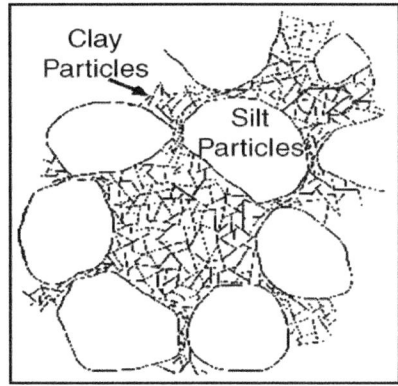

First of all, consider soil as a engineering material — it is not a coherent solid material like steel and concrete, but is a particulate material. It is important to understand the significance of particle size, shape and composition, and of a soil's internal structure or fabric.

Soil as an Engineering Material

The term "soil" means different things to different people : To a geologist it represents the products of past surface processes. To a pedologist it represents currently occurring physical and chemical processes.

To an engineer it is a material that can be :

Built on : foundations to buildings, bridges.

Built in : tunnels, culverts, basements.

Built with : roads, runways, embankments, dams.

Supported : retaining walls, quays.

Soils may be described in different ways by different people for their different purposes. Engineers' descriptions give engineering terms that will convey some sense of a soil's current state and probable susceptibility to future changes (*e.g.* in loading, drainage, structure, surface level). Engineers are primarily interested in a soil's mechanical properties : strength, stiffness, permeability. These depend primarily on the nature of the soil grains, the current stress, the water content and unit weight.

Size Range of Grains

Clay	Silt		Sand		Gravel		Cobble	Boulder
	0.006	0.02	6	20	6	20		
0.002mm		0.06mm		2mm			60mm	200mm

The range of particle sizes encountered in soil is very large : from boulders with a controlling dimension of over 200 mm down to clay particles less than 0.002 mm (2 µm). Some clays contain particles less than 1 µµ in size which behave as colloids, *i.e.* do not settle in water due solely to gravity. In the British Soil Classification System, soils are classified into named Basic Soil Type groups according to size, and the groups further divided into coarse, medium and fine sub-groups :

Very coarse soils	Boulders		> 200 mm
	Cobbles		60 - 200 mm
Coarse soils	G Gravel	Coarse	20 - 60 mm
		Medium	6 - 20 mm
		Fine	2 - 6 mm
	S Sand	Coarse	0.6 - 2.0 mm
		Medium	0.2 - 0.6 mm
		Fine	0.06 - 0.2 mm
Fine soils	M Silt	Coarse	0.02 - 0.06 mm
		Medium	0.006 - 0.02 mm
		Fine	0.002 - 0.006 mm
	C Clay		< 0.002 mm

Aids to Size Identification

Soils possess a number of physical characteristics which can be used as aids to size identification in the field.

A handful of soil rubbed through the fingers can yield the following :
- Sand (and coarser) particles are visible to the naked eye.
- Silt particles become dusty when dry and are easily brushed off hands and boots.
- Clay particles are greasy and sticky when wet and hard when dry, and have to be scraped or washed off hands and boots.

Shape of Grains

The majority of soils may be regarded as either Sands or Clays :
- Sands include gravelly sands and gravel-sands. Sand grains are generally broken rock particles that have been formed by physical weathering, or they are the resistant components of rocks broken down by chemical weathering. Sand grains generally have a rotund shape.
- Clays include silty clays and clay-silts; there are few pure silts (*e.g.* areas formed by windblown Löess). Clay grains are usually the product of chemical

weathering or rocks and soils. Clay particles have a flaky shape.

There are major differences in engineering behaviour between SANDS and CLAYS (*e.g.* in permeability, compressibility, shrinking/swelling potential). The shape and size of the soil grains has an important bearing on these differences.

Shape Characteristics of Sand Grains

Sand and larger-sised grains are rotund. Coarse soil grains (silt-sised, sand-sised and larger) have different shape characteristics and surface roughness depending on the amount of wear during transportation (by water, wind or ice), or after crushing in manufactured aggregates. They have a relatively low specific surface (surface area).

Shape Characteristics of Clay Grains

Clay particles are flaky. Their thickness is very small relative to their length and breadth, in some cases as thin as 1/100th of the length. They therefore have high to very high specific surface values. These surfaces carry a small negative electrical charge, that will attract the positive end of water molecules. This charge depends on the soil mineral and may be affected by an electrolite in the pore water. This causes some additional forces between the soil grains which are proportional to the specific surface. Thus a lot of water may be held asadsorbed water within a clay mass.

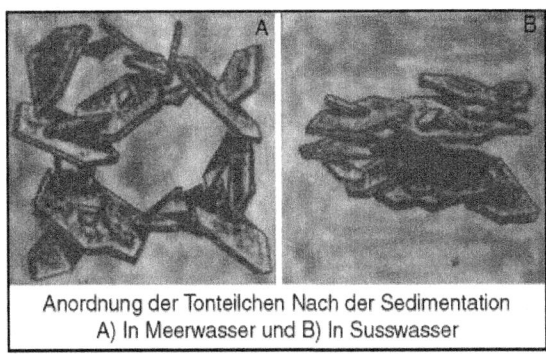

Anordnung der Tonteilchen Nach der Sedimentation
A) In Meerwasser und B) In Susswasser

Structure or Fabric

Natural soils are rarely the same from one point in the ground to another. The content and nature of grains varies, but more importantly, so does the arrangement of these. The arrangement and organisation of particles and other features within a soil mass is termed its structure or fabric. This includes bedding orientation, stratification, layer thickness, the occurrence of joints and fissures, the occurrence of voids, artefacts, tree roots and nodules, the presence of cementing or bonding agents between grains.

Structural features can have a major influence on in situ properties :

- Vertical and horizontal permeabilities will be different in alternating layers of fine and coarse soils.
- The presence of fissures affects some aspects of strength.
- The presence of layers or lenses of different stiffness can affect stability.
- The presence of cementing or bonding influences strength and stiffness.

Origins, Formation and Mineralogy

Soils are the results of geological events. The nature and structure of a given soil depends on the geological processes that formed it :

- Breakdown of parent rock : weathering, decomposition, erosion.
- Transportation to site of final deposition : gravity, flowing water, ice, wind.
- Environment of final deposition : flood plain, river terrace, glacial moraine, lacustrine or marine.
- Subsequent conditions of loading and drainage : little or no surcharge, heavy surcharge due to ice or overlying deposits, change from saline to freshwater, leaching, contamination.

Origins of Soils from Rocks

All soils originate, directly or indirectly, from solid rocks in the Earth's crust :

- *Igneous Rocks* : Crystalline bodies of cooled magma *e.g.* granite, basalt, dolerite, gabbro, syenite, porphyry
- *Sedimentary Rocks* : Layers of consolidated and cemented sediments, mostly formed in bodies of water (seas, lakes, etc.) *e.g.* limestone, sandstones, mudstone, shale, conglomerate
- *Metamorphic Rocks* : Formed by the alteration of existing rocks due to heat from igneous intrusions (*e.g.* marble, quartzite, hornfels) or pressure due to crustal movement (*e.g.* slate, schist, gneiss).

Weathering of Rocks

Physical Weathering

Physical or mechanical processes taking place on the Earth's surface, including the actions of water, frost, temperature changes, wind and ice; cause disintegration and wearing. The products are mainly coarse soils (silts, sands and gravels). Physical weathering produces Very Coarse soils and Gravels consisting of broken rock particles, but Sands and Silts will be mainly consists of mineral grains.

Chemical Weathering

Chemical weathering occurs in wet and warm conditions and consists of degradation by decomposition and/or alteration. The results of chemical weathering are generally fine soils with separate mineral grains, such as Clays and Clay-Silts. The type of clay mineral depends on the parent rock and on local drainage. Some minerals, such as quartz, are resistant to the chemical weathering and remain unchanged.

Quartz

A resistant and enduring mineral found in many rocks (*e.g.* granite, sandstone). It is the principal constituent of sands and silts, and the most abundant soil mineral. It occurs as equidimensional hard grains.

Haematite

A red iron (ferric) oxide : resistant to change, results from extreme weathering. It is responsible for the widespread red or pink colouration in rocks and soils. It can form a cement in rocks, or a duricrust in soils in arid climates.

Micas

Flaky minerals present in many igneous rocks. Some are resistant, *e.g.* muscovite; some are broken down, *e.g.* biotite.

Clay Minerals

These result mainly from the breakdown of feldspar minerals. They are very flaky and therefore have very large surface areas. They are major constituents of clay soils, although clay soil also contains silt sised particles.

Clay Minerals

Clay minerals are produced mainly from the chemical weathering and decomposition of feldspars, such as orthoclase and plagioclase, and some micas. They are small in size and very flaky in shape. The key to some of the properties of clay soils, *e.g.* plasticity, compressibility, swelling/shrinkage potential, lies in the structure of clay minerals.

There are three main groups of clay minerals :

- *Kaolinites* : (Include kaolinite, dickite and nacrite) formed by the decomposition of orthoclase feldspar (*e.g.* in granite); kaolin is the principal constituent in china clay and ball clay.
- *Illites* : (Include illite and glauconite) are the commonest clay minerals; formed by the decomposition of some micas and feldspars; predominant in marine clays and shales (*e.g.* London clay, Oxford clay).
- *Montmorillonites* : (Also called smectites or fullers' earth minerals) (include calcium and sodium momtmorillonites, bentonite and vermiculite) formed by the alteration of basic igneous rocks containing silicates rich in Ca and Mg; weak linkage by cations (*e.g.* Na^+, Ca^{++}) results in high swelling/shrinking potential.

Transportation and Deposition

The effects of weathering and transportation largely determine the basic nature of the soil (*i.e.* the size, shape, composition and distribution of the grains).

The environment into which deposition takes place, and subsequent geological events that take place there, largely determine the state of the soil, (*i.e.* density, moisture content) and the structure or fabric of the soil (*i.e.* bedding, stratification, occurrence of joints or fissures, tree roots, voids, etc.)

Transportation

Due to combinations of gravity, flowing water or air, and moving ice. In water or air : grains become sub-rounded or rounded, grain sizes are sorted, producing poorly-graded deposits. In moving ice : grinding and crushing occur, size distribution becomes wider, deposits are well-graded, ranging from rock flour to boulders.

Deposition

In flowing water, larger particles are deposited as velocity drops, *e.g.* gravels in river terraces, sands in floodplains and estuaries, silts and clays in lakes and seas. In still water : horizontal layers of successive sediments are formed, which may change with time, even seasonally or daily.

- Deltaic and shelf deposits,often vary both horizontally and vertically.
- From glaciers, deposition varies from well-graded basal tills and boulder clays to poorly-graded deposits in moraines and outwash fans.
- In arid conditions : scree material is usually poorly-graded and lies on slopes.
- Wind-blown Löess is generally uniformly-graded and false-bedded.

Loading and Drainage History

The current state (*i.e.* density and consistency) of a soil will have been profoundly influenced by the history of loading and unloading since it was deposited. Changes in drainage conditions may also have occurred which may have brought about changes in water content.

LOADING /UNLOADING HISTORY

Initial Loading

During deposition the load applied to a layer of soil increases as more layers are deposited over it; thus, it is compressed and water is squeezed out; as deposition continues, the soil becomes stiffer and stronger.

Unloading

The principal natural mechanism of unloading is erosion of overlying layers. Unloading can also occur as overlying ice-sheets and glaciers retreat, or due to large excavations made by man. Soil expands when it is unloaded, but not as much as it was initially compressed; thus it stays compressed and is said to be over consolidated. The degree of over consolidation depends on the history of loading and unloading.

DRAINAGE HISTORY

Chemical Changes

Some soils initially deposited loosely in saline water and then inundated with fresh water develop weak collapsing structure. In arid climates with intermittent rainy periods, cycles of wetting and drying can bring minerals to the surface to form a cemented soil.

Climate Changes

Some clays (*e.g.* montmorillonite clays) are prone to large volume changes due to wetting and drying; thus, seasonal changes in surface level occur, often causing foundation damage, especially after exceptionally dry summers.

Trees extract water from soil in the process of evapotranspiration; The soil near to trees can therefore either shrink as trees grow larger, or expand following the removal of large trees.

SOIL CLASSIFICATION AND CHARACTERISATION

Development of sustainable agricultural systems such as alley farming is an attempt to reduce degradation of natural resources and to find environmentally compatible ways of increasing production and promoting broad-scale development. Intensification of agriculture on land currently used for traditional farming requires a thorough knowledge of the soil as a resource and attributes of the land.

Information on distribution, potential and constraints of major soils is needed, so that the most appropriate soil management systems can be designed. In addition knowledge on land capability and suitability is also essential to determine the best land use for sustained crop production.

This chapter reviews current systems used to classify soils and land capabilities. It also provides an introduction to the management requirements of the major soils in the humid and sub-humid zones of tropical Africa.

Soils and their Classification

Soil is the thin layer covering the entire earth's surface, except for open water surfaces and rock outcrops. The properties of soil are determined by environmental factors.

Five dominant factors are often considered in the development of the various soils :

1. The climate,

2. Parent materials (rocks and physical and chemical derivatives of same),

3. Relief,
4. Organisms (fauna and flora), and
5. The time factor.

There are a large number of different soils, reflecting different kinds and degrees of soil forming factors and their combinations.

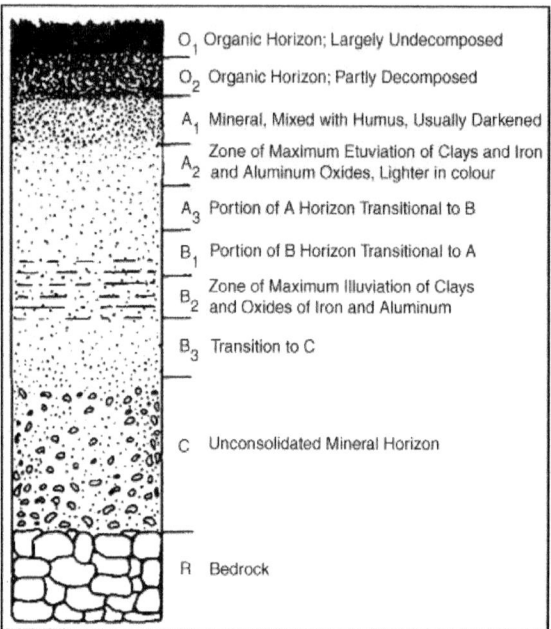

Fig. A hypothetical soil profile.

Scientists have developed different systems of soil classification to group soils of similar properties in one class, allowing them to exchange information on soils found in different areas. Soil classification also helps in determining the best possible use and management of soils. Soil classification is however a controversial subject at both national and international levels. There is lack of agreement for a common classification system, because soil scientists do not agree on the characteristics for differentiating and classifying soils.

Although many soil classification systems exist; however, two system are widely used : The USDA Soil Taxonomy and the FAO/UNESCO legend. The French system (ORSTROM) is also commonly used in France and in Francophone Africa. The classification of soils starts with examination of soil profiles. Morphologically, soils are composed of a series of horizons.

Soil horizons are layers of different appearance, thickness, and properties which have arisen by the action of various soil-forming processes. The horizons are normally parallel to the surface. Collectively, the horizons make up what is called the soil profile or soil "pedon".

A soil profile is defined as a vertical section of the soil to expose layering. Figure sketches a hypothetical soil profile having all the principal horizons, with a brief description of the characteristics of each horizon. Individual soils have one or more of these horizons. Very young soils may not yet have started the soil horizonisation process. In soil classification, the item to be classified is the soil profile.

The classification or study of the entire profile consists of recognising and naming the horizons which make up the profile. In the study of soil profiles, sub-soil horizons are given greater emphasis than surface horizons which are frequently changed by human activity to such an extent that they bear hardly any relationship with genetic process.

The USDA Soil Taxonomy

The Soil Taxonomy developed since the early 1950's is the most comprehensive soil classification system in the world, developed with international co-operation it is sometimes described as the best system so far. However, for use with the soils of the tropics, the system would need continuous improvement.

Hierarchy of Categories in the Soil Taxonomy

There are six levels in the hierarchy of categories : Orders (the highest category), sub-orders, great groups, sub-groups, families and series (the lowest category).

Orders

There are ten orders, differentiated on gross morphological features by the presence or absence of diagnostic horizons or features which show the dominant set of soil-forming processes that have taken place. The ten orders and their major characteristics. The occurrence of the major soils in the humid and sub-humid tropics.

Sub-orders

It is the next level of generalisation. It permits more statements to be made about a given soil. In addition to morphological characteristics other soil properties are used to classify the soil. The sub-order focusses on genetic homogeneity like wetness or other climatic factors.

There are 47 sub-orders within the 10 orders. The names of the sub-orders consist of two syllables. The first connotes the diagnostics properties; the second is the formative element from the soil order name. For example, an Ustalf is an alfisol with an ustic moisture regime (associated with sub-humid climates).

Great Groups

The great group permits more specific statements about a given soil as it notes the arrangement of the soil horizons. A total of 230 great groups (140 of which occur in the tropics) have been defined for the 47 sub-orders.

The name of a great group consists of the name of the sub-order and a prefix suggesting diagnostic properties. For example, a Plinthustalf is an ustalf that has developed plinthite in the profile. Plinthite development is selected as the important property and so forms the prefix for the great group name.

Table. Brief Descriptions of the Ten Soil Orders According to Soil Taxonomy.

Soil Orders	Description
Alfisols	- Soils with a clayey B horizon and exchangeable cation (Ca + Mg + K + Na) saturation greater than 50% calculated from NH_4OAc-CEC at pH7.
Ultisols	- Soils with a clayey B horizon and base saturation less than 50%. They are acidic, leached soils from humid areas of the tropics and subtropics.
Oxisols	- Oxisols are strongly weathered soils but have very little variation in texture with depth. Some strongly weathered, red, deep, porous oxisols contain large amounts of clay-sized Fe and Al oxides.
Vertisols	- Dark clay soils containing large amounts of swelling clay minerals (smectite). The soils crack widely during the dry season and become very sticky in the wet season.
Mollisols	- Prairie soils formed from colluvial materials with dark surface horizon and base saturation greater than 50%, dominating in exchangeable Ca.
Inceptisols	- Young soils with limited profile development. They are mostly formed from colluvial and alluvial materials. Soils derived from volcanic ash are considered a special group of Inceptisols, presently classified under the Andept suborder (also known as Andosols).
Entisols	- Soils with little or no horizon development in the profile. They are mostly derived from alluvial materials.
Aridisols	- Soils of arid region, such as desert soils. Some are saline.
Spodosols	- Soils with a bleached surface layer (A2 horizon) and an alluvial accumulation of sesquioxides and organic matter in the B horizon. These soils are mostly formed under humid conditions and coniferous forest in the temperate region.
Histosols	- Soils rich in organic matter such as peat and muck.

Table. Occurrence of Major Soils in the Huld and Sub-humid Tropics.

Classification (USDA)	Occurrence
1. Alfisols	Savanna and drier forest zones
2. Hydromorphic Soils	Valley bottom of a rolling topography
3. Vertisols	Alluvial plains in savanna
4. Ultisols	Rain forest zone and derived savanna
5. Oxisols	Rain forest and savanna
6. Inceptisols	All regions
7. Andepts (suborder of Inceptisols)	Limited and localized distribution relating to present and past volcanic activities

Subgroups

There are three kinds of sub-groups :

1. The typical subgroup which represents the central concept of the great group, for example Typic Paleustalfs.
2. Intergrades are transitional forms to other orders, sub-orders or great groups, for example Aridic Paleustalfs or Oxic Paleustalfs.
3. Extragrades have some properties which are not representative of the great group but do not indicate transitions, for example, Petrocalcic Paleustalf.

Families of Soil

The grouping of soils within families is based on the presence or absence of physical and chemical properties important for plant growth and may not be indicative of any particular process. The properties include particle size distribution and mineralogy beneath the plough layer, temperature regime, and thickness of rooting zone. Typical family names are clayey, kaolinitic, isohyperthermic, etc. There are thousands of families.

Series of Soil

The soil series is the lowest category. It is a grouping of soil individuals on the basis of narrowly defined properties, relating to kind and arrangement of horizons; colour, texture, structure, consistence and reaction of horizons; chemical and mineralogical properties of the horizons. The soil series are given local place names following the earlier practice in the old systems in naming soil series. There are tens of thousands of series.

Classified Soils in the Tropics

According to the USDA Soil Taxonomy, Oxisols are the most abundant soils in the humid and perhumid tropics covering about 35 per cent of the land area. Ultisols are the second most abundant, covering an estimated 28 per cent of the region. About half of the Ultisols and 60 per cent of the Oxisols are located in humid and per-humid tropical Africa and Asia. In tropical Africa, they are abundant in the eastern Congo basin bordering the lake region; in the forested zones of Sierra Leone; in Ivory Coast; in parts of Liberia; and in the forested coastal strip from Ivory Coast to Cameroon.

The Alfisols, which have high to moderate fertility, cover a smaller area of the humid tropics. In west Africa they are found in Ivory Coast, Ghana, Togo, Benin, Nigeria and Cameroon. They are, however, the most abundant soils in Africa's sub-humid and semi-arid zones, covering about one third of these regions. The Alfisols are widely distributed in the sub-humid and semiarid tropical regions of Africa, including large areas in western, eastern, central, and south-eastern Africa.

Soil Order	Tropical Africa	Tropical Asia	Tropical America	Total	Percent
Humid Tropics					
Oxisols	179	14	332	525	35
Ultisols	69	131	213	413	28
Alfisols	21	15	18	54	4
Others	176	219	103	498	33
Total	445	379	666	1490	100
Semi-arid Tropics[2]					
Alfisols	466	121	107	694	33
Ultisols	24	20	8	52	1
Others	972	178	198	1348	66
Total	1462	319	313	2094	100

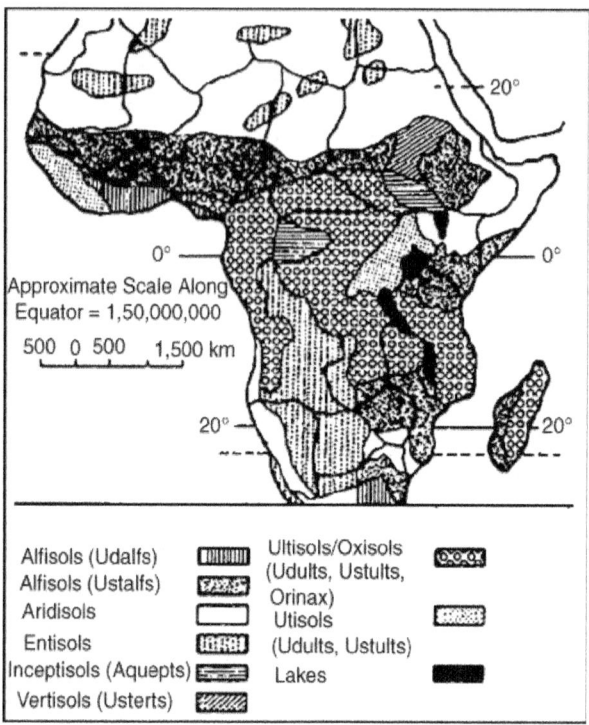

Fig. Soils of Tropical Africa; According to the USDA Soil Taxonomy.

The Fao/Unesco System of A Small Scale Soil

The FAO/UNESCO system was devised more as a tool for the preparation of a small-scale soil map of the world than a comprehensive system of soil classification. The map shows only the presence of major soils, being associations of many soils combined in general units. The legend of the soil map of the world lists 106 units classified into 26 groupings. The soil units correspond roughly to great groups from the USDA Soil Taxonomy, while larger main grouping are similar to the USDA soil sub-order.

In 1986 FAO published a soil map of Africa following the FAO/UNESCO system of soil classification. In this map, all the soils of Africa have been grouped into 10 soil associations. Though it is not very precise, the map provides an overview of the soil resources of the continent of the ten major associations, the desert and shallow soil associations (comprising Yermosols, Xerosols and Luvisols) occupy about one-third of Africa's land area. However, only a part of the area occupied by these associations falls in the tropics.

The French System of Classifying Soils

The so-called French System of classifying soils is based on principles of soil evolution and degree of evolution of soil profiles. It also takes into account humus type, structure, and the degree of hydromorphism. The system was developed by the Office de la recherché scientifique *et* technique d'outremer (ORSTROM, now Institut français de recherche scientifique pour le développement en co-opération).

Table. The Soil Correlation Between Systems of Soil Classification : Taxonomy, FAO/UNESCO Legend and the INRA System.

FAO/UNESCO	Soil Taxonomy*	INRA System
Acrisols	Ultisols	Sols Lessive
Andosols	Andepts	Andosols
Arenosols	Psamments	Sols mineraux bruts
Cambisols	Tropepts	Sols bruns eutrophes tropicaux
Ferralsols	Oxisols (Latosols)	Sols Ferraltique
Fluvisols	Fluvents (Alluvial soils)	Sols mineraux bruts
Gleysols	Aquepts and Aquents (Aquic great groups of Entisols, Inceptisols)	Sols a gley peu profond peu bumiferes
Histosols	Histosols	Sols hydromorphes organiques
Litbosols	Lithic subgroups	Lithosols
Luvisols	Alfisols	Sols lessives modaux
Nitosols	Tropics, Rhodic great groups of Alfisols and Ultisols	
Podzols	Spodosols	Podsols
Regosols	Orthents, Psamments	Sols mineraux bruts d'apport; eolien ou volcamque; sols peu evolves regosolique d'erosion etc.
Vertisols	Vertisols	Vertisols

Note : * = Name in old USDA system.

Characteristics of The Major of Tropical Soils

The following sections provide additional information on the properties and management of the most important soils in the humid and sub-humid zones of tropical Africa.

Alfisols

The Alfisols are less leached and have lower acidity than Ultisols and Oxisols, but they exhibit high base saturation and their fertility is low to moderate. The Alfisols and associated soils support a wide variety of cereal crops (maise, rice, sorghum, millet), root and tuber crops (yam, cassava, cocoyam, sweet potato), and grain legumes (soybean, cowpeas, groundnuts, pigeon peas, chick peas). Distribution of the Alfisols, Ultisols, and Oxisols is shown in the Soil Taxonomy. Examples of chemical characteristics of Alfisols and Ultisols.

The productivity of the Alfisols is limited mainly by their physical characteristics :
- They have low structural stability and are susceptible to surface crusting, soil compaction and erosion.
- They have low water retention capacity and are subject to drought.
- Deficiencies of N and P are common while deficiencies of K, Mg, S. Fe, and Zn occur under intensive cultivation.
- Because of their low buffering capacity, Alfisols acidify rapidly under continuous cultivation, particularly with the use of high rates of nitrogenous fertilizers.

Some of the chemical properties of an Alfisols profile from Southwest Nigeria, where the soil is slightly acidic with high base saturation even in the lower soil horizons. Benefits from N. P. and K application for continuous crop production on the Alfisols have been well documented.

With intensive cropping, N is the primary limiting nutrient, followed by P. Potassium is generally needed with long-term continuous cropping, particularly on soils derived from sedimentary rocks. The Alfisols and associated soils have low P-fixation and high residual effects from applied P. In addition, mycorrhiza symbiosis is common and effective on these soils particularly with root crops, resulting in a low P requirement for crop production.

Continuous cultivation and fertilizer application can significantly affect the properties of Alfisols and associated soils. Cropping, and in particular fertilizer application, reduces soil pH, soil organic matter, and extractable cations. Lowering of soil pH on the Alfisols can result in increased toxic levels of Al and Mn (Kang and Spain, 1986).

Ultisols and Oxisols

The Oxisols and especially the Ultisols are acidic, with low base saturation. Both soil orders commonly have multiple nutrient deficiencies

(N. P. K, Ca and Zn). Oxisols are highly weathered and leached, while Ultisols are susceptible to erosion and compaction. The poor productivity of these soils is due to their low capacity to provide nutrients to crops as well as their Al and Mn toxicity. Soils have medium to high P fixation. Chemical characteristics of some Nigerian Ultisols.

The Ultisols and Oxisols support a lesser variety of food crops than Alfisols, being more suitable for tree crop production. Crops that do well on the Ultisols and Oxisols include some cereal crops (*e.g.*, rice), root and tuber crops (cassava, yam, cocoyam, sweet potato), grain legumes (cowpeas, groundnuts). Plantains and bananas also do well. In traditional system, maise is grown only on newly cleared and burnt plots.

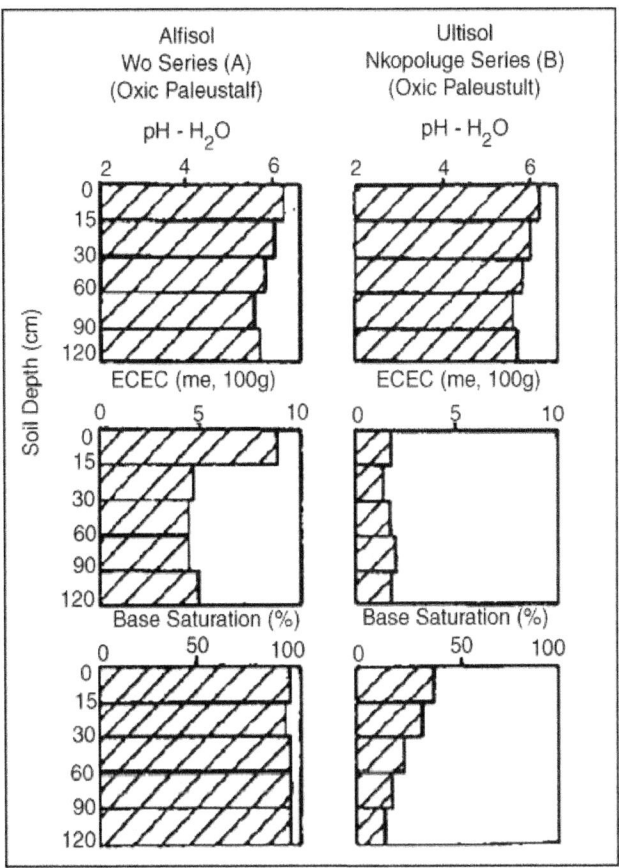

Fig. Soil pH, Effective Cation Exchange Capacity (ECEC), and Degree of Base Saturation.

In many early studies, acid soils in the humid tropics were limed to neutral pH, with generally poor results due to nutrient imbalance. Following the finding in the 1950s that acid soils contain more exchangeable Al^{3+} than H^+, primary consideration has been given to removal of toxic factors which limit plant growth. Research on acid soils in West Africa has confirmed these findings. Low lime rates are needed to reduce toxic levels of Al^{3+} and application of 0.5 to 1.0 tons of lime per hectare was found to be adequate for highly acid soils (IITA, 1984). These soils are usually deficient in P as well. Rock phosphates can be used on unlimed acid soils as an inexpensive and efficient way of supplying P to acid-tolerant crops.

Management of Low-Activity Clays (LAC) Soils

For purpose of management, the majority of the upland soils in the humid and sub-humid tropics is grouped as low activity clays (LAC) soils. A LAC soils has a low effective cation exchange capacity (ECEC) of ≤ 16 meq/ 100 g clay in the sub-soil. The LAC soils are predominantly Alfisols, Ultisols, Oxisols, and associated soils. Vast areas of the rainfed uplands in the humid and sub-humid tropics currently used for traditional food crop production are dominated by these "fragile" soils. Observations have shown that the majority of the LAC soils in West Africa have an especially low ECEC of < 8 meq. As the clay fraction of these soils are composed mainly of kaolinite, halloysite, and oxides of Fe and Al, the soil ECEC depends mainly on the soil organic matter level, which controls nutrient absorption and release.

Problems in Fertility Management of LAC Soils

One of the major problems associated with extended cultivation of LAC soils is the maintenance of favourable soil physical conditions and the control of soil erosion. Significant changes in soil chemical and biological properties also occur following forest or bush fallow clearing and cropping. Soil organic matter declines sharply during the first few years under cropping and the effect is more pronounced with intensive continuous cropping. The loss of organic matter and acidification resulted in a decrease in the effective cation exchange capacity (ECEC) and the loss of Ca and Mg. The arbitrary application of exotic, high-input food crop production technologies on these fragile soils therefore often leads to rapid chemical, physical, and biological degradation of the soil. Although soil fertility problems on the LAC soils can be corrected by liming and appropriate fertilization, socio-economic constraints often limit the application of these crop production technologies in many areas of tropical Africa. Currently, sub-Saharan Africa's per capita and per hectare fertilizer use is very low compared with that of other regions. There is a need to develop integrated soil fertility management systems for the region based on better utilisation of local nutrient sources. Such systems should be supplemented with external inputs wherever that is feasible and affordable. For sustained crop production in addition to adequate supply of plant nutrients, the LAC soils also require continuous addition of organic matter.

Integrated Nutrient Management Options

Integrated soil fertility management for LAC soils can be achieved by various methods including :

- Promoting maximum recycling and more efficient use of nutrients from plant residues,
- Increasing contribution of biological nitrogen fixation,
- Improving efficiency of use of mineral nitrogen fertilizers and local sources of phosphate fertilizers,

- Using organic residues to reduce soil acidity problems, and
- Using acid-tolerant cultivars.

Use of low levels of chemical inputs in combination with fallowing and agroforestry systems has shown varying degrees of success. Fallowing and addition of organic mulches may correct chemical soil degradation resulting from continuous cultivation; at the same time, it may also increase efficiency of fertilizer use.

Crop residue management and seed bed preparation methods can play an important role in sustaining the productivity of these soils for crop production. This can be achieved in reduced tillage systems through the use of crop residue mulches, *in situ* mulches from cover crops, and/or hedgerow prunings from alley farming. The presence of adequate amounts of mulch cover helps maintain high soil nutrient status and high biological activity. Mulch also protects the soil against high temperatures, soil erosion, and run-off, thereby preventing the breakdown of soil structure and the resultant soil compaction and decreased permeability. Furthermore, mulching increases soil moisture retention and reduces run-off and soil erosion.

Results of long-term field experiments carried out on Alfisols have also shown that with judicious fertilizer use and crop rotation, high and sustained crop yields can be obtained. Similar principles also apply for managing the Ultisols/Oxisols. For sustained crop production, the Ultisols and Oxisols additionally require judicious liming.

Performance of Woody Species on Alfisols and Ultisols/Oxisols

The integration of food crops and forages with multi-purpose tree species (MPTs) in agroforestry and alley farming systems have received much attention in recent years as an alternative, low chemical input management possibility for LAC soils. However, little information is available on the soil requirements for growing the MPTs.

As with crops, the capacities of MPTs for biomass production and nutrient recycling are affected by soil and climatic conditions. Under the same climatic regime, growth and biomass production of MPTs is expected to be higher on the more productive Alfisols than on the less productive Ultisols/Oxisols. Additions of nutrients may be needed for good growth of MPTs.

MPTs for alley farming such as Leucaena leucocephala and Gliricidia sepium do well on non-acid or slightly acid Alfisols. Both species perform poorly on acid soils. On the low pH soils, MPTs such as Acioa barteri, Calliandra calothyrsus, and Flemingia macrophylla perform well.

Land Capability Classification System

The technique which allows determination of the most suitable use for any area of land is called land classification. A great number of systems of land classification are in use, varying mainly according to the purpose for which the land is classified. Land may be classified according to its present land use, its suitability for a specific crop under the existing forms of management, its capability for producing crops or combinations of crops under optimum management, or its suitability for non-agricultural types of land use. A good knowledge of the land capability and suitability combined with good understanding of the soil characteristics and management aspects are the keys to more productive and sustainable agriculture. The purpose of land capability classification systems is to study and record all data relevant to finding the combination of agricultural and conservation measures which would permit the most intensive and appropriate agricultural use of the land without undue danger of soil degradation.

The USDA Land Capability Classification System

The best known of these systems is the United States Department of Agriculture system. The USDA land classification system is interpretative, using the USDA soil survey map as a basis and classifying the individual soil map units in groups that have similar management requirements.

At the highest of categorisation, eight soil classes are distinguished, namely :

- *Class I soils* : have few limitations restricting their use. Erosion hazards on these soils are low; they are deep, productive and easily worked. For optimum production, these soils need ordinary management practices to maintain productivity, as regards both soil fertility and favourable physical soil properties.
- *Class II soils* : have some limitations that reduce the choice of plants or require moderate conservation practices. Limitations of soils in Class II include (singly or in combination) the effect of gentle slopes, moderate susceptibility to erosion, less than ideal soil depth, somewhat unfavourable soil structure, slight to moderate correctable salinity, occasional damaging overflow, wetness correctable by drainage, slight climatic limitation. Soils in this class require more than ordinary management practices for obtaining optimum production and for maintaining productivity.
- *Class III soils* : have severe limitations that reduce the choice of plants or require special conservation practices. The limitation of soils in this class are those of Class II, but in higher degree; including additional limitations such as shallow depth, low moisture-holding capacity, and low fertility that is not easily corrected. Class III soils require considerable management inputs, but even so, choice of crops or cropping systems remains restricted because of inherent limiting factors.
- *Class IV soils* : have very severe limitations that restrict the choice of plants and or require very careful management. Restrictions, both in terms of choice of plants and or management and conservation practices are greater than in Class III to such an extent that production is often marginal in relation to the inputs required. Limiting factors re of the same nature as in the previous classes but more severe and difficult to overcome. Several

limitations such as steep slopes are a permanent feature of the land. Some of the limitations due to sloppiness and erosion hazards in classes II to IV can be reduced by biological terracing as practiced in agroforestry and alley

- *Cropping* : In the USDA system, soils of classes V to VIII are generally not suited for cultivation, although certain of them may be made suitable for agricultural use with costly measures.
- *Class V soils* : have few or no erosion hazards but have other limitations, impracticable to remove, that restrict their use to pasture, range, woodland, or wildlife food and cover. Although they may be level or nearly level, many of these soils are subject to inundation or are stony or rocky.
- *Class VI soils* : have severe limitations — that make them generally unsuited to cultivation and limit their use largely to pasture or range, woodland, or wildlife food cover. This class is a continuation of Class IV, with very severe limitations that cannot be corrected. They may serve for some kinds of crops, such as tree crops, provided unusually intensive management is practiced.
- *Class VII soils* : have very severe limitations that make them unsuited to cultivation and also, restrict their use largely to grasing, woodland, or wildlife. The limitations are such that these soils are not suited for any of the common crops.
- *Class VIII soils* : and land forms have limitations that preclude their use for commercial plant production.

In the second level of generalisation of the USDA land capability classification system, sub-classes specify the kind of limitations. Four kinds of limitations are recognised at this level, namely, risk of erosion; wetness, drainage or overflow; rooting zone limitations, and climatic limitation. The third level, that of the capability unit, provides more specific and detailed information for application to specific fields on a farm. A new standard framework for land evaluation by means of land suitability classification has been developed by FAO (1983). As in other systems, the land suitability component of land evaluation is based on the survey of the physical attributes of the land (soils, climate, vegetation, topography, hydrology, etc.), and consequently requires interpretation of these attributes. The proposed FAO land suitability classification integrates relevant social and economic factors with the technical suitability classification. At the present stage, the system mainly concentrates on the classification of land based on technical suitability.

CLASSIFICATION OF SOILS IN INDIA

The Indian Council of Agricultural Research (ICAR) has made an authentic and standardised classification of soils and divided the soils of India into the following 8 groups :

1   Alluvial soils
2   Black soils

3 Red and Yellow soils
4 Laterite soils
5 Arid soils
6 Saline soils
7 Peaty and Organic soils
8 Forest soils.

ALLUVIAL SOILS

Nature of Soil

The alluvial soils are of many shades depending upon depth, (deep and shallow), deposition conditions (coarse and finer) and time (older and recent). In the western Ganga plains, Punjab and Haryana, the quantity of loam and clay loam increases while in the middle Ganga plain sand decreases and loam increases. In Punjab and other plains, the excess of irrigation has made the soils water-logged and saline crusts have been formed. This has also caused the formation of heavy soils in low lying areas. In the eastern parts finer particles predominate and loams and fine silty clays are formed. Due to heavy rainfall, the alluvial soils have been laterised. In the river valley plain of northern India floods result in deposition of silt. This new alluvium is known as Khadar. The higher areas, where floods do not reach has old alluvium and is known as Bangar. Under the bangar deposits, beds of lime modules are found and are known as Kankar and these are usually found in Haryana and are a good source of raw material for cement plants. Along the coast of the Peninsula, where sea water enters the delta, saltish soils are the result. The salt encrustations in Kutchch (Gujarat) shine with a glaze on sunny days. In the north-west, the drier parts draw sodium, potassium, calcium, magnesium, etc., from the depth and create a powdery layer on the top of the soil because of evaporation of water.

This layer is called Reh (Kallar). No vegetation grows on the Kallar land.

Alluvial soils arc found in two different and distinct regions in India :

- *Northern Plains* : The whole of the northern plains; from Punjab, Haryana, U.P., Bihar to West Bengal are included in this region. The river courses and the deltas form alluvial soil regime.
- *Southern Coastal Area* : Starting from the Eastern Coast along a narrow belt passing through the flood plains, terraces, deltaic and lagoon areas of the rivers like the Mahanadi, Godavari, Krishna, Cauvery, etc.

Narrow areas along the coast and lower portions of the Narmada, the Tapti, the Sabarmati and the Mahi also have such type of soils. The alluvial soils are fertile and are responsible for making the northern plains, the granary of India. Agricultural activities and crop productivity are attributed to these soils.

Black Soils

Black soils are said to have developed in the Deccan Trap area on basalt rocks in semi-arid conditions. They are of three kinds- shallow, medium and deep. These are also called Regur. They are deep black in colour but there is almost complete absence of humus. Water can remain stored in the soils for a long period and this can continue to provide water to the roots of the plants in the dry period.

That is why these soils are used for the cultivation of cotton even in those areas where irrigation is not available. Their black colour is due to certain salts. On getting dry, the soils develop cracks.

The soils have some deficiency of potash, nitrogen and phosphorus but have lime, magnesium, aluminium, etc. Thus they are fertile soils. Such soils are in eastern Gujarat, south-western M.P (Narmada, parts of Vindhya and Satpura), almost the whole of Maharashtra, northern Karnataka, north-western T.N., western A.P, etc. They are about 6 metres deep in the lower parts of the Narmada, the Tapti, the upper parts of Godavari and the Krishna rivers.

Black soils are very conducive to cotton cultivation. The Deccan Trap area has become famous for cotton cultivation because of these soils. In fact this region is called cotton bowl of India. Wherever soils arc less deep as on the slopes of the plateaus and hills, and unable to hold water, there instead of cotton, barley, millets, pulses etc., are grown. In areas of deep soils, besides cotton, a host of other crops like tobacco, groundnuts etc. are successfully raised.

Red and Yellow Soils

The reddish-yellow colour is due to the presence of iron oxide. These soils are formed where the rainfall is low and there is a little leaching lesser than that in the laterite soils. Red soils are as such usually developed on old crystalline and metamorphic rocks. These are sandier and comparatively less clayey. These soils cannot retain moisture for a long-time. Use of manures increases their fertility. The soils are deficient in phosphorus, nitrogen and humus.

They are acidic in nature and have iron, aluminium and lime in sufficient quantities. If the soils are fine grained, they are fertile. These are porous soils. They are not fertile on higher dry lands.

These soils are found in three regions :

- Central— From Bundelkhand, Baghelkhand to the south, from Orissa, eastern A.P and T.N., these soils occupy large areas.
- Western— The eastern and south-eastern narrow belt to the eastern side of the Aravalis.
- Eastern— Parts of Meghalaya, Nagaland, Manipur, Mizoram, etc.

If the soils are not fertile, millets are grown. Where they are deep, deep red and fertile, the main crops grown are wheat, cotton, potato, rough grains and others.

Laterite Soils

These soils are in those areas which are hot and get seasonal rainfall. Due to higher temperatures the bacteria eat away humus and the rainfall leaches silica and lime. As a result the soils are acidic and are rich in aluminium and iron oxides. At places where aluminium compounds dominate, the laterites are called bauxite. On account of presence of iron oxides in them the soils appear red.

These soils are classified into three types on the basis of their particles :

- Deep Red Laterite. They have excess of iron oxide and potash but are short of Kaolin. The soils are not fertile.
- White Laterite. The colour of the soil is due to excess of Kaolin. Soils lose fertility at a faster rate.
- Underground Laterite. The upper parts are dissolved especially in iron which settles down below the upper layer. This makes the soils fertile.

Laterite soils do not retain moisture. The use of manure is necessary for increasing soil fertility. Their occurrence is not spread on large areas but they occur in patches, however, continuous also in some areas.

Bihar and Jharkhand Plateau has laterite soils. They are in patches on the Eastern Ghats through Orissa, A.P and T.N. In the western parts of India such soils are in a narrow belt from the north to the south through Maharashtra, Karnataka and Kerala extending more or less continuously. Shillong Plateau has a laterite soil belt which extends towards Sadiya in Assam.

Soils are useful for making bricks because of presence of lot of iron in them. Its form in-which aluminium is in excess is called Bauxite and is used for extracting aluminium. Soils become fertile with the addition of fertilizers and manures. These soils are devoted to the cultivation of cotton, rice, wheat, pulses, tea, coffee, etc. These are intensively cultivated in south India. Tapioca and cashew nuts are also grown in these soils. The latter is a cash crop.

Arid Soils

These soils are usually shallow. They have sandy texture. They have low clay and salt content; usually below 10 per cent. The colour ranges from red to brown and light brown.

Due to high evaporation in arid regions, the soils suffer from deficiency of humus and moisture but wherever water is provided through irrigation, soils become fertile. Though these soils are poor in nitrogen yet they are somewhat rich in plant food. The entire area, west of the Aravalis has arid soils.

This is the part of the Thar Desert and it continues into neighbouring Pakistan. The strong desert winds remove fine particles of sand to far off places causing infertile barren lands.

The use of manures and provision of irrigation facilities to such soils result in fairly good crop yields. Afforestation can help stabilising shifting sand dunes.

Indira Gandhi canal has proven to be a boon for the region by way of converting dry desert lands into blooming landscape full of greenery and economic prosperity.

Saline Soils

Saline soils are foun d in various climatic regimes – dry, semi-dry and swampy. These soils possess sodium, potassium and magnesium salts. Salts reach these areas by defective drainage and dry climate. The salts in their solutions move up and are found lying over the surface like a white sheet. Its encrustation is very hard and inpenetrable. It does not allow any vegetation to grow. The S.W Monsoons which cross Rann of Kutchch bring with them salt particles and form a layer in the Gujarat state. In the swampy areas and in the coastal tidal areas, the swamps are saturated with salts. These soils are deficient in nitrogen and calcium. The western Gujarat area (Kutchch) is known for these type of soils. These soils are known as Khar, Khanjan, etc.

In the Cauvery and Mahanadi deltas, the sea water makes the soil saline. In West Bengal the Sunderbans are well known for such soils. In Punjab, Haryana, U.P and Bihar, Saline soils are encompassing more and more agricultural areas. Same is the position in the southern Indian states. However, the fertility of soils can be regained by way of putting gypsum in the soils and improving drainage.

Peaty and Organic Soils

A large amount of dead organic matter accumulates in areas which have heavy rainfall and high humidity. As a result these soils are saline, rich in organic matter (40 per cent) but deficient in potash and phosphorus.

These are alkaline, heavy and black in colour. Such soils are found in the coastal areas of W. Bengal, Orissa and Tamil Nadu, northern Bihar and Almora area of U.P.

Forest Soils

These soils are found in the hilly areas, covered with forests. The main characteristic of these soils is the accumulation of organic matter derived from forest cover. The soils are not uniform everywhere but there are variations in their distribution. The soils are loamy and have silt in the valley areas and are coarse grained, kankar etc., in the higher areas.

There are some important types of soils which have been spread over areas described below :

- *Fine Textured Soil* : Usually the outwash and river valleys develop these type of soils. For example, in many areas of upper Himalayas (Lahul-Spiti, Kinnaur and even in Ladakh), soils have not fully developed as such stone, kankar and shallow soils are met with.
- *Alpine Soil* : In the higher areas about 3,000 metres high, the climate is cold. As a result, the soils have undecomposed vegetative matter derived from grasses resulting in immature soils.

- *Podzol* : The area where Podzol soils are found varies in height from 2,000 to 3,000 metres. The soil consists of partly decomposed vegetation derived from the coniferous forests that grow at this height.

Heavy rainfall results in leaching of the soils and turns it acidic. Its colour is greyish brown.

Soils are not much fertile. Lower Forest Soil. The height of the mountainous area where these soils develop lies between 1000 to 2000 metres. The forest cover is mostly of deciduous trees. The soils are brown in colour, deep and slightly acidic. The soils have humus and are thus fertile.

UNITED STATES SOIL CLASSIFICATION SYSTEM

The first formal system of soil classification was introduced in the United States by Curtis F. Marbut in the 1930s. This system, however, had some serious limitations, and by the early 1950s the United States Soil Conservation Service began the development of a new method of soil classification. The process of development of the new system took nearly a decade to complete. By 1960, the review process was completed and the Seventh Approximation Soil Classification System was introduced. Since 1960, this soil classification system has undergone numerous minor modifications and is now under the control of Natural Resources Conservation Service (NRCS), which is a branch of the Department of Agriculture. The current version of the system has six levels of classification in its hierarchical structure.

The major divisions in this classification system, from general to specific, are: orders, sub-orders, great groups, sub-groups, families, and series. At its lowest level of organisation, the U.S. system of soil classification recognises approximately 15,000 different soil series. The most general category of the NRCS Soil Classification System recognises eleven distinct soil orders : oxisols, aridsols, mollisols, alfisols, ultisols, spodsols, entisols, inceptisols, vertisols, histosols, and andisols.

- *Oxisols* : develop in tropical and sub-tropical latitudes that experience an environment with high precipitation and temperature. The profiles of oxisols contain mixtures of quartz, kaolin clay, iron and aluminum oxides, and organic matter. For the most part they have a nearly featureless soil profile without clearly marked horizons. The abundance of iron and aluminum oxides found in these soils results from strong chemical weathering and heavy leaching. Many oxisols contain laterite layers because of a seasonally fluctuating water table.
- *Aridsols* : are soils that develop in very dry environments. The main characteristic of this soil is poor and shallow soil horizon development. Aridsols also tend to be light coloured because of limited humus additions from vegetation. The hot climate under which these soils develop tends to restrict vegetation growth. Because of limited rain and high temperatures soil water tends to migrate in these soils in an upward direction. This condition causes the deposition of salts carried by the water at or near

the ground surface because of evaporation. This soil process is of course called salinisation.
- *Mollisols* : are soils common to grassland environments. In the United States most of the natural grasslands have been converted into agricultural fields for crop growth. Mollisols have a dark coloured surface horizon, tend to be base rich, and are quite fertile. The dark colour of the A horizon is the result of humus enrichment from the decomposition of litterfall. Mollisols found in more arid environments often exhibitcalcification.
- *Alfisols* : form under forest vegetation where the parent material has undergone significant weathering. These soils are quite widespread in their distribution and are found from southern Florida to northern Minnesota. The most distinguishing characteristics of this soil type are the illuviation of clay in the B horizon, moderate to high concentrations of base cations, and light coloured surface horizons.
- *Ultisols* : are soils common to the southeastern United States. This region receives high amounts of precipitation because of summer thunderstorms and the winter dominance of the mid-latitude cyclone. Warm temperatures and the abundant availability of moisture enhances the weathering process and increases the rate of leaching in these soils. Enhanced weathering causes mineral alteration and the dominance of iron and aluminum oxides. The presence of the iron oxides causes the A horizon of these soils to be stained red. Leaching causes these soils to have low quantities of base cations.
- *Spodsols* : are soils that develop under coniferous vegetation and as a result are modified by podzolisation. Parent materials of these soils tend to be rich in sand. The litter of the coniferous vegetation is low in base cations and contributes to acid accumulations in the soil. In these soils, mixtures of organic matter and aluminum, with or without iron, accumulate in the B horizon. The A horizon of these soils normally has an eluvial layer that has the colour of more or less quartz sand. Most spodosols have little silicate clay and only small quantities of humus in their A horizon.
- *Entisols* : are immature soils that lack the vertical development of horizons. These soils are often associated with recently deposited sediments from wind, water, or ice erosion. Given more time, these soils will develop into another soil type.
- *Inceptisols* : are young soils that are more developed than entisols. These soils are found in arctic tundra environments, glacial deposits, and relatively recent deposits of stream alluvium. Common characteristics of recognition include immature development of eluviation in the A horizon and illuviation in the B horizon, and evidence of the beginning of weathering processes on parent material sediments.
- *Vertisols* : are heavy clay soils that show significant expansion and contraction due to the presence or absence of moisture. Vertisols are common in

areas that have shale parent material and heavy precipitation. The location of these soils in the United States is primarily found in Texas where they are used to grow cotton.

- *Histosols* : are organic soils that form in areas of poor drainage. Their profile consists of thick accumulations of organic matter at various stages of decomposition.
- *Andisols* : develop from volcanic parent materials. Volcanic deposits have a unique process of weathering that causes the accumulation of allophane and oxides of iron and aluminum in developing soils.

Canadian System of Soil Classification

Canada's first independent taxonomic system of soil classification was first introduced in 1955. Prior to 1955, systems of classification used in Canada were strongly based on methods being applied in the United States. However, the U.S. system was based on environmental conditions common to the United States. Canadian soil scientists required a new method of soil classification that focused on pedogenic processes in cool climatic environments.

Like the US system, the Canadian System of Soil Classification differentiates soil types on the basis of measured properties of the profile and uses a hierarchical scheme to classify soils from general to specific. The most recent version of the classification system has five categories in its hierarchical structure. From general to specific, the major categories in this system are : orders, great groups, sub-groups, families, and series. At its most general level, the Canadian System recognises nine different soil orders :

Brunisol is a normally immature soil commonly found under forested ecosystems. The most identifying trait of these soils is the presence of a B horizon that is brownish in colour. The soils under the dry pine forests of south-central British Columbia are typically brunisols.

Brunisolic Pine Landscape Brunisol Profile (Central British Columbia)
Fig. Associated Surface Environment and Profile of a Brunisol Soil.

Chernozem is a soil common to grassland ecosystems. This soil is dark in colour (brown to black) and has an A horizon that is rich in organic matter. Chernozems are common in the Canadian prairies. The images below are from the eastern prairies where higher seasonal rainfalls produce black chernozemic soils.

Fig. Chernozemic Landscape (Prairies).

Chernozen Profile
Fig. Associated Surface Environment and Profile of a Chernozem Soil.

Cryosol is a high latitudes soil common in the tundra. This soil has a layer of permafrost within one meter of the soil surface. The image on the left is of tundra landscape dominated by moss and lichen vegetation. The soil profile has a permanently frozen ice wedge beneath its surface.

Tundra Cryosolic Landscape (N.W.T.) Organic Cryosol Profile

Fig. Associated Surface Environment and Profile of a Cryosol Soil.

Gleysol is a soil found in an ecosystem that is frequently flooded or permanently waterlogged. Its soil horizons show the chemical signs of oxidation and reduction.

Flooded Gleysolic Landscape (Atlantic Coast) Gleysol Profile

Fig. Associated Surface Environment and Profile of a Gleysol Soil.

Luvisol is another type of soil that develops under forested conditions. This soil, however, has a calcareous parent material which results in a high pH and strong eluviation of clay from the A horizon.

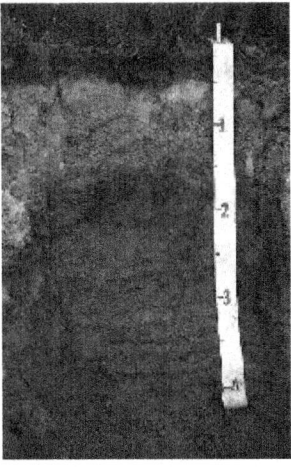

Luvisolic Sub-Boreal Forest Landscape Luvisol Profile (Northern British Columbia)

Fig. Associated Surface Environment and Profile of a Luvisol Soil.

Organic this soil is mainly composed of organic matter in various stages of decomposition. Organic soils are common in fens and bogs. The profiles of these soils have an obvious absence of mineral soil particles.

Organic Soil Landscape (British Columbia) Organic Soil Profile
Fig : Associated Surface Environment and Profile of an Organic Soil.

Podzol is a soil commonly found under coniferous forests. Its main identifying traits are a poorly decomposed organic layer, an eluviated A horizon, and a B horizon with illuviated organic matter, aluminum, and iron. The forested regions of southern Ontario and the temperate rainforests of British Columbia normally have podzolic soils.

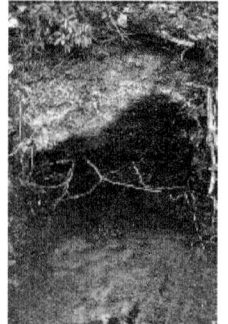

Forested Podzolic Landscape (Ontario) Podzol Profile
Fig. Associated Surface Environment and Profile of a Podzol Soil.

Regosol is any young underdeveloped soil. Immature soils are common in geomorphically dynamic environments. Many mountain river valleys in British Columbia have floodplains with surface deposits that are less than 3000 years old. The soils in these environments tend to be regosols.

Immature Regosolic Landscape Regosol Profile (Floodplain British Columbia)

Fig. Associated Surface Environment and Profile of a Regosol Soil.

Solsonetzic is a grassland soil where high levels of evapotranspiration cause the deposition of salts at or near the soil surface. Solonetzic soils are common in the dry regions of the prairies where evapotranspiration greatly exceeds precipitation input. The movement of water to the earth's surface because of capillary action, transpiration, and evaporation causes the deposition of salts when the water evaporates into the atmosphere.

Saline Solonetzic Landscape Solonetzic Profile (Saskatchewan)
Fig. Associated Surface Environment and Profile of a Solonetzic Soil.

SOIL CLASSIFICATION FOR THE NEEDS OF REGIONAL PLANNING

Soil is an important natural resource and common wealth of each nation. Its use, protection and disposal are regulated by the law in the Republic of Croatia. The problem of soil conservation and protection from non-agricultural use is present in all regions of Croatia. Hence, the conservation and protection of soil constitutes an unavoidable requirement of rational land management.

About 6,700 ha of agricultural land is lost in Croatia every year, which is a lot with respect to the overall area of the country, its population and proportion of agricultural soils. Statutory regulations highlight the need of conserving good agricultural land and preventing its permanent loss by construction of towns, roads, airports, harbours, sports facilities, exploitation of sand, stone, brick and pottery clay, surface excavations, and solid and liquid waste dumps.

Regulations pertaining to soil conservation and protection from unappropriated use have undergone many changes in the Republic Croatia. The latest Agricultural Land Act (Official Gazette 66/2001) stipulates that any switch of the purpose of agricultural land into non-agricultural must comply with the regional planning documents. Cadastre capability evaluation of soil for this purpose has been abandoned in Croatia.

Instead, soil classification for land documents is based on three categories that should be protected : S1 — highly valuable agricultural soils, S2 — valuable agricultural soils, and S3 — other agricultural soils. Other agricultural soils (pastures, meadows, low-quality plough-land, karst orchards and vineyards, etc.) are allotted to the category of other agricultural soils and forest soils (SF). Unfortunately, the guidelines do not provide the criteria for this classification. Further, forest soils are evaluated on the basis of the forestry management and ecological or protective role of forests into categories : F1 soils for forestry management purposes, F2 — forest soils where forests have a protective role in terms of conservation and prevention of erosion, and F3 — forest soils for special purposes, such as national parks, nature reserves, endemic forests, etc. Since there are no statutory regulations on the method of classification for land evaluation (Bogunovic *et. al.*, 1999, 2000 and 2001, Bogunovic and Husnjak, 2000a and b, and Husnjak and Bogunovic, 2001), the authors applied the basic elements of the FAO land classification, which was accepted as a possible solution by the regional planners.

For this reason, two examples of land evaluation are given in further text : at the regional (county) level and at the municipality level (as the smallest spatial-political entity in our administrative system).

Material and Methods of Soil

Classification is based on the data from the Basic Soil Map of the Republic Croatia, scale 1 : 50,000. The map is digitalised and is the basis for targeted interpretation and classification of soils. Forests were separated pursuant to military maps, scale 1 : 50,000 and 1 : 25,000. Manual digitalisation with the CalComp digitaliser was applied initially, while maps scanned by Microstation programme have been used lately. Criteria of agricultural land classification are applied in accordance with the FAO classification (1976), based on recommendations given by Brinkman and Smith (1972).

Investigation Results

Investigation results and land classification for the needs of regional planning are based on land information provided by the GIS technology. Pedological data were digitally processed using the ArcInfo programme, and the data are suitable for detailed processing, quick retrieval and interpretation. Digital processing offers multiple options of using these data for various purposes Thus, besides direct data for the needs of regional planning, other data are also provided for their direct use for soil improvement and for future land use.

According to the original data and the results of the analysis of the Lika-Senj County, agricultural land, including degraded soils and pastures subject to overgrowing, accounts for 207,189 ha or 39 per cent of the total county area. Forests cover 327,536 ha or 61 per cent of the total county area. According to quality evaluation, agricultural soils have been classified into four categories for the needs of regional planning. Their distribution is given in the enclosed reduced map of land suitability for agriculture, intended for regional planning.

S1 category, the best soils classified by the Regional Planning Act as highly valuable agricultural soils, covers only 5,656 ha or 2.7 per cent of total agricultural land. This includes the best soils of this County, which cover only a part of Krbavsko Polje with eutric cambisols in association with rendzinas and luvisols. Under the FAO land classification (1986); Brinkman and Smith (1972), these soils belong to the S1 land suitability class. Pursuant to the regional planning standards, these soils should be strictly protected from being converted to other purposes.

S2 category, valuable agricultural soils, includes 16,322 ha or 7.9 per cent of total area. These are mainly soils of other Karst fields, such as Gacko, the largest parts of Licko and Krbavsko, Korenicko, Lapacko, Otocacko, Plaško, Gracacko fields and the wider area of Sadilovac and Dre•nik. This also includes the Isle of Pag syncline around the old Novalja. Several soil types are found here : from medium deep eutric cambisols to rendzina, luvisols, colluvial and other soils on quaternary and fluvioglacial accumulations. This category is the second important one in terms of prevention of soil conversion to non-agricultural purposes.

S3 category of other agricultural soils includes 58,153 ha or 28 per cent of total agricultural land. These soils can be tilled despite their highest constraints in terms of acidity, depth, skeletal structure, slope, occurrence of erosion, drainage and stagnation of precipitation water. Typologically, the soils are : dystric cambisols, luvisols, calcocambisols, pseudogley and skeletal rendzina.

The fourth or SF category, a combination of other agricultural and forest soils, accounts for the largest part of the Lika-Senj County. It covers a total of 127,058 ha, or 61 per cent of agricultural land. This fact is one of the reasons why the Lika-Senj County belongs to the poorest regions of our country. These are mountain soils covered with pastures, meadows, and less frequently abandoned plough-fields and small wild-growing forests. These soils are suitable for natural pasturing of bovine cattle and small stock.

Forest areas prevail in this County and occupy 60 per cent of its total area. The largest part of forests or 83 per cent are used for forestry management purposes (F1), 5.1 per cent of forests have a protective function (F2), while special-purpose forests (F3) in national parks of the Plitvice Lakes, Paklenica and North Velebit cover 12 per cent of the area.

Table : Area Categories of the Lika-Senj County According to Land Use.

Area Category					
Agriculture			Forest		
Category	ha	%	Category	ha	%
S 1	5 656.3	2.73	F 1	271 421.8	82.87
S 2	16 322.1	7.88	F 2	16 772.5	5.13
S 3	58 152.8	28.07	F 3	39 341.8	12.0
Total	80 131.2	38.68	Total	327 536 1	100
SF	127 058.3	61.32			
Grand Total	207 189.5	100			

The second example addresses land classification for the needs of regional planning at the municipality level. The example involves the classification of soils in the municipality of Lepoglava. It is mandatory that soil classification should be based on a pedological sketch at a scale of 1 : 25,000, where system units are presented independently or in associations.

Naturally, also pedological investigations done for this purpose must comply with the standards for making soil maps at this scale. The map for this purpose presents the distribution of soils in the municipality of Lepoglava at a reduced scale. Forests are separated on the map because for the regional planning purposes they are, as already mentioned, divided into three categories according to their actual function.

Table : Area Categories of the Municiplity of Lepoglava According to Land use.

Area Category					
Agriculture			Forest		
Category	ha	%	Category	ha	%
S 1	59.0	2.11	F 1	3 505.0	96.63
S 2	348.0	12.42	F 2	64.0	1.78
S 3	998.0	35.63	F 3	58.0	1.59
Total	1 405.0	50.16	Total	3 627.0	100
SF	1 396.0	49.84			
Grand Total	2 801.0	100			

To conclude this part, it should be stressed that soil separation at the semi-detailed reconnaissance level (scale 1 : 100,000) does not provide sufficiently detailed data for the explicit need of defining the soil purpose. Therefore, the objection is addressed to regional planners at the county level that the obligations arising from the primary regional plans are not adequate for the needs of detailed defining of land use at the municipality level.

Chapter 7

SOIL SURVEY

SOIL AND SOIL SURVEY

A soil survey describes the characteristics of the soils in a given area, classifies the soils according to a standard system of classification, plots the boundaries of the soils on a map, and makes predictions about the behaviour of soils. The different uses of the soils and how the response of management affects them are considered. The information collected in a soil survey helps in the development of land-use plans and evaluates and predicts the effects of land use on the environment.

Soil surveys were first authorised in the United States in 1896. Although extensive writings on husbandry by L.J.M. Columella were published in the first century A.D., practical experience was the teacher of most farmers until the advent of agricultural chemistry in the nineteenth century. By the end of the nineteenth century the knowledge about soils that had been gained from farming, agricultural chemistry, biology, and geology grew into a unified concept of the soil itself.

Early Concepts and Study of Soil

The first scholar to study soils in the United States was Edmund Ruffin of Virginia. He worked diligently to find the secret of liming and discovered what we now call exchangeable calcium. After writing a brief essay in the American Farmer in 1822, he published the first edition of An Essay on Calcareous Manures in 1832. Much of what Ruffin had learned about soils had to be rediscovered because his writings circulated only in the South.

E.W. Hilgard was one of the first true pedologists in the United States, but he never received the credit that his accomplishments deserved during his lifetime. The early concepts of soil were based on ideas developed by a German chemist, Justus von Liebig, and modified and refined by agricultural scientists who worked on samples of soil in laboratories, greenhouses, and on small field plots. The soils were rarely examined below the depth of normal tillage. These chemists held the "balance-sheet" theory of plant nutrition. Soil was considered a more or less static storage bin for plant nutrients – the soils could be used and replaced. This concept still has value when applied within the framework of modern soil science, although a useful understanding of soils goes beyond the removal of nutrients from soil by harvested crops and their return in manure, lime and fertilizer.

The early geologists generally accepted the balance-sheet theory of soil fertility and applied it within the framework of their own discipline. They described soil as disintegrated rock of various sorts – granite, sandstone, glacial till, and the like. They went further, however, and described how the weathering processes modified this material and how geologic processes shaped it into landforms such as glacial moraines, alluvial plains, loess plains, and marine terraces. N.S. Shaler's monograph on the origin and nature of soils summarised the late 19th century geological concept of soils.

Near the end of the nineteenth century, Professor Milton Whitney inaugurated the National Soil Survey Programme. Professor Whitney and his co-workers in the newly organised soil research unit of the U.S. Department of Agriculture became impressed by the great variations among natural soils – persistent variations that were in no way related to the effects of agricultural use. Whitney and his co-workers emphasised soil texture and the capacity of the soil to furnish plants with moisture as well as nutrients. Professor F.H. King of the University of Wisconsin was also reporting the importance of the physical properties of soils about this time.

Early soil surveys were made to help farmers locate soils responsive to different management practices and to help them decide what crops and management practices were most suitable for the particular kinds of soil on their farms. Many of the early workers were geologists because only geologists were skilled in the necessary field methods and in scientific correlation appropriate to the study of soils. They conceived soils as mainly the weathering products of geologic formations, defined by landform and lithologic composition. Most of the soil surveys published before 1910 were strongly influenced by these concepts. Those published from 1910 to 1920 gradually added greater refinements and recognised more soil features but retained fundamentally geological concepts.

Early field workers soon learned that many important soil properties were not necessarily related to either landform or kind of rock. They noted that soils with poor natural drainage had different properties from soils with good natural drainage and that many sloping soils were unlike level ones. Topography was clearly related to soil profile differences. As early as 1902, soil structure was described in the soil survey of Dubuque County, Iowa. The 1904 soil survey of Tama County, Iowa, reported that, on similar parent material, soils that had formed under forest contrasted markedly with soils that had formed under grass.

The balance-sheet theory of plant nutrition dominated the laboratory and the geological concept dominated field work. Both approaches were taught in many classrooms until the late 1920s. Although broader and more generally useful concepts of soil were being developed by some soil scientists, especially the United States and soil scientists in Russia, the necessary data for formulating these broader concepts came from the field work of the soil survey during the first decade of its operations in the United States. After the work of Hilgard, the longest step towards a more satisfactory concept of soil was made by G.N. Coffey, who determined the ideal classification to be a hierarchical system that was based on the unique characteristics of soil as "a natural body having a definite genesis and distinct nature of its own and occupying an independent position in the formations con stituting the surface of the earth".

Beginning in 1870, the Russian school of soil science under the leadership of V.V. Dokuchaiev and N.M. Sibertsev was developing a new concept of soil. The Russian workers conceived of soils as independent natural bodies, each with unique properties resulting from a unique combination of climate, living matter, parent material, relief, and time. They hypothesised that properties of each soil reflected the combined effects of the particular set of genetic factors responsible for the soil's formation. Hans Jenny later emphasised the functionally relatedness of soil properties and soil formation. The results of this work became generally available to Americans through.

The Russian concepts were revolutionary. Properties of soils no longer were based wholly on inferences from the nature of the rocks or from climate or other environmental factors, considered singly or collectively; rather, by going directly to the soil itself, the integrated expression of all these factors could be seen in the morphology of the soils. This concept required that all properties of soils be considered collectively in terms of a completely integrated natural body. In short, it made possible a science of soil.

The early enthusiasm for the new concept and for the rising new discipline of soil science led some to suggest the study of soil could proceed without regard to the older concepts derived from geology and agricultural chemistry. Certainly the reverse is true. Besides laying the foundation for a soil science with its own principles, the new concept makes the other sciences even more useful. Soil morphology provides a firm basis on which to group the results of observation, experiments, and practical experience and to develop integrated principles that predict the behaviour of the soils.

Under the leadership of Marbut, the Russian concept was broadened and adapted to conditions in the United States. This concept emphasised individual soil profiles to the subordination of external soil features and surface geology. By emphasising soil profiles, however, soil scientists at first tended to overlook the natural variability of soils which can be substantial even within a small area. Overlooking the variability of soils seriously reduced the value of the maps which showed the location of the soils. This weakness soon became evident in the United States, perhaps because of the emphasis here on making detailed soil maps for their practical, predictive value. Progress in transforming the profile concept into a more reliable predictive tool was rapid because a large body of important field data had already been accumulated. By 1925, a large amount of morphological and chemical work was being done on soils throughout the country. The data available by 1930 were summarised and interpreted in accordance with this concept, as viewed by Marbut in his work on the soils of the United States.

Furthermore, early emphasis on genetic soil profiles was so great as to suggest that material lacking a genetic profile, such as recent alluvium, was not soil. A sharp distinction was drawn between rock weathering and soil formation. Although a distinction between these sets of processes is useful for some purposes, rock and mineral weathering and soil formation are commonly indistinguishable.

The concept of soil was gradually broadened and extended during the years following 1930, essentially through consolidation and balance. The major emphasis had been on the soil profile. After 1930, morphological studies were extended from single pits to long trenches or a series of pits in an area of a soil. The morphology of a soil came to be described by ranges of properties deviating from a central concept instead of by a single "typical" profile. The development of techniques for mineralogical studies of clays also emphasised the need for laboratory studies.

Marbut emphasised strongly that classification of soils should be based on morphology instead of on theories of soil genesis, because theories are both ephemeral and dynamic. He perhaps over-emphasised this point to offset other workers who assumed that soils had certain characteristics without examining the soils. Marbut tried to make clear that examination of the soils themselves was essential in developing a system of Soil Classification and in making usable soil maps. In spite of this, Marbut's work reveals his personal understanding of the contributions of geology to soil science. His soil classification of 1935 depends heavily on the concept of a "normal soil," the product of equilibrium on a landscape where downward erosion keeps pace with soil formation.

Clarification and broadening of the concept of a soil science also grew out of the increasing emphasis on detailed soil mapping. Concepts changed with increased emphasis on predicting crop yields for each kind of soil shown on the maps. Many of the older descriptions of soils had not been quantitative enough and the units of classification had been too heterogeneous for making yield and management predictions needed for planning the management of individual farms or fields.

During the 1930s, soil formation was explained in terms of loosely conceived processes, such as "podzolisation," "laterisation," and "calcification." These were presumed to be unique processes responsible for the observed common properties of the soils of a region.

In 1941 Hans Jenny's Factors of Soil Formation, a system of quantitative pedology, concisely summarised and illustrated many of the basic principles of modern soil science to that date. Since 1940, time has assumed much greater significance among the factors of soil formation, and geomorphological studies have become important in determining the time that soil material at any place has been subjected to soil-forming processes. Meanwhile, advances in soil chemistry, soil physics, soil mineralogy, and soil biology, as well as in the basic sciences that underlie them, have added new tools and new dimensions to the study of soil formation. As a consequence, the formation of soil has come to be treated as the aggregate of many interrelated physical, chemical, and biological processes. These processes are subject to quantitative study in soil physics, soil chemistry, soil mineralogy, and soil biology. The focus of attention also has shifted from the study of gross attributes of the whole soil to the co-varying detail of individual parts, including grain-tograin relationships.

In both the classification of Marbut and the 1938 classification developed by the U.S. Department of Agriculture, the classes were described mainly in qualitative terms. Classes were not defined in quantitative terms that would permit consistent application of the system by different scientists. Neither system definitely linked the classes of its higher categories, largely influenced by genetic concepts initiated by the Russian soil scientists, to the soil series and their sub-divisions that were used in soil mapping in the United States. Both systems reflected the concepts and theories of soil genesis of the time, which were themselves predominantly qualitative in character. Modification of the 1938 system in 1949 corrected some of its deficiencies but also illustrated the need for a reappraisal of concepts and principles. More than 15 years of work under the leadership of Guy Smith culminated in a new soil classification system. This became the official classification system of the U.S. National Co-operative Soil Survey in 1965 and was published in 1975 as Soil Taxonomy : A Basic System of Soil Classification for Making and Interpreting Soil Surveys.

Categories and classes of the new taxonomy are direct consequences of new and revised concepts and theories. The system of soil classification discussed in Soil Taxonomy is dynamic and will change as new knowledge is obtained. Its most significant contribution comes from defining class limits quantitatively. The theories on which the system is based are tested every time the taxonomy is applied. For soil survey, the application of quantitatively defined classes to bodies of soil produces quantitatively defined mapping units. This permits the soil maps to be interpreted with more precision than was formerly achieved. Furthermore, this soil-classification system simplifies and accelerates the process of soil correlation.

In addition to the new soil classification system, several other techniques have contributed to the increased precision of soil survey. The use of aerial photographs as mapping bases became almost universal in detailed soil mapping during the late 1930s and early 1940s. Using aerial photographs has greatly increased the precision with which soil boundaries can be delineated on maps. At the same time, the scale of published maps was increased from about 1 : 63,360 to 1 : 24,000 to 1 : 15,840. The smallest area that can be delineated legibly at a scale of 1 : 63,360 is about 15.8 ha; areas of 1 ha can be delineated legibly at a scale of 1 : 15,840.

Another factor has had an immense impact on soil survey, especially during the 1960s. Before 1950, the primary applications of soil surveys were farming, ranching, and forestry. Applications for highway planning were recognised in some States as early as the late 1920s, and soil interpretations were placed in field manuals for highway engineers of some States during the 1930s and 1940s. Nevertheless, the changes in soil surveys during this period were mainly responses to the needs of farming, ranching, and forestry. During the 1950s and 1960s non-farm uses of the soil increased rapidly. This created a great need for information about the effects of soils on those non-farm uses. Beginning about 1950, co-operative research with the Bureau of Public Roads and State highway departments established a firm basis for applying soil surveys to road construction. Soil scientists, engineers, and others have worked together to develop interpretations of soils for roads and other non-farm uses. These interpretations, which have become standard parts of published soil surveys, require different information about soils. Some soil properties that are not important for growth of plants are very important in evaluating soils for building sites, sewage disposal systems, highways, pipelines, and recreation. Many of these uses of soil require very large capital investments per unit area; errors can be extremely costly. Consequently, the location of soil boundaries, the identification of the areas delineated, and the quantitative definition of map units have assumed great importance.

Modern Concept of Soil

Soil is "the collection of natural bodies in the earth's [sic] surface, in places modified or even made by man of earthy materials, containing living matter and supporting or capable of supporting plants out-of-doors. Its upper limit is air or shallow water. At its margins it grades to deep water or to barren areas of rock or ice. Its lower limit to the not-soil beneath is perhaps the most difficult to define. Soil includes the horizons near the surface that differ from the underlying rock material as a result of interactions, through time, of climate, living organisms, parent materials, and relief. In the few places where it contains thin cemented horizons that are impermeable to roots, soil is as deep as the deepest horizon. More commonly soil grades at its lower margin to hard rock or to earthy materials virtually devoid of roots, animals, or marks of other biologic activity. The lower limit of soil, therefore, is normally the lower limit of biologic activity, which generally coincides with the common rooting depth of native perennial plants".

The "natural bodies" of this definition include all genetically related parts of the soil. A given part, such as a cemented layer, may not contain living matter or be capable of supporting plants. It is, however, still a part of the soil if it is genetically related to the other parts and if the body as a unit contains living matter and is capable of supporting plants.

The definition includes as soil all natural bodies that contain living matter and are capable of supporting plants even though they do not have genetically differentiated parts. A fresh deposit of alluvium or earthy constructed fill is soil if it can support plants. To be soil, a natural body must contain living matter. This excludes former soils now buried below the effects of organisms. This is not to say that buried soils may not be characterised by reference to taxonomic classes. It merely means that they are not now members of the collection of natural bodies called soil; they are buried paleosols.

Not everything "capable of supporting plants out-of-doors" is soil. Bodies of water that support floating plants, such as algae, are not soil, but the sediment below shallow water is soil if it can support bottom-rooting plants such as cattails or reeds. The above-ground parts of plants are also not soil, although they may support parasitic plants. Rock that mainly supports lichens on the surface or plants only in widely spaced cracks is also excluded.

The time transition from not-soil to soil can be illustrated by recent lava flows in warm regions under heavy and very frequent rainfall. Plants become established very quickly in such climates on the basaltic lava, even through there is very little earthy material. The plants are supported by the porous rock filled with water containing plant nutrients. Organic matter soon accumulates; but, before it does, the dominantly porous broken lava in which plant roots grow is soil.

More than 50 years ago, Marbut's definition of soil as "the outer layer" of the Earth's crust implied a concept of soil as a continuum. The current definition refers to soil as a collection of natural bodies on the surface of the Earth, which divides Marbut's continuum into discrete, defined parts that can be treated as members of a population. The perspective of soil has changed from one in which the whole was emphasised and its parts were loosely defined to one in which the parts are sharply defined and the whole is an organised collection of these parts.

Factors that Control the Distribution of Soils

The properties of soil vary from place to place, but this variation is not random. Natural soil bodies are the result of climate and living organisms acting on parent material, with topography or local relief exerting a modifying influence and with time required for soil-forming processes to act. For the most part, soils are the same wherever all elements of the five factors are the same. Under similar environments in different places, soils are similar. This regularity permits prediction of the location of many different kinds of soil.

When soils are studied in small areas, the effects of topography or local relief, parent material, and time on soil becomes apparent. In the humid region, for example, wet soils and the properties associated with wetness are common in low-lying places; better drained soils form in most instances in higher lying areas. The correct conclusion is that topography or relief is important. In arid regions, the differences associated with relief may be salinity or sodicity, but the conclusion is the same. In a local environment, different soils are associated with contrasting parent materials, such as residuum from shale and from sandstone, and the correct conclusion is that parent material is important. Soils on a flood plain differ from soils on higher and older terraces where there is no longer deposition of parent material on the surface. The correct conclusion is that time is important. The influence of topography, parent material, and time on the formation of soil is observed repeatedly while studying the soils of an area.

With the notable exception of the contrasting patterns of vegetation in transition zones, local differences in vegetation are closely associated with differences in relief, parent material, or time. The effects of micro-climate on vegetation may be reflected in the soil, but such effects are likely associated with differences in local relief.

Regional climate and vegetation influence the soil as well as topography or relief, parent material, and time. In spite of local differences, most of the soils in an area typically have some properties in common. The low-base status of many soils in humid or naturally acid rock or sediment regions stands in marked contrast to the typical, high-base status in arid or calcareous sandstone or limestone regions. To one who has studied soils only on old landscapes of humid regions, however, low base status is so commonplace that little significance is attached to it. Regional patterns of climate, vegetation, and parent material can be used to predict the kinds of soil in large areas. The local patterns of topography or relief, parent material, and time, and their relationships to vegetation and micro-climate, can be used to predict the kinds of soil in small areas. Soil surveyors learn to use local features, especially topography and associated vegetation, as marks of unique combinations of all five factors. These features are used to predict boundaries of different kinds of soil and to predict some of the properties of the soil within those boundaries.

Soil-Landscape Relationships

Geographic order suggests natural relationships. Running water, with weathering and gravitation, commonly sculptures landforms within a landscape. Over the ages, earthy material has been removed from some landforms and deposited on others. Landforms are interrelated. An entire area has unity through the interrelationships of its landforms.

Each distinguishable landform may have one kind of soil or several. Climate, including its change with time, commonly will have been about the same throughout the extent of a minor landform. The kinds of vegetation associated with climate also likely will have been fairly uniform. Relief varies within some limits that are characteristic of the landform. The time that the material has been subjected to soil formation has probably been about the same throughout the landform. The surface of the landform may extend through one kind of parent material and into another. Of course, position on the landform may have influenced soil-water relationships, micro-climate, and vegetation.

Just as different kinds of soil are commonly associated in a landscape, several landscapes are commonly associated in still larger areas. These areas cover thousands or tens of thousands of square kilometers. Many can be identified on photographs taken from satellites. From this vantage point, broad geomorphic units — the East Gulf Coastal Plain, the Allegheny Plateau, the Laramie Basin, and the Great Valley of California — are apparent. These broad units usually have some unity of landscape, which is characterised by such terms as "plain," "plateau," and "mountain." These physiographic units are composed of many kinds of soil.

The main relief features of a physiographic unit are usually the joint products of deep-seated forces and a complex set of surface processes that have acted over long spans of time. Within a physiographic unit, groups of minor landforms are shaped principally by climate-controlled processes. The climate and biological factors, however, vary much less within a geomorphic unit than across a continent.

Still broader than the geomorphic units are great morphogenetic regions having distinctive climates. For example, one classification recognises glacial, periglacial, arid, semi-arid-sub-humid, humid-temperate, and humid-tropical climatic regions associated with distinctive sets of geomorphic processes. Other major regions characterised by seasonal climatic variation are also recognised. These geomorphic-climatic regions are related to soil moisture and soil temperature regimes. Thus, the great climatic regions are divided into major geomorphic units. Landforms and associated soil landscapes are small parts of these units and are commonly of relatively recent origin. The landforms of concern in soil mapping may include constructional units, such as glacial moraines, and elements of local sequences of graded erosional and constructional land surfaces. These bear the imprint of local, base-level controls under climate-induced processes. Most surfaces that have formed within the last 10,000 years have been subject to climatic and base-level controls similar to those of the present. Older surfaces may retain the imprint of climatic conditions and related vegetation of the distant past. Most landforms of the present started to form during the Quaternary Period; some started in late Tertiary time. In many places conditions of the past differed significantly from those of the present. Understanding climatic changes locally and worldwide far into the past contributes to understanding the attributes of landforms in the present.

Geomorphic processes are important in mapping soils. Soil scientists need a working knowledge of local geomorphic relationships in areas where they map and should understand the interpretations of landforms and land surfaces made by geomorphologists. The intricate inter-relationships of soil and landscape are best studied by a collaboration between soil scientists and geomorphologists.

Development of the Soil Survey

Soil surveys were authorised in the United States by the U.S. Department of Agriculture Appropriations Act for fiscal year 1896, which provided funds for an investigation "of the relation of soils to climate and organic life" and "of the texture and composition of soils in field and laboratory."

In 1899 the U.S. Department of Agriculture completed field investigations and soil mapping of portions of Utah, Colorado, New Mexico, and Connecticut. Reports of these soil surveys and similar works were published by legislative directive. At the same time, the State of Maryland, using similar procedures and State funds, completed a soil survey of Cecil County. Since then many soil surveys have been initiated, completed, and published co-operatively by the Department of Agriculture, State agencies, and other Federal agencies. The total effort is the National Co-operative Soil Survey (NCSS).

The early soil surveys investigated the use of soils for farming, ranching, and forestry. As experience was acquired in the use of soil surveys, predictions were made about other uses, such as highways, airfields, and residential and industrial developments. As the making and the use of soil surveys expanded, the knowledge about soils – about their nature, occurrence, and behaviour for defined uses and management – also increased. The Highway Department of Michigan was applying soil survey experience to assist in planning highway construction in the late 1920s. At about the same time soil surveys in North Dakota were used in tax assessment.

Soil surveys published between 1920 and 1930 reveal a marked transition from earlier concepts to give emphasis to soil profiles and soils as independent bodies. The maps retained significant geologic boundaries as soil maps do today. Many of the surveys of that period provide excellent general maps for evaluating engineering properties of geologic material. In addition, maps and texts of the period show more recognition of other soil properties significant to farming and forestry than do earlier surveys and have value for broad generalisations about farming practices in large areas. The use of aerial photographs for soil mapping, which began during the late 1920s and early 1930s, greatly increased the precision of plotting soil boundaries. To meet the needs for planning the management of individual fields and farms, greater precision of interpretation was required. The changing objectives of soil surveys initiated changes in methods and techniques that made surveys more useful and forced reconsideration of the concept of soil itself.

Beginning in the 1930s, the Soil Conservation Service (SCS) emphasised the control of soil erosion as it used soil surveys for the resource conservation planning of farms and ranches. In the 1950s, extensive use was made of soil survey information in urban land development in Fairfax County, Virginia, and in the sub-division design of sub-urban areas of Chicago, Illinois. Soil surveys were an important base for resource information in regional land-use planning in southeastern Wisconsin. Rural land zoning has also relied on soil surveys.

Soil surveys necessarily involve thousands of different kinds of soils – as many as there are significantly different combinations of genetic factors. The history of a soil and evidence of its potential for use are contained in the properties soil scientists are able to identify through observation and research in the field and laboratory. These properties determine the limitations, suitability, and potential for rural and urban land use of soils. Soil surveys are particularly valuable because they identify specific soil properties and help soil scientists make broad generalisations significant to farming and forestry practices.

The programme of the NCSS can be divided into soil mapping, description of the mapping concepts, and the prediction of the behaviour of these mapping concepts for various uses. Soil behaviour prediction relies on the evaluated and named soil properties to interpret the concept of map units.

Soil Survey and the Soil Map

The different kinds of soil used to name soil map units have sets of interrelated properties that are characteristic of soil as a natural body. This definition is intended to exclude maps showing the distribution of a single soil property such as texture, slope, or depth, alone or in limited combinations; maps that show the distribution of soil qualities such as productivity or erodibility; and maps of soil-forming factors, such as climate, topography, vegetation, or geologic material. A soil map delineates areas occupied by different kinds of soil, each of which has a unique set of interrelated properties characteristic of the material from which it formed, its environment, and its history. The soils mapped by the NCSS are identified by names that serve as references to a national system of soil taxonomy.

The geographic distribution of many individual soil properties or soil qualities can be extracted from soil maps and shown on separate maps for special purposes, such as showing predicted soil behaviour for a particular use. The number of such interpretative maps that can be derived from a soil map is large, and each such map would differ from the others according to its purpose. A map made for one specific interpretation can rarely serve a different purpose.

Maps to show one or more soil properties can be made directly from field observations without making a basic soil map. Such maps serve their specific purposes but have few other applications. Predictions of soil behaviour can also be mapped directly; however, most such interpretations need to be changed with changes in land use and in the cultural and economic environment. A map showing the productivity of crops on soils that are wet and undrained, for example, has little value after drainage systems have been installed. If the basic soil map is made accurately, interpretative maps can be revised as needed without doing additional fieldwork. In planning soil surveys, this point needs to be emphasised. Occasionally, "short-cut" inventories are made for some narrow objective, perhaps at a cost lower than that of a soil survey. Such maps quickly become obsolete. They cannot be revised without fieldwork because vital data are missing, facts are mixed with interpretations, or boundaries between significantly different soil units have been omitted.

The basic objective of soil surveys is the same for all kinds of land, although the number of mapping units, their composition, and the detail of mapping vary with the complexity of the soil patterns and the specific needs of the users. Thus a soil survey is matched to the soils and the soil-related problems of the area. Soil surveys increase our general knowledge about soils and serve practical purposes. They satisfy a need for soils information about specific geographic areas for State, county, and community land-use plans. These plans include resource conservation plans for farms and ranches, development of reclamation projects, forest management, engineering projects, as well as other purposes. The storage and retrieval of soil survey data are possible through the use of Automatic Data Processing (ADP). ADP helps develop important interpretations and policy decisions for both the present and the future.

SOIL SURVEY : EXAMINATION AND DESCRIPTION OF SOILS

Erosion

Erosion is the detachment and movement of soil material. The process may be natural or accelerated by human activity. Depending on the local landscape and weather conditions, erosion may be very slow or very rapid. Natural erosion has sculptured landforms on the uplands and built landforms on the lowlands. Its rate and distribution in time controls the age of land surfaces and many of the internal properties of soils on the surfaces. The formation of Channel Scablands in the state of Washington is an example of extremely rapid natural, or geologic, erosion. The broad, nearly level interstream divides on the Coastal Plain of the Southeastern United States are examples of areas with very slow or no natural erosion.

Landscapes and their soils are evaluated from the perspective of their natural erosional history. Buried soils, stone lines, deposits of wind-blown material, and other evidence that material has been moved and redeposited is helpful in understanding natural erosion history. Thick weathered zones that developed under earlier climatic conditions may have been exposed to become the material in which new soils formed. In landscapes of the most recently glaciated areas, the consequences of natural erosion, or lack of it, are less obvious than where the surface and the landscape are of an early Pleistocene or even Tertiary age. Even on the landscapes of most recent glaciation, however, postglacial natural erosion may have redistributed soil materials on the local landscape. Natural erosion is an important process that affects soil formation and, like man-induced erosion, may remove all or part of soils formed in the natural landscape.

Accelerated erosion is largely the consequence of human activity. The primary causes are tillage, grasing, and cutting of timber. The rate of erosion can be increased by activities other than those of humans. Fire that destroys vegetation and triggers erosion has the same effect. The spectacular episodes of erosion, such as the soil blowing on the Great Plains of the Central United States in the 1930s, have not all been due to human habitation. Frequent dust storms were recorded on the Great Plains before the region became a grain-producing area. "Natural" erosion is not easily distinguished from "accelerated" erosion on every soil. A distinction can be made by studying and understanding the sequence of sediments and surfaces on the local landscape, as well as by studying soil properties.

Landslip Erosion

Landslip erosion refers to the mass movement of soil. Slides and flows are two kinds of landslip erosion. In the slide process, shear takes place along one or a limited number of surfaces. Slide movement may be categorised as slightly or highly deformed, depending on the extent of re-arrangement from the original organisation. In flow movement the soil mass acts as a viscous fluid. Failure is not restricted to a surface or a small set of surfaces. Classes of landslip erosion are not provided. Location of the mass movement relevant to landscape features generally and the size of the mass movement in terms of area parallel to the land surface and the depth may be indicated. Information about the time since the mass movement took place may be very useful.

Water Erosion

Water erosion results from the removal of soil material by flowing water. A part of the process is the detachment of soil material by the impact of raindrops. The soil material is suspended in run-off water and carried away. Four kinds of accelerated water erosion are commonly recognised : sheet, rill, gully, and tunnel (piping).

Sheet erosion is the more or less uniform removal of soil from an area without the development of conspicuous water channels. The channels are tiny or tortuous, exceedingly numerous, and unstable; they enlarge and straighten as the volume of run-off increases. Sheet erosion is less apparent, particularly in its early stages, than other types of erosion. It can be serious on soils that have a slope gradient of only 1 or 2 per cent; however, it is generally more serious as slope gradient increases.

Rill erosion is the removal of soil through the cutting of many small, but conspicuous, channels where run-off concentrates. Rill erosion is intermediate between sheet and gully erosion. The channels are shallow enough that they are easily obliterated by tillage; thus, after an eroded field has been cultivated, determining whether the soil losses resulted from sheet or rill erosion is generally impossible.

Gully erosion is the consequence of water that cuts down into the soil along the line of flow. Gullies form in exposed natural drainage-ways, in plow furrows, in animal trails, in vehicle ruts, between rows of crop plants, and below broken man-made terraces. In contrast to rills, they cannot be obliterated by ordinary tillage. Deep gullies cannot be crossed with common types of farm equipment.

Gullies and gully patterns vary widely. V-shaped gullies form in material that is equally or increasingly resistant to erosion with depth. U-shaped gullies form in material that is equally or decreasingly resistant to erosion with depth. As the substratum is washed away, the overlying material loses its support and falls into the gully to be washed away. Most-U-shaped gullies become modified towards a V shape once the channel stabilises and the banks start to spall and slump. The maximum depth to which gullies are cut is governed by resistant layers in the soil, by bedrock, or by the local base level. Many gullies develop headward; that is, they extend up the slope as the gully deepens in the lower part.

Tunnel erosion may occur in soils with subsurface horizons or layers that are more subject to entrainment in moving free water than is the surface horizon or layer. The free water enters the soil through ponded infiltration into surface-connected macropores. Desiccation cracks and rodent burrows are examples of macropores that may initiate the process. The soil material entrained in the moving water moves downward within the soil and may move out of the soil completely if there is an outlet. The result is the formation of tunnels (also referred to as pipes) which enlarge and coalesce. The portion of the tunnel near the inlet may enlarge disproportionately to form a funnel-shaped feature often referred to as a "jug." Hence, the term "piping" and "jugging." The phenomenon is favoured by the presence of appreciable exchangeable sodium.

Deposition of sediment carried by water is likely anywhere that the velocity of running water is reduced—at the mouth of gullies, at the base of slopes, along stream banks, on alluvial plains, in reservoirs, and at the mouth of streams. Rapidly moving water, when slowed, drops stones, then cobbles, pebbles, sand, and finally silt and clay. Sediment transport slope length has been defined as the distance from the highest point on the slope where run-off may start to where the sediment in the run-off would be deposited.

Wind Erosion

Wind Erosion in regions of low rainfall, can be widespread, especially during periods of drought. Unlike water erosion, wind erosion is generally not related to slope gradient. The hazard of wind erosion is increased by removing or reducing the vegetation. When winds are strong, coarser particles are rolled or swept along on or near the soil surface, kicking finer particles into the air. The particles are deposited in places sheltered from the wind. When wind erosion is severe, the sand particles may drift back and forth locally with changes in wind direction while the silt and clay are carried away. Small areas from which the surface layer has blown away may be associated with areas of deposition in such an intricate pattern that the two cannot be identified separately on soil maps.

Estimating the Degree of Erosion

The degree to which accelerated erosion has modified the soil may be estimated during soil examinations. The conditions of eroded soil are based on a comparison of the suitability for use and the management needs of the eroded soil with those of the uneroded soil. The eroded soil is identified and classified on the basis of the properties of the soil that remains. An estimate of the soil lost is described. Eroded soils are defined so that the boundaries on the soil maps separate soil areas of unlike use suitabilities and unlike management needs.

The depth to a reference horizon or soil characteristic of the soil under a use that has minimised accelerated erosion are compared to the same properties under uses that have favoured accelerated erosion. For example, a soil that supports native grass or large trees with no evidence of cultivation would be the basis for comparison of the same or similar soil that has been cleared and cultivated for a relatively long time. The depth to reference layers is measured from the top of the mineral soil because organic horizons at the surface of mineral soils are destroyed by cultivation.

The depths to a reference layer must be interpreted in terms of recent soil use or history. Cultivation may cause differences in thickness of layers. The upper parts of many forested soils have roots that make up as much as one-half of the soil volume. When these roots decay, the soil settles. Rock fragment removal can also lower the surface. The thickness of surficial zones that have been bulked by tillage should be adjusted downward to what they would be if water had compacted them.

The thickness of a plowed layer of a specific soil cannot be used as a standard for either losses or additions of material because, as a soil erodes, the plow cuts progressively deeper. Nor can the thickness of the uncultivated and uneroded A horizon be used as a standard for all cultivated soil, unless the A horizon is much thicker than the plow layer. If the horizon immediately below the plowed layer of an uneroded soil is distinctly higher in clay than the A horizon, the plow layer becomes progressively more clayey under continued cultivation as erosion progresses; the texture of the plow layer may then be a criterion of erosion.

Comparisons must be made on comparable slopes. Near the upper limit of the range of slope gradient for a soil, horizons may normally be thinner than near the lower limit of the range for the same soil. Roadsides, cemeteries, fence rows, and similar uncultivated areas that are a small part of the landscape as a whole or are subject to unusual cultural histories must be used cautiously for setting standards, because the reference standards for surface-layer thickness are generally set too high. In naturally treeless areas or in areas cleared of trees, dust may collect in fence rows, along roadsides, and in other small uncultivated areas that are covered with grass or other stabilising plants. The dust thus accumulated may cause the surface horizon to become several centimeters thicker in a short time.

For soils having clearly defined horizons, differences due to erosion can be accurately determined by comparison of the undisturbed or uncultivated norms within the limitations discussed. Guides for soils having a thin A horizon and little or no other horizon are more difficult to establish. After the thin surface layer is gone or has been mixed with underlying material, few clues remain for estimating the degree of erosion. The physical conditions of the material in the plowed layer, the appearance and amount of rock fragments on the surface, the number and shape of gullies, and similar evidence are relied on. For many soils having almost no horizon expression, attempting to estimate the degree of erosion serves little useful purpose.

SOIL SURVEY INFORMATION AND HYDRIC SOILS

A soil survey is the systematic examination, description, classification, and mapping of the soils in a survey area and contains published maps, descriptions, and interpretations of the soils in the survey area. The National Co-operative Soil Survey programme was established as a partnership between the Soil Conservation Service (now the Natural Resources Conservation Service), land-grant universities, and federal, state, and county governments to produce soil surveys for private landowners. All private agricultural land with cropland history or potential in the United States has had a modern soil survey as of 1990. No similar programme has been established for federal land, however, except in a few areas where the owner agency agreed to a contract with the Natural Resource Conservation Service, which surveyed the soils in the survey area and ensured quality control.

National Co-operative Soil Survey standard, multiple-purpose soil surveys for farm— and ranch-scale planning have scales of 1 : 12,000, 1 : 15,840, 1 : 20,000, or 1 : 24,000. The scale of the map determines the smallest delineation that can be identified, generally 4 ac (~1.6 ha) or more for 1 : 20,000 and 1.5 ac (~0.6 ha) or more for 1 : 12,000 scale. Landscape units such as summits, shoulders, backslopes, footslopes, terraces, or flood plains are identified and delineated using topographic map or aerial photo references. The soil scientists attempt to sub-divide the landscape units into soil map units with a composition as pure and as predictable as possible for the scale of their field maps.

Where possible, soils are identified at the soil series level, the lowest category of the U.S. system of soil taxonomy. A soil series is a conceptualised class of soils that have the same classification and that are similar in major characteristics that influence their behaviour or potential uses (Soil Survey Division Staff, 1993). Extremely variable soils, such as those on flood plains subject to frequent flooding, are often classified at the suborder level (*e.g.*, Aquepts) or the great group level (*e.g.*, Fluvaqents) in Soil Taxonomy (Soil Survey Staff, 1999). These soils are called higher taxa in soil surveys.

Map units are named for the most extensive soil or soils; other soils that are of lesser extent are called inclusions. Consociations are mapped on landscape units with relatively pure composition of one extensive soil and similar soils. Included dissimilar soils generally comprise 25 per cent or less of the map unit. The included dissimilar soils that limit land use more than the most extensive soil and similar soils generally comprise 15 per cent or less of the map unit. Complexes of two or three soils are mapped on landscape units with repeatable and predictable composition of soils in areas too small or scattered to be mapped separately. Undifferentiated map units are used to identify areas where the most extensive soils do not occur predictably in all delineations. The named soil or soils in consociations or complexes are found in all correctly mapped delineations, but the named soils of undifferentiated groups are not necessarily present in all delineation, or the composition of the named soils varies widely. Other small, but important features such as rock outcrop or small wet areas that do not occur in a regular predictable pattern are indicated on maps by special spot symbols.

Hydric Soils and Soil Survey

Detailed soil surveys (scales of 1 : 100 to 1 : 400) or onsite investigations must be conducted for the single purpose of delineating hydric soils. The standard soil surveys were drawn with a minimum delineation size, which is not detailed enough to identify wetlands as small as 5,000 ft2 (~0.05 ha), as required by some regulations. Therefore, standard soil surveys are best used with hydric soils lists for offsite planning and resource evaluation.

The most important information in a printed or digital soil survey is the recorded knowledge of the relationships between soils, landscape positions, hydrology, and vegetation in the survey area. The location of a suspected wetland on a detailed soil map provides the names of soils to check against the hydric soils list and to record as part of the Delineation Manual delineation process. However, the information in standard soil surveys is also useful for prediction of hydric soils and for reconnaissance of soil resources prior to onsite investigation. Several spot symbols, such as those for springs, seeps, wet spots, and marshes, can also be reliable indicators of the presence of hydric soils.

Precautions must be taken in using standard soil surveys for gathering hydric soil information because map units commonly contain both hydric and non-hydric soils. The location of hydric soils within any map unit delineation is generally unknown unless detailed landscape position descriptions were given for major and minor components, and because the range of a soil series may include hydric and non-hydric soils. Knowledge of the limitations of soil surveys is important to appreciate the possible and appropriate uses of soil surveys for hydric soil identification. Both printed and digital standard soil surveys have information concerning hydric soils, although each is slightly different.

Printed Soil Surveys

The following sections of printed or scanned soil survey reports are useful in identifying hydric soil resources. Not all soil surveys have the same tables, but most soil surveys published after the mid-1970s should contain similar information.

These are listed in order of decreasing reliability :

- Information in the climate tables (Temperature and Precipitation, Freeze Dates in Spring and Fall, and Growing Season) can be used to identify the growing season as part of the Hydric Soil definition.
- Some newer soil surveys contain hydric soil lists. In others the national, state, or county hydric soils list (Soil Survey Division, 2003c) can be compared with higher taxa, series, and phase names in the Soil Legend or the Index to Map Units.
- In other cases the Classification of the Soils table can lend information by comparison against the hydric soils criteria.
- The Soil and Water Features table includes information about flooding or ponding frequency to be matched against hydric soils criteria 3 and 4. Also, water table depth and months may be matched against hydric soil criteria 2a and 2b to provide general information, but not to confirm meeting the hydric soil definition.
- The Soil Series and Their Morphology section gives a detailed description of representative soil morphology, and these may be detailed enough to match against current Field Indicators.
- The section Detailed Soil Map Units gives hydrology indicators such as drainage class or water table depth and a relative description of landscape position. Soils that are poorly drained or very poorly drained and that are in flooded, ponded, low-lying, or nearly level landscape positions are likely to be hydric soils.

- Other supportive interpretative information can be gathered from the Woodland Management and Productivity table that lists trees that may be wetland indicators.
- The Wildlife Habitat table gives information about potential habitat for wetland plants and shallow water areas.

Web-based Soil Survey

Web Soil Survey (WSS) is an on-line version of Soil Survey Geographic Database (SSURGO). WSS is operated by the USDA NRCS and has soil maps and data available online for more than 95 per cent of the nation's counties. The site is updated and maintained online as the single authoritative source of soil survey information. WSS is simple to use and provides spatial soil survey data with a legend on a background satellite imagery similar to that in Google Maps™ (Google, Mountain View, CA), and an interface that is similar to Soil Data Viewer (Soil Survey Division, 2007) for producing thematic maps. Web Soil Survey is currently limited to producing soil maps of 10,000 acres or less, and does not have an ability to export files that can be loaded into a geographic information system software. However, WSS does produce a colour map and full report output option and a soil primer for introductory users of soil information.

Digital Soil Surveys – Tabular Data

Many standard soil surveys have been converted to a digital version called the Soil Survey Geographic Database (SSURGO). The following information exists in the database tables downloaded with the spatial files. The tables are related by fields (columns) that provide soil survey information such as map unit symbol in the musym field, sequence of the named soil map unit components in the seqnum field, the name of the map unit component in the compname field, the soil series identification number in the s5id field, and the map unit composition percentage in the comppct field. Fields are filled with code values that are identified in the codes table.

Table. Example of Data from Montgomery county, Virginia, SSURGO comp (map unit composition) table.

Stssaid	Muid	Seqnum	Musym	Compname	S5id	Comppct
VA121	12110	1	10	Craigsville	VA0072	100
VA121	12111B	1	11B	Duffield	PA0013	45
VA121	12111B	2	11B	Ernest	WV0018	35

The comp (map unit component) table identifies the named map unit components that are hydric soils in the hydric field; annual flooding in the anflood, anflodur, anflobeg, and anfloend fields; flooding and ponding during the growing season in the pnddepl, pnddeph, pnddur, pndbeg, pndend, gsflobeg, gsflodur, gsfloend, and gsflood fields; soil classification in the clascode field; drainage class in the drainage field; and water table information in the wtdepl, wtdeph, wtkind, wtbeg, and wtend fields.

Table. Example of data from Montgomery County, Virginia, SSUEGO comp (map unit composition) table.

Muid	Seqnum	Musym	Compname	S5id	Comppct	Hydric	Anflood	Anflodur	Anflobeg	Anfloend
12124D	1	24D	Jefferson	KY0113	100	N				
12125	1	25	Mcgary	IN0112	40	N	Occas	Brief	JAN	MAY
12125	2	25	Purdy	WV0034	35	Y				

Muid	Seqnum	Musym	Gsflood	Gsflodur	Gsflobeg	Gsfloend	Pnddepl	Pnddeph	Pnddur	Pndbeg	Pndend
12124D	1	24D					0	0			
12125	1	25					0	0			
12125	2	25					0	0			

Muid	Seqnum	Musym	Clascode	Drainage	Wtdepl	Wtdeph	Wtkind	Wtbeg	Wtend
12124D	1	24D	UUDHAAA 09602460202021602	W	0	0			
12125	1	25	AAQEPAE 12602340202021602	SP	1	3	APPAR	JAN	APR
12125	2	25	UAQENAA 11402340202021602	P	0	0	APPAR	NOV	JUN

The inclusn (hydric soil inclusions) table identifies the named map unit components that are hydric soils in the hydric field, the soil that is considered hydric in the inclsoil field, the map unit per cent composition of the included hydric soil in the inclpct field, the landform location in the landfmlo field, the criteria under which the soil met the hydric soil criteria in the hydcrit field and the hydric soil criteria that must be identified onsite in the onsite field.

Table. Example of data from Montgomery County, Virginia, SSUEGO comp (hydric soil inclusions) table.

Muid	Seqnum	Inclsoil	Inclpct	Hydric	Landfmlo	Condition	Hydcrit	Onsite
12121C	1	Purdy		Y	FP		2B3	3
12125	1	Ponded Soils			DP		2B3	3
12125	2	Flooded Soils			FP		2B3	4

The hydric condition (hydcond) field is available in the comp (map unit component) and inclusn (hydric soil inclusions) tables downloaded with the spatial files. The hydric condition field identifies each map unit component and all map unit inclusions that meet current hydric soil criteria and Field Indicators. The mapunit (map unit) table identifies the name of the map unit components and any special varieties (phases) in the muname field for comparison with national, state, or county hydric soils.

Table. Example of data from Montgomery County, Virginia, SSUEGO mapunit (map unit) table.

State	Ssaid	Musym	Muname
VA	121	10	Craigsville soils
VA	121	11B	Duffield-Ernest complex, 2 to 7 percent slopes
VA	121	11C	Duffield-Ernest complex, 7 to 15 percent slopes

The taxclass (classification) table identifies the classification in the clascode field for comparison with hydric soil criteria 1, 3, and 4.

Table. Example of data from Montgomery County, Virginia, SSUEGO taxclass (classification) table.

Clascode:c	Class:c
AAQEPAE 12602340202021602	Fine, mixed, mesic Aeric Epiaqualfs
AUDHAAA 12602340202021602	Fine, mixed, mesic Typic Hapludalfs
AUDHAAQ0612602340202021602	Fine, mixed, mesic Aquic Hapludalfs

The wlhabit (wildlife habitat) table identifies suitability of the soil to produce the habitat (wetland) requirements for wetland wildlife in the wlwet field, the wildlife (wetland plants) habitat element in the wlwetplt field, and the habitat (shallow water) element in the wlshlwat field.

Table. Example of data from Montgomery County, Virginia, SSUEGO wlhabit (wildlife habitat) table.

Stssaid	Muid	Seqnum	Wlwetplt	Wlshlwat	Wlwet
VA121	12124D	1	Very poor	Very poor	Very poor
VA121	12125	1	Fair	Fair	Fair
VA121	12125	2	Good	Good	Good

DIGITAL SOIL SURVEYS – SPATIAL DATA

Soil Data Viewer is available as an extension tool to ArcMap™ (Environmental Systems Research Institute, Redlands, CA), and allows users to easily create hydric soil maps. This tool is designed to make maps from SSURGO spatial data and data tables.

Outside of using Soil Data Viewer, advanced users of geographic information systems software may generate hydric soil maps using the hydric soil hydric field in the comp (component) and inclusn (hydric soil inclusion) tables. Maps that show delineations with hydric soils reveal spatial relationships that may indicate the likely occurrence of hydric soils.

For example, the map units with the following characteristics are likely to have hydric soils within most delineations :

- Flood plains, especially between and adjacent to hydric units;
- Level and nearly level areas with shallow, perching layers such as fragipans or somewhat poorly drained soils;
- Spot symbols, such as springs, seeps, wet spots, and marshes; and
- Shallow water tables during the growing season.

USING SOIL SURVEY TO IDENTIFY HYDRIC SOILS

A soil survey is the systematic examination, description, classification, and mapping of soils in an area that produces published maps, descriptions, and interpretations of the soils in an area. The National Co-operative Soil Survey (NCSS) programme was established as a partnership between USDANRCS, land-grant universities, federal, state, and county governments to produce soil surveys for private landowners. All private agricultural land with cropland history or potential in the United States has had a modern soil survey as of 1990. No similar programme has been established for federal land, however, except in a few areas where the owner agency agreed upon a contract with USDA-Natural Resources Conservation Service (USDA-NRCS) to do the work and oversee the quality control. Standard multiple-purpose soil surveys were conducted by NCSS for farm- and ranch-scale planning on maps of 1 : 12,000, 1 : 15,840, 1 : 20,000, or 1 : 24,000 scale. The scale of the map determined the smallest delineation that could be identified, usually four acres or more. Landscape units such as summits, shoulders, backslopes, footslopes, terraces, or floodplains were identified and delineated using topographic map or aerial photo references. The objective of the soil scientist was to sub-divide the landscape units into soil map units with as pure and predictable soil composition as possible for the scale of their field maps.

When possible, soils were identified to the soil series level, the lowest category of the U.S. system of soil taxonomy. A soil series is a conceptualised class of soils that have the same classification and are similar in major characteristics that influence their behaviour or potential uses. When the soil was extremely variable, such as in frequently-flooded floodplains, soils were classified only down to the Suborder (*i.e.*, Aquepts) or Great Group level (*i.e.*, Fluvaqents) in Soil Taxonomy. These soils are higher taxa in the soil survey.

Map units were named for the most extensive soil(s); other soils of lesser extent were called inclusions. Consociations were mapped on landscape units with relatively pure (≥ 75 per cent) composition of one extensive soil. The included soils must not have limited the land use more than the most extensive soil. Complexes of two or three soils were mapped on landscape units with repeatable and predictable composition of soils that were too small or infrequent to be mapped separately. Undifferentiated map units were used to identify areas where the most extensive soils did not occur predictably in every delineation. The named soil(s) in consociations and complexes occur in every mapped delineation, but inclusions and the named soil(s) of undifferentiated map units are not present in every delineation. Other small but important features that could not be identified to a soil series were indicated on maps by special "spot" symbols placed at a single point on the map.

Hydric Soils

Hydric soils are soils that satisfy all requirements of the Hydric Soil Definition. The Hydric Soil Definition is : "A hydric soil is a soil that formed under conditions of saturation, flooding, or ponding long enough during the growing season to develop anaerobic conditions in the upper part." Hydric Soil Lists contain a listing of soils that have a probability of meeting the hydric soil definition. Hydric Soil Criteria are used to generate Hydric Soil Lists.

Hydric Soil Lists are created by comparing the range of soil series properties in the Official Soil Series Description with the Hydric Soil Criteria. Soil series that have any overlap with the set of criteria are interpreted to meet the hydric soil definition. Thus, the presence of a soil on a hydric soil list is only an interpretive rating and does not mean that the soil in question is in fact hydric.

Hydric Soil Criteria and Hydric Soil Lists are used as offsite assessment tools along with standard soil surveys for planning and management. Onsite evaluations are conducted to confirm that the soil meets the hydric soil definition through the positive morphological evidence outlined in the 1987 Corps of Engineers Wetland Delineation Manual plus updates and in the latest version of the Field Indicators of Hydric Soils of the United States (USDA-NRCS, 2002). Field evidence that Hydric Soil Criteria 1, 3, or 4 is met is also evidence that the hydric soil definition is met.

Hydric Soils and Soil Survey

Detailed large scale soil surveys (scales of 1 : 100 to 1 : 400) and onsite investigations must be conducted for the single-purpose of delineating hydric soils. The standard soil surveys were drawn with minimum delineation size of four to eight acres (1.6 to 3.2 ha), which is not detailed enough to identify wetlands as small as 5000 ft^2 (~0.05 ha) as required by some regulations. Therefore, standard soil surveys are best used with Hydric Soils Lists for offsite planning and resource evaluation. The most important information in a printed or digital soil survey is the recorded knowledge of the relationships between the soils, landscape positions, hydrology, and vegetation in the survey area. The location of a suspected wetland on the soil survey map provides the name of soils that can be checked against the Hydric Soils List and recorded as part of the U.S. ACOE delineation process. However, the information in standard soil surveys is also useful for prediction of hydric soils and for reconnaissance of soil resources prior to onsite investigation. Several spot symbols are also reliable indicators of hydric soil presence, such as springs, seeps, wet spots, and marshes.

Precautions must be taken in using standard soil survey for gathering hydric soil information because map units often contain both hydric and non-hydric soils. The location of hydric soils within any map unit delineation is unknown unless detailed landscape position descriptions were given, and because soil series ranges may contain hydric and non-hydric soils. Knowing the limitations is important to appreciate the possible and appropriate uses of soil survey for hydric soil identification. Both printed and digital standard soil surveys have information concerning hydric soils, although each is slightly different.

Chapter 8

SOIL FERTILITY

INTRODUCTION

A successful soil fertility programme for wheat requires knowledge of a field's yield potential and a recent soil test. The soil test will provide current levels of phosphorus and potassium in the soil and the soil pH. Soil pH will assist in determining the need for micro-nutrients and other soil amendments most importantly lime. When the proper soil pH is maintained, adequate levels of micro-nutrients and secondary nutrients (*e.g.* sulfur) should be released by the soil organic matter. The proper soil pH for western Ohio (sub-soils derived from limestone) should be above 6.0 and below 7.0, and above 6.5 and below 7.0 for eastern Ohio (sub-soils derived from shale and sandstone). The lime test index or buffer pH on the soil test should be used for lime recommendations. Lime recommendations are available from the Ohio State University Extension fact sheet AGF-505-07 Soil Acidity and Liming for Agronomic Production or bulletin 472 Ohio Agronomy Guide 14th Edition. These recommendations are for mineral soils with adequate drainage containing 1 to 5 per cent organic matter. Organic soils (organic matter > 20 per cent) and sandy soils (CEC < 6) will require different recommendations.

Nitrogen : rates are based on yield potential and not on soil analysis. Total nitrogen recommendations are given in Table or may be calculated by the following equation :

$$40 + [1.75 \times (\text{yield potential} - 50)]$$

For the corresponding rate, part of it should be applied in the fall and the rest after greenup. Generally, 20 to 30 pounds of fall applied nitrogen should be adequate for early fall and spring growth. Spring recommendations should be the total nitrogen required less the amount applied in the fall.

No credits are given for previous crops. For example, a wheat crop with a 90 bu/A yield goal would require 110 pounds of nitrogen. If the grower applied 30 pounds in the fall, the remaining 80 pounds should be applied in the spring.

Yield Potential (bu/A)	Nitrogen Rate (lb/A)
60	60
70	75
80	90
90	110
100	130

Yields are generally not affected when the initial spring nitrogen is applied between greenup and early stem eslongation. Nitrogen losses may be severe on applications prior to greenup and may cause significant yield reductions, regardless of nitrogen source. Significant yield losses may also occur if initial spring applications are delayed until after early stem elongation.

Table. Grain Yield Response to Spring N (70-80 lb/ A) at different Application Times — Custar (2000 2006) and S. Charleston, Ohio (2005 - 2006).

Application Time	Location	
	Custar	S. Charleston
	bu/acre	
Greenup	82.2	93.5
Early Stem Elongation (Feekes GS 6)	85.5	94.5
Average	83.8	94.0

Table. Grain yield response to spring nitrogen (70 lb/A) applied at different growth stages - Custar, Ohio (Lentz, 2003)

Application Time	Nitrogen Source			Average
	Urea	Urea-ammonium Nitrate Solution	Ammonium Sulfate	
	bu/acre			
Pregreenup (2 wks before greenup)	55.0	54.0	62.8	57.3
Greenup	76.3	74.1	83.9	78.1
Early Stem Elongation	82.3	76.0	77.9	78.7
Late Stem elongation	70.2	68.7	73.0	70.6
Average	70.9	68.2	74.4	

SPLIT APPLICATIONS AND NITROGEN SOURCE

In most years, yield gains from a split application have not been large enough to offset the cost of a second trip across a field. A split spring application programme may be a benefit in poorly drained fields that are prone to nitrogen loss, and also in years that the potential for nitrogen loss is great.

Years that have a potential for nitrogen loss generally have a warmer than normal winter followed by a warm and wet April. Delaying initial nitrogen application until closer to early stem elongation would have the same effect as a split application without sacrificing yields.

Nitrogen source is not a concern unless conditions are conducive for nitrogen loss. In general, urea-ammonium nitrate solutions have the greatest potential for loss, then urea, and ammonium sulfate the least. Risk for nitrogen loss potential is the greatest for early applications and decreases as plants approach early stem elongation. Fields prone to wet conditions would also be susceptible to nitrogen loss. If nitrogen loss is not a concern, economics should determine nitrogen source.

In summary, initial spring application should be applied between greenup and early stem elongation. Waiting until early stem elongation may increase yields slightly but the small gain is offset by the risk of wet conditions at elongation time.

If these wet conditions delay application until late stem elongation or later, a yield decrease may occur. Nitrogen source should be dependent upon the risk of nitrogen loss conditions and cost.

Phosphorus: should be applied before planting when the soil test level is below 50 ppm. Actual phosphorus recommendations are determined by the yield goal and soil test level. Phosphorus and fall applied nitrogen are often applied as diammonium phosphate (DAP) or mono-ammonium phosphate

Table. Phosphorus Recommendations for Wheat at Various Yield Potentials and Soil Test Levels.

Yield Potential (bu/A)	Soil Test (ppm)				
	15	20	25-40	45	50
	lb P_2O_5/A				
60	90	65	40	20	0
70	95	70	45	20	0
80	100	75	50	25	0
90	105	80	55	30	0
100	115	90	65	30	0

Potash: recommendations are based upon the yield goal, soil cation exchange capacity (CEC) and the soil test level.

Soils with larger CEC values have a greater chance of potassium becoming unavailable to the crop, and require more potash than low CEC soils. Recommendations should be greater in fields where the straw may be baled and removed.

Table. Potash Recommendations for Wheat at Various Yield Potentials, CEC, and Soil Test Levels - only Grain Removed not Straw.

Yield Potential (bu/A)	Soil Test (ppm)				
	15	20	25-40	45	50
	lb P_2O_5/A				
60	90	65	40	20	0
70	95	70	45	20	0
80	100	75	50	25	0
90	105	80	55	30	0
100	115	90	65	30	0

Table. Potash Recommendations for Wheat at Various Yield Potentials, CEC, and Soil Test Levels - only Grain and Straw Removed.

Yield Potential (bu/acre)	Soil CEC	Soil test K (ppm)						
		25	50	75	100	125	150	175
60		lb K$_2$O/acre						
	10	210	170	135	100	100	0	0
	15	250	205	160	120	100	60	0
	20	300	250	200	150	100	100	0
80		lb K$_2$O/acre						
	10	235	200	160	120	120	0	0
	15	275	230	190	145	120	80	0
	20	320	270	220	170	120	120	0
100		lb K$_2$O/acre						
	10	260	225	185	150	150	0	0
	15	300	260	215	170	150	95	0
	20	350	300	250	200	150	150	0

Sulfur : historically data has not supported the need for sulfur on medium to fine textured soils with adequate organic matter. However, atmospheric depositions have decreased over past decades as sulfur emissions from manufacturing processes have diminished, casting doubt whether Ohio soils still contain adequate sulfur levels for optimum wheat production. Studies were repeated in 2005 and 2006 to determine if wheat yields may respond to supplemental sulfur. Yields were similar to non-sulfur and sulfur treatments confirming that typical Ohio soils would have minimal response to sulfur fertilizer.

Table. Grain Yield Response to Spring N (80 lb/A) and Sulfur (20 or 40 lb/A) — Lentz and Mullen, 2005 - 2006.

Fertilizer Source	Location	
	Custar	S. Charleston
	bu/acre	
Urea	82.2	94.4
Urea-gypsum 20 lb S/A	81.3	95.6
Urea-gypsum 40 lb S/A	83.7	93.7
Average	82.4	94.6

Manganese : Wheat is almost as sensitive to manganese deficiencies as is soybean and occurs in the same areas of fields. Deficiency symptoms are usually not severe enough to be seen but will reduce yield. Maintaining soil pH between 6.0 and 7.0 will usually eliminate the problem. For fields that have a history of manganese problems, manganese sulfate can be applied in a band or in contact with seeds at planting or mix 4 pounds per acre with urea-ammonium nitrate solution at spring topdress.

MAINTENANCE OF SOIL FERTILITY

No two soils are alike either in respect of their nature or in respect of quantities of plant nutrients they contain. Under a given situation, the system of farming, soil management and manuring practices, etc., influence the productiovity of soils and crop yields obtained from them. It is estimated that the different agricultural crops in India remove about 4.27 million tonnes of nitrogen, 2.13 million tonnes of phosphoric acid, 7.42 million tonnes of potash and 4.88 million tonnes of lime per year. The production of larger yields through improved varities of crops and intensive cultivation will increase the depletion of nutrients still further. But erosion and leaching cause additional losses.

The present production of synthetic nitrogenous fertilizers in the country reached 1.5 million tonnes of nitrogen, and the bulky organic manures might supply another 1.5 million tonnes of total nitrogen. The amounts of phosphorous, potash, etc., added to the soil are very small. It is thus obvious that the current huge drain on nutrient supplies will continue to impoverish the soils unless these supplies are replenished by natural or by artificial means.

The principal methods of supplementing natural recuperation and for improving the productive capacity of the soils are :
- To add organic matter to the soil, so that through decay, it may furnish a more or less continuous supply of nuttrients for crops, to restore or increase the amount of defficient nutrients by the application of fertilizers.

The urgent need of the constantly expanding agriculture production to meet the requirements of the continually increasing human and cattle populations in India makes the supply of additional plant nutrients through fertilizers and organic manures, a problem of supreme importance.

Types of Manures and Fertilizers

Indian soils are usually very poor in organin matter as well as in nitrogen. Pohsphate deficiency is less wide-spread and potash deficiency generally occurs in compact areas. In acid soils the addition of lome steps up production.Materials which are commonly used to maintain and improve soil fertility.

- *Manures* : These are relatively bulky materials, such as animal or green manures, which are added mainly to improve the physical condition of the soil, to replenish and keep up its humus status, to maintain the optimum conditions for the activities of soil micro-organisms and make good a small part of the plant nutrients removed by crops or otherwise lost through leaching and soil erosion.
- They, thus, supply practically all the elements of fertility which crops require, though not in adequate proportions. The plant-food elements contained in a manure are released in an available form after it is applied to the soil and is decomposed by soil micro-organisms. Similarly, the green manures add not only substantial amounts of organic matter but also nitrogen.
- *Fertilizers* : Fertilizers are inorganic materials of a concentrated nature; they are applied mainly to increase the supply of one or more of the essential nutrients,

- *e.g.* nitrogen, phosphorous and potash. Fertilizers contain these elements in the form of soluble of readily available chemical compounds. This distinction is, however, not very rigid. In common parlance the fertilizers are sometimes called 'chemical','artificial' or 'inorganic' manures.
- *Concentrated Organic Manures* : Some of the concentrated materials, such as oil-cakes, bone-meal, urine and blood are of organic origin. The use of manures and fertilizers is complementary and not as a substitute for each other.
- *Bulky Organic Manures* : The properties and role of organic matter and humus in the soil have been explained already.

Farmyard Manure

Good-quality farmyard manure is perhaps the most valuable organic matter applied to a soil. It is the most commonly used organic manure in India. It consists of a mixture of cattle dung, the bedding, used in the stable and of any ramnants of straw and plant stalks fed to cattle. Though its crop-increasing value has been recognised from time immemorial, more than 50 per cent of the cattle dung produced in the country today is burnt as fuel and is thus lost to agriculture. Not only this tremendous waste, but also the tradition method of preparing and storing the farmyard manure is generally faulty. The cattle-dung, together with stable-waste and house sweeping, is first collected in the open backyard, and when a cartload has been collected, it is removed to another heap or to an uncovered pit in a common plot outside the village. The loose heaps lie exposed to the sun, with the result that the raw organic matter dries up quickly and does not rot properly. Very often, a part of the dry dung is blown off by wind or washed away by rain. Cattle urine is either not conserved or is stored in a defective manner. American studies on the distribution of soil derived elements between urine and faeces of dry cows have shown that 95 per cent of Potassium, 63 per cent of nitrogen and 50 per cent of sulphur are contained in the urine. The wastage of nitrogen-rich urine, the loss of nitrogen (in the form of ammonia) due to the fermentation of exposed cattle dung, and the washing away of soluble mineral elements be leaching reduce its manurial value in India to a great extent. Its average content of plant nutrients under Indian and European conditions is shown below for comparison :

Percentage Content

	N	P_2O_5	K_2O
India	0.3	0.15	0.3
European Countries	1.0	0.30	1.0

About half of this nitrogen, one-sixth of phosphorous and more than half of potash are readily soluble and subject to dissipation. However, the loss of nitrogen and minerals elements caused by careless handling can be reduced greatly by using absorbent bedding for cattle, storing dung in stone or brick-line pits, mixing large quantioties of straw and other vegitable matter with cattle dung, and keeping the heap compact and moist. Thus, if urine is properly conserved, the loss of soluble mineral elements through seepage is prevented, bacterial decomposition of raw organic matter is encouraged, plant nutrients are made soluble, and nitrogen losses are minimised.

Table. The Relative Absorbent Capacity of Different Materials used as Bedding for Cattle

Material by one kg of the 24 hours of soaking	Quantity of water (in kg) retained following materials after
Wheat straw	2.20
Sawdust	4.35
Peat straw	2.80
Dry leaves	2.00
Peat	6.00
Soil	0.50
Sand	0.25

If urine is not conserved in the bedding used for cattle it must be collected in covered pucca cistern and then, added to the dung in the manure pit. Nitrogen in the urine is mainly in the form of urea, which readily changes into the highly volatile ammonium carbonate through bacterial action, and quickly loses ammonia thereafter by evapouration. This loss can be reduced to a great deal if the manure and the urine-soaked absorptive itter for bedding are kept compacted in a pit.

The pit may be 1 m in depth, 1.3 to 1.5 m in width and 4.5 to 6 m in length, depending upon the no. of cattle on a farm. The filling of the pit should be 'sectional' and when each section of three or 1.3 m in length is filled to about 45 cm above the ground level, it should be clustered with 2.5 cm layer of a mixture of mud and dung in equal proportions. Before plastering, 4 to 5 buckets of water should be added to the manure in the pit. Plastering conserves moisture and nitrogen and also prevents housefly nuisance. The manure becomes ready for use in about 4 to 5 months after plastering.

The quality of manure is also improved by the concentrated feeds given to cattle. Cotton-seed, cotton-seed cake, linseed-meal, wheat bran, grain husk, groundnut cake, gram, horse-gram, etc., are rich in nitrogen, phosphorous, potassium, magnesium and sulphur. It has been found that in the case of adult working-cattle about 80 per cent of nitrogen and the other mineral elements contained in the feed is recovered in urine, faeces and other animal by-products. Accordingly, manure from cattle fed on cereal straws and grass hay is much less valuable than that from animals fed on legume hays, grains and concentrates. In foreign countries, considerable attention has been given to the use of preservatives on manure. Calcium sulphate or gypsum and superphosphate have proved most promising in preventing the escape of ammonia. Gypsum has been found specially effective as an ammonia-absorbing agent. Superphosphate, besides absorbing ammonia, supplies additional phosphorous and, thus, improves the crop producing capacity of the manure.

Partially rotten farmyard manure should generally be applied to the soil about three to four weeks before the sowing of a crop. In case there is sufficient moisture in the soil, there sill be enough time for its decomposition and for improving the soil structure. Its application too long before sowing a crop will either cause a drying up of the rotted manure or too quick decomppostion, depending on the incidence of rains. But in each case, there will be a serious loss of ammonia and nitrogen. If the manure is already well rotted, it is advisable to apply it just before sowing to a crop.

This procedure is perticularly essential in the case of light soils. In any case, after the manure is carted to the field, it should be evenly spread and worked into the soil soon to avoid the loss of nitrogen. The existing practice of leaving the manure in small heaps scattered in the field for several days, before spreading it in the field and incorporating it into the soil, results in the serious deterioration of its quality, particularly if strong winds blow.

In vagitable and fruit cultivation, the application of well-rotten manure, in conjunction with fertilizers to young plants individually, has been founnd to give the best results. In Egypt, even the cotton crop, which is invariably grown on ridges, is given a top-dressing a handful of the rotten manure being applied to the soil at the base of each plant before an irrigation, and is worked into the soil with a hand hoe.

The practice of penning cattle, sheep and goats in the fields in summer is common in some parts of the country. Folding 7,000 sheep for one night is said to add the equivalent of 149.3 quintals (14.93 tonnes) of cattle dung. The fresh dung left in the field in such cases rapidly dries up. This drying checks ammonification and loss of nitrogen. With the first fall of rain, the dung is worked into the soil.

It, therefore, does not lose much of its fertilizing value. Further more, its beneficia effects on the physical condition of the soil are undeniable. However, sheep-folding is said to make the land more weedy.

There must be adequate moisture in the soil for the proper decomposition of organic matter. Farmyard manure can, therefore, be applied to all crops grown in the rainy season or grown under irrigation. The quantity of manure to be applied to unirrigated crops varies from 1.5 to 2 cartloads per hectare in areas of heavier rainfall. If sufficient farmy manure is not available, it may be applied at the usual rate to a part of the land, say, to one-third or one-forth of the area, in rotation every year, so that all parts of the field receive the manure regularly once in three or four years.

For irrigated field crops, the rate varies from 10 to 20 cartloads. Sugarcane, maise and garden crops, such as potatoes, turmeric, ginger, vegetables and fruits receive still higher doses, amounting sometimes to 15 to 25 cartloads. A cartload of manure, measuring 9 cubic metres, weighs about half a tonne.

SOIL FERTILITY AND PRODUCTIVITY

Soil fertility is the capacity/ability of the soil to supply the plant nutrients required by the crop plants in available and balanced forms or it is the capacity of soil to produce crops of economic value to man and maintain the health of the soil for future use or the soil is said to be fertile when it contains all the required nutrients in the right proportion for luxuriant plant growth. Plants like animals and human beings require food for growth and development.

This food is composed of certain chemical elements often referred to as plant nutrients or plant food elements. These nutrients are obtained from soil through roots. Plants need 16 elements for their growth and completion of life cycle. In addition to these, 4 more elements *viz.* sodium, vanadium, cobalt and silicon are absorbed by some plants for special purposes.

Classification and source of nutrients :

Class	Nutrient	Source
Basic	C, H, O	Air and water
Macro	N, P, K, Ca, Mg, S	Soil
Micro	Fe, Mn, Zn, Cu, B, Mo & Cl	Soil

Four more recognised nutrients are NA, Co, VA & SI.

Basic nutrients (C, H, and O) constitute 96 per cent of total dry matter of plants. Macro (Major) nutrients (primary-N, P, K, and secondary-Ca, Mg, S) are required in large quantities while Micro-nutrients (Trace elements-Fe, Zn, Cu, B, Mo, Cl, and Mn) are required in small quantities. These trace elements are very efficient and minute quantities produce optimum effect. On the other hand, even a slight deficiency or excess is harmful to plants.

Function of the plant :
- Elements that provide basic structure to the plant – C, H, O.
- Elements useful in energy storage, transfer and bonding – N, S & P. these are accessory structural elements which are more active and vital for living tissues.
- Elements necessary for change balance – K, Ca & Mg, act as regulators and carrier.
- Elements involved in enzyme activation and electron transports. Fe, Mg, Cu, Zn, B, Mo & Cl are catalysers and activators.

Criteria of Essentlailty : Armon and Stout (1939) proposed criteria of essentiality which was refined by Arnon (1954) as :
- The plant must be unable to grow normally or complete its life cycle in the absence of the element.
- The element is specific and cannot be replaced by another.
- The element plays a direct role in metabolism and

- The deficiency symptoms of the element can be corrected or prevented by application of that element only.

In general, an element is considered as essential, when plants can't complete vegetative or reproductive stage of life cycle due to its deficiency when this deficiency can be corrected or prevented only by supplying this element and when the element is directly involved in the metabolism of the plant. The term functional nutrient for any mineral nutrient that functions in plant metabolism whether or not its action is specific. *E.g.* : Na, Co, Va and Si.

Soil fertility denotes the capacity of the soil to produce crops of economic value and maintain the health of the soil for future use or it is the capacity of soil to supply essential nutrients to normal plants in adequate amounts and in a balanced proportion or it is better to cultivate small piece of fertile land than large nutrient needs of the crop or the soil is said to be fertile when it contains all sixteen of the required nutrients in the right proportion for luxuriant plant growth.

Preparation of Soil Test Summaries and Soil Fertility Maps

Besides their use in making individual fertilizer recommendation to the farmers, soil test data usually are summarised for a respective block and district and on an all India level. Such soil fertility summaries are useful to administrators and planner in deciding the kind and amount of fertilizer most suitable in each area or district and determining the policy of fertilizer, distribution and consumption in different region. The data also are of use to fertilizer association, fertilizer industries and extension workers in promoting their respective programme and to research workers, particularly from the point of view of changes in fertility levels, conditioned by different fertilizer use or by different soil and crop management practices. Its satisfactory soil test summaries of any unit have been sufficiently large and the samples are representative of the whole area. For district summaries there should be at least one sample for one every 500 acres for block summaries one sample for every 50 acres and for village summaries one sample for every 5 acres. Many soil test summaries include tables giving percentage of the total number of samples falling in different categories of classification. *E.g.* Low, medium and high.

FACTORS OF FERTILIZER APPLICATION

Factors for Determining the Fertilizer Schedule are :
- Soil supplying power.
- Total uptake by crops.
- Residual effect of fertilizers.
- Nutrients added by legume crops.
- Crop residues left on the soil.
- Efficiency of crops in utilising the soil and applied nutrients.

Soil Supplying Power

Growing different crops during different seasons alters the soil nutrient status, estimated by soil analysis at the beginning of the season. The soil supplying power increases with legume in rotation. Fertilizer application and addition of crop residues. The available nitrogen and potassium in soil after groundnut are higher to initial status of the soil. But after pearl millet, only potassium status in the soil is improved and no changes in P.

Nutrient Uptake by Crops

The total amount of nutrients taken by the crops in one sequence gives an indication of the fertilizer requirement of the system. The balance is obtained by subtracting the fertilize applied to crops that nutrient taken by the crops.

Residual Effect of Fertilizers

The extent of residues left over in the soil depends on the type of fertilizer used. Phosphatic fertilizer and FYM have considerable residue in the soil, which is useful for subsequent crops. The residues left by potassium fertilizers are marginal. legume effect : Legumes add nitrogen to the soil in the range of 15 to 20 kg/ha. The amount of nitrogen added depends on the purpose. Green gram grown for grain, contributes 24 and 30 kg N respectively to the succeeding crop. Inclusion of leguminous green manures in the system add 40 kg to 120 kg N/ha. The availability of phosphorous is also increased by incorporation of green manure crops. Potassium availability to subsequent crop is also increased by groundnut crop residues : crop residues add considerable quantity of nutrients to the soil, cotton planted in finger millet stubbles benefits by 20 to 30 kg/ha due to decomposition of stubbles. Deep rooted crops-cotton, red gram absorbs nutrients from deeper laers. Leaf fall and decomposing add phosphorus to op layers crop residue contain high C : N ratio like stubbles of sorghum, pearl millet temporarily immobile nitrogen. Residue of legume's crop contains low C : N ratio and they decompose quickly and release nutrients.

Efficiency of crops : jute is more efficient crop for utilising of nitrogen followed by summer rice, maise, potato and groundnut in that order. Phosphorus efficient crops, jute > summer rice> Kharif rice> potato > groundnut > maise. Groundnut is more efficient in potassium utilisation followed by maise, jute, summer rice, Kharif rice, and potato. Fertilizer recommendation should be based on cropping system *e.g.* in wheat based cropping system an extra dose of 25 per cent nitrogen is recommended for wheat when it is grown after sorghum, pearl millet. When wheat, after pulse crop needs 20 to 30 kg less nitrogen per hectare. Phosphatic fertilizers are added through green manure crops, not to apply phosphates to succeeding wheat crop. In rice based cropping system consisting of rice- rice in Kharif and rabbi and sorghum, maise, finger millet, soybean in summer it is sufficient to apply phosphorus and potassium to summer crops only while nitrogen is applied to all the crops. Thus, following system approach in fertilizer recommendation can save lot of fertilizer.

Water Management

There is no carry over effect of irrigation as in case of fertilizer, rice-rice is efficient cropping system for total yield, but it consume large amount of water especially in summer. If water is scare in summer instead of rice, groundnut is used in cropping system.

Method of irrigation : the layout should be so planned that most of the crops can be suitable, in rice-rice-groundnut system; rice is irrigated by flood method, while groundnut by boarder strips. In cotton-sorghum-finger millet system, cotton, sorghum by furrow method while finger millet checks-basin method is adopted.

More remunerative and less water consuming crop rotation have standardised have been standardised at different location of India. Rice-mustard-green gram, rice- potato- green gram rotation were found more water efficient systems at memari in Memari in W.B under high level of irrigation in tarai region of U.P, rice-lentil and rice-wheat cropping system were found better. Pre monsoon groundnut-rabi sorghum sequence was highly remunerative with high water use efficiency compared to sugarcane alone in M.S when irrigation water is not limiting. Under limited water supply, however, rice-chickpea-green gram and rice- mustard-green gram are more remunerative with high water use efficiency.

Weed management : weed problems observed individual crop, and weed shift occur and their carry over effect of weed control method on succeeding crop. Weeds are dynamic in nature, generally broad- leaved weeds occur in wheat occur in wheat at later stage and 2, 4 D is applied as post emergence herbicide to control them. In rice- wheat system, canary grass (phallaris minor) is a menace for wheat crop. Other weed seed species are decomposed and loss viability, but Phataris minor seed do not loss viability. When sown in rice stubble is heavily infested with Phalaris minor. In cotton-sorghum-finger millet sequences cropping with zero tillage weeds are controlled by herbicide in two rotations.

Herbicide applied to the previous crop may be toxic to the succeeding crop. Higher dose of Atrazine applied to sorghum crop affect germination of succeeding pulse crops. Herbicide recommendation should be depends on succeeding crops, ploughing before the planting season helps to kill most of the weeds.

Pest and Diseases

Pest and diseases infestation more in sequence cropping due to continuous cropping, carry over effect of insecticides is not observed.

Harvesting

In sequences cropping crop can be harvested at physiological maturity stage instead of harvesting maturity. The field can vacate one week earlier. Because of continuous cropping the harvesting time may coincide with heavy rains and special post harvest operations, like artificial drying, treating the crop with common salt etc. are practices to save the produce. Integrated farming system, components and its advantages. India with 2.2 per cent of global geographical area support more than 15 per cent of the total world population, 70 per cent of who depends on agriculture. It is also support nearly 15 per cent of the total livestock population of the world.

As of now, out of 328.73 ha of geographical areas approximately 18 per cent is forest only 13.5 per cent is not available for cultivation. Total problems areas constitute 173.65 million ha, which include areas subjected to wind and water erosion (145 million ha) water – logged areas (8.53 million ha), alkali soils (3.58 million ha), saline and coastal sandy areas (5.50 million ha), ravines and gullies (3.97 million ha), shifting cultivation (4.91 million ha) and reverie torrents 2.73 (million ha). Besides 40 million ha are prone to flood and 260 million ha. Is drought prone. Thus the net sown area is 136.18 million ha. Is drought prone. Thus, the net sworne is 136.18 million ha (41.42 per cent of the total geographical area).

Unlike industries, agriculture is practiced by 105 million farm families who live in 0.6 million village. More than 40 per cent of them are below the poverty line. Nearly 85 million farm families belong to small and marginal categories. Only 25 to 30 per cent of the modern agricultural technology has reached the farmers. This is often because the technology has not been consistent with the condition with condition of the farm situations. Since there is no further scope for horizontal expansion of land for cultivation, the only alternative left is for vertical expansion on space and time particularly for small and marginal farmers (constituting 76 per cent farming community) who do not have much of reasons, especially in rain fed areas. The new farming system research strategy should, therefore to develop technology with participatory approach of farmers.

PRINCIPLES OF SOIL MANURING

In their natural condition all soils not absolutely barren are capable of supporting a certain amount of vegetation, and they continue to do so for an unlimited period, because the whole of the substances extracted from them are again restored, either directly by the decay of the plants, or indirectly by the droppings of the wild animals which have browzed upon them. Under these circumstances, a soil yields what may be called its normal produce, which varies within comparatively narrow limits, according to the nature of the season, temperature, and other climatic conditions. But the case is completely altered if the crop, in place of being allowed to decay on the soil, is removed from it, for, though the air will continue to afford an undiminished supply of those elements of the food of plants which may be derived from it, the fixed substances, which can only be obtained from the soil, decrease in quantity, and are at length entirely exhausted.

In this way a gradual diminution of the fertility of the soil takes place, until, after the lapse of a period, longer or shorter, according to its natural resources, it will become entirely incapable of maintaining a crop, and fall into absolute infertility unless the substances removed from it are restored from some other source in the form of manure. When this is done, the fertility of the soil may not only be sustained but greatly increased, and, in point of fact, all cultivated soils, by the use of manure, are made to yield a much larger crop than they can do in their natural condition.

The fundamental principle upon which a manure is employed is that of adding to the soil an abundant supply of the elements removed from it by plants in the condition best fitted for absorption by their roots; but looked at in its broadest point of view, it acts not merely in this way, but also by promoting the decomposition of the already partially disintegrated rocks of which the soil is composed, setting free those substances it already contains, and facilitating their absorption by the plants. In considering the practical applications of the broad general principle just stated, it might be assumed that a manure ought invariably to contain all the elements of plants in the quantities in which they are removed by the crops, and that when this has been accurately ascertained by analysis, it would only be necessary to use the various substances in the pro this, though a very important, and no doubt in many cases essential condition, is by no means the only matter which requires to be taken into consideration in the economical application of manures. And this becomes sufficiently obvious when the circumstances attending the exhaustion of the soil are minutely examined.

When a soil is cropped during a succession of years with the same plant, and at length becomes incapable of longer maintaining it, the exhaustion is rarely, if ever, due to the simultaneous consumption of all its different constituents, but generally depends upon that of one individual substance, which, from its having originally existed in the soil in comparatively small quantity, is removed in a shorter time than the others. To restore the fertility of a soil in this condition, it is by no means necessary to supply all the different substances required by the plant, for it will suffice to add that which has been entirely removed. On the other hand, if an ordinary soil be supplied with a manure containing a very small quantity of one of the elements of plant food, along with abundance of all the others, the amount of increase which it yields must obviously be measured, not by those which are abundant, but by that which is deficient; for the crop which grows luxuriantly so long as it obtains a supply of all its constituents, is arrested as effectually by the want of one as of all, as has been proved by the experiments of Prince Salm Horstmar and others; and hence, in order to obtain a good crop, it would be necessary to use the manure in such abundance as to supply a sufficiency of the deficient element for that purpose.

If this course were persevered in for a succession of years, the other substances which would have been used in much more than the quantity required by the crops, must either have been entirely lost or have accumulated in the soil. In the latter case it is sufficiently obvious that the soil must have been gradually acquiring an amount of resources which must remain dormant until the system of manuring is changed. To render them available, it is only necessary to add to it a quantity of the particular substance in which the manure hitherto employed has been deficient, so as to restore the lost balance, and enable the plant to make use of those which have been stored up within it.

The substance so used is called a special manure; that containing all the constituents of the crop is a general manure. The distinction of these two classes of manures is very important in a practical point of view, because a special manure is not by itself capable of maintaining the life of plants, but is only a means of bringing into use the natural and acquired resources of the soil. In place of preventing or retarding its exhaustion, it rather accelerates it by causing the increased crops to consume more portions thus indicated. But this, though a very important, and no doubt in many cases essential condition, is by no means the only matter which requires to be taken into consideration in the economical application of manures. And this becomes sufficiently obvious when the circumstances attending the exhaustion of the soil are minutely examined. When a soil is cropped during a succession of years with the same plant, and at length becomes incapable of longer maintaining it, the exhaustion is rarely, if ever, due to the simultaneous consumption of all its different constituents, but generally depends upon that of one individual substance, which, from its having originally existed in the soil in comparatively small quantity, is removed in a shorter time than the others. To restore the fertility of a soil in this condition, it is by no means necessary to supply all the different substances required by the plant, for it will suffice to add that which has been entirely removed. On the other hand, if an ordinary soil be supplied with a manure containing a very small quantity of one of the elements of plant food, along with abundance of all the others, the amount of increase which it yields must obviously be measured, not by those which are abundant, but by that which is deficient; for the crop which grows luxuriantly so long as it obtains a supply of all its constituents, is arrested as effectually by the want of one as of all, as has been proved by the experiments of Prince Salm Horstmar and others; and hence, in order to obtain a good crop, it would be necessary to use the manure in such abundance as to supply a sufficiency of the deficient element for that purpose.

If this course were persevered in for a succession of years, the other substances which would have been used in much more than the quantity required by the crops, must either have been entirely lost or have accumulated in the soil. In the latter case it is sufficiently obvious that the soil must have been gradually acquiring an amount of resources which must remain dormant until the system of manuring is changed. To render them available, it is only necessary to add to it a quantity of the particular substance in which the manure hitherto employed has been deficient, so as to restore the lost balance, and enable the plant to make use of those which have been stored up within it. The substance so used is called a special manure; that containing all the constituents of the crop is a general manure.

The distinction of these two classes of manures is very important in a practical point of view, because a special manure is not by itself capable of maintaining the life of plants, but is only a means of bringing into use the natural and acquired resources of the soil. In place of preventing or retarding its exhaustion, it rather accelerates it by causing the increased crops to consume more abundantly, and within a shorter period of time, those substances which it contains. On the other hand, a general manure prevents or diminishes the consumption of the elements of plant-food contained in the soil, and if added in sufficient abundance, may cause them to accumulate in it, and even enable an almost absolutely barren soil to yield a tolerable crop.

General manures must therefore always be the most important and essential, and no others would be used if it were possible to obtain them of a composition exactly suited to the requirements of the crop to be raised. Practically, however, this condition cannot be fulfilled, because all the substances available for the purpose, and particularly farm-yard manure, are refuse matters, the exact composition of which is not under our control, and they do not necessarily contain their constituents either in the most suitable proportions, or the most available forms, and consequently when they are used during a succession of years, certain of their constituents may accumulate in the soil, and it is under such circumstances that special manures are both necessary and advantageous. Several different substances, but more especially farm-yard manure, fulfil in a very remarkable manner the conditions of a general manure, and supply abundantly, not merely the mineral, but also the carbonaceous and nitrogenous matters necessary for building up the organic part of the plant; and hence its use is governed by principles of comparative simplicity, and really resolves itself into determining the best mode of managing it so as effectually to preserve its useful constituents, and, at the same time, to bring them into those forms of combination in which they are most available to the plant. But the employment of a special manure opens up nice questions as to the relative importance of the different elements of plants which have given rise to much controversy and difference of opinion. In treating of the food of plants, it has been already observed that the fixed or mineral constituents which are contained in their ash, are necessarily derived exclusively from the soil, but that the carbon, hydrogen, nitrogen, and oxygen, of which their organic part is composed, may be obtained either from that source or from the air.

The important distinction which thus exists between these two classes of substances, has given rise to two different views regarding the theory of manures. Basing his views on the presence of the organic elements in the air, Liebig has maintained that it is unnecessary to supply them in the manure, while others, among whom Messrs. Lawes and Gilbert have taken a prominent position, hold that, as a rule, fertile soils, cultivated in the ordinary manner, contain a sufficient supply of mineral matters for the production of the largest possible crops, but that the quantity of ammonia and nitric acid which the plants are capable of extracting from the air is insufficient, and must be supplemented by manures containing them.

A large number of experiments have been made in support of these views, but the inferences which can be drawn from them are not absolutely conclusive on either side, and it is necessary to consider the matter in a general point of view. Setting out from the proposition already so frequently referred to, that the plant cannot grow unless it receives a supply of all its elements, it must be obvious that if, to a soil containing a sufficiency of mineral matters to raise a given number of crops, a supply of ammonia be added, its total productive capacity cannot be thus increased; and though it may yield larger crops than it would have done without that substance, this can only be accomplished by a proportionate diminution of their number.

In either case, the same quantity of vegetable matter will be produced, but the time within which it is obtained will be regulated by the supply of ammonia. That substance differs in no respect from any other element of plant-food, and used in this way is to all intents and purposes a special manure, and acts merely by bringing into play those substances which the soil already contains. Its effect may not be apparent until after the lapse of a very long period of time, but it ultimately leads to the exhaustion of the soil. If, on the other hand, a soil be continuously cropped until it ceases to yield any produce, it is manifest that the exhaustion must in this instance be entirely due to the removal of its available mineral nutriment, because the superincumbent air constantly changed by the winds must continue to afford the same unvarying supply of the organic elements, and the power of supporting vegetation would be restored to it, by adding the necessary inorganic matters. Hence when a soil, which in its natural condition is capable of yielding a certain amount of vegetable matter, is rendered barren by the removal of the crop, it may be laid down as an incontrovertible position, that its infertility is due to the loss of mineral matters, and that it may be restored to its pristine condition by the use of them, and of them only. But the case is materially altered when we come to consider the course of events in a cultivated soil.

The object of agriculture is to cause the soil, by appropriate treatment, to yield much more than its normal produce, and the question is, how this can be best and most economically effected in practice. According to Liebig, it is attained by adding to the soil a liberal supply of those mineral substances required by the plant, and that it is unnecessary to use any of the organic elements, because they are supplied by the air in sufficient quantity to meet the requirements of the most abundant crops. Other chemists and vegetable physiologists again hold that though a certain increase may be obtained in this way, a point is soon reached beyond which mineral matters will not cause the plant to absorb more ammonia from the air, although a further increase may be obtained by the addition of nitrogen in that or some other available form. It is admitted on both sides, that all the elements of plant food are equally essential, and the controversy really lies in determining what practically limits the crop producible on any soil.

The point at issue may be put in a clear point of view by considering the course of events on a soil altogether devoid of the elements of plants. If a small quantity of mineral matters be added to such a soil, it immediately becomes capable of supporting a certain amount of vegetation, deriving from the air the organic elements necessary for this purpose, and with every increase of the former, the air will be laid under a larger contribution of the latter, to support the increased growth, and this must proceed until the limit of supply from the atmosphere is reached. At this point a further supply of mineral matters alone must obviously be incapable of again increasing the crop, and it would thus be absolutely necessary to conjoin them with a proportionate quantity of organic substances. Liebig maintains that this limit is never attained in practice, but that the air affords ammonia and the other organic elements in excess of the requirements of the largest crop, while mineral matters are generally though not invariably present in the soil in insufficient quantity. *Messrs.*

Lawes and Gilbert, on the other hand, believe that the soil generally contains an excess of mineral matters, and that a manure which is to bring out their full effect must contain ammonia, or some other nitrogenous substance fitted to supplement the deficient supply afforded by the atmosphere. In short, the question at issue is, whether there is or is not a sufficiency of atmospheric food to meet the demands of the largest crop which can practically be produced. An absolutely conclusive reply to this question is by no means easy. The experiments by which it is to be resolved are complicated by the fact, that all soils capable of supporting anything like a crop, contain not only the mineral, but the organic elements of its food in large and generally in greatly superabundant quantity, and it is impossible satisfactorily to ascertain how much is derived from this source, and how much from the atmosphere.

Purely Mineral Soil

There are in fact no experiments in which the effects of a purely mineral soil have been ascertained. The important and carefully performed researches of Messrs. Lawes and Gilbert were made upon a soil which had been long under cultivation, and contained decaying vegetable matters in sufficient abundance to supply nitrogen to many successive crops, and it would be most unreasonable to assert that the produce they did obtain by means of mineral manures, drew the whole of its nitrogen from the air. On the contrary, it may be fairly assumed that the soil did yield a certain quantity of its nitrogenous compounds, but to what extent this occurs, it is impossible to determine. This difficulty is encountered more or less in all the other experiments, and precludes absolute conclusions. The same fallacy also besets the arguments of Liebig when he holds that the crop, increased by means of mineral manures alone, must derive the whole of the additional quantity of nitrogen which it contains from the air, as appears to be tacitly assumed throughout the whole discussion.

So far from this being the case, it is just as likely that the mineral matters should cause the plants to take it from the soil, if it is there, as from the atmosphere. Taking a general view of the whole question, it is evident that a certain amount of vegetation may always be produced by means of mineral manures, and the quantity obtained is generally much beyond the normal produce of the soil. But it is still open to doubt whether the largest possible crop can be thus obtained, although the balance of evidence is against it, and in favour of the addition of ammonia, and other nitrogenous and organic substances, to the soil.

In actual practice manures containing nitrogen are more important, and more extensively applied than any others, and the quantity of that element thus used is very much larger than is generally supposed. Twenty tons of farm-yard manure, a quantity commonly applied, and often exceeded on well cultivated land, contain a sufficiency of organic matters to yield about 2-1/2 cwt. of nitrogen. A complete rotation, according to the six-course shift, contains almost exactly the same quantity of nitrogen, when we assume average crops throughout the whole, and it is thus made up.

The supply is therefore quite sufficient for the requirements of the crop; and when it is borne in mind that a considerable quantity of ammonia and nitric acid is annually carried down by the rain, and that during a long rotation other substances are very generally used in addition to farm-yard manure, it is obvious that the crop need not depend to any extent upon what it derives from the air. What is true of the nitrogenous matters applies with still greater force to the mineral constituents of the manure. Twenty tons of farm-yard manure contain 32 cwt. of mineral matters, while the average crops of a six course-shift contain only 1088 lbs., or less than one-third of this quantity. It is obvious, therefore, that in well manured land there must be a gradual increase of all the constituents of plants, but that of the mineral matters is relatively much greater than that of the nitrogenous.

If therefore from any cause the crop produced on a soil to which farmyard manure had been applied were greatly to exceed the average, the amount of produce, so far as the soil is concerned, would be limited not by deficiency of mineral, but of nitrogenous food. Hence also when farm-yard manure is liberally applied, there is a gradual accumulation of valuable matters, and a progressive improvement of the productive capacity of the soil. It is far otherwise, however, if a special manure is employed, because in that case the crop is thrown upon the resources of the soil itself for all its constituents except those contained in the substance employed, and by persisting in its exclusive use exhaustion is the inevitable result. It would be wrong, however, to infer from this, that special manures are to be avoided. On the contrary, great benefits are derived from their judicious employment, and the circumstances under which they are admissible may be readily gathered from what has already been said.

They are agents which bring into useful activity the dormant resources of the soil, they restore the proper balance between its different constituents, and supply the excessive demand of some particular elements.

Soil Containing and Mineral Matters

Thus, for instance, in a soil containing an abundant supply of mineral matters, a salt of ammonia or nitric acid increases the crop, by promoting the absorption of the substances already present. So likewise a soil on which young cattle and milch cows have been long pastured has its fertility restored by phosphate of lime, because that substance is removed in the bones and milk in relatively much larger proportion than any others.

The choice of a special manure is necessarily dependent on a great variety of circumstances, and is governed partly by the nature of the soil, and partly by that of the crop. It is obvious that cases may occur in which any individual element of the plant may be deficient, and ought to be supplied, but experience has shown that, as a rule, nitrogen and phosphoric acid are the substances which it is most necessary to furnish in this way, and which in all but exceptional cases produce a marked effect on the crop.

The other substances, such as potash, soda, magnesia, etc., occasionally act beneficially, but the results obtained from them are very uncertain, and frequently entirely negative. It has been commonly asserted that phosphates are specially adapted to root crops, and ammonia or nitrates to the cereals, and this statement is so far true, that the former are used with advantage on the turnip, while the latter act with great benefit on grain crops and more especially on oats and barley.

The effect of the latter, however, is more or less apparent in all crops and on all soils, because it promotes the assimilation of the mineral matters already present. But its peculiar importance lies in the power which it has of promoting the rapid development of the young plant, causing it to send its roots out into the soil, and to spread its leaves into the air, thus enabling it to take from those two sources, abundance of the useful substances existing in them.

But it ought to be distinctly understood, that the statement that particular manures are specially suited to particular crops must be assumed with some reservation, because everything depends upon the nature of the food contained in the soil. It is well known that there are many soils in which ammonia acts more favourably on the turnip than phosphates, and *vice versa*, and the difference is often due to the previous treatment. In many cases in which ammonia when first used proved most beneficial, it now begins to lose its effect, and the reason no doubt is, that by its means the phosphates existing in these soils have been reduced in amount, while the ammonia has accumulated, so that a change in the system of manuring becomes necessary. A general manure may be used year after year in a perfectly routine manner, but where a special manure is employed, the importance of watching its effects, and altering it as circumstances indicate, cannot be overestimated. The length of time during which special manures have been extensively used has not been sufficient to bring this prominently before the agriculturist, but its importance must sooner or later force itself upon him, and he will then see the necessity for studying the succession of manures as well as that of crops.

Hitherto we have considered a manure merely as a source from which plants derive their food, but it exercises a scarcely less important action on the chemical and physical properties of the soil. Farm-yard manure, which, as we shall afterwards see, contains a large amount of decomposing vegetable and animal matters, yields a supply of carbonic acid, which operates on the mineral constituents, promotes their further disintegration, and thus liberates their useful elements. It affects also their physical properties, for it diminishes the tenacity of heavy clays; each straw as it decomposes forming a channel through which the roots of plants, air, and moisture can penetrate more readily than through the stiff clay itself.

On the other hand, it diminishes the porosity of light sandy soils, causes them to retain moisture, and generally makes their texture more suitable to the plant. Special manures probably act to some extent chemically on the soil, but the nature of the changes they produce is as yet imperfectly understood. Superphosphates which are highly acid in all probability act powerfully on the mineral substances, and common salt, which, though of little importance to the plant, occasionally produces very striking effects, appears to exercise some decomposing action on the soil. It is difficult, however, to trace the mode in which they operate on a substance of such complexity as the soil.

Lime, as we shall afterwards see, acts by promoting the decomposition of the vegetable matters on the soil, and possibly some other substances may have a similar effect. In the application of manures to the soil there are several circumstances which must be taken into consideration. It is generally stated that they ought to be distributed as uniformly as possible, but this is not always necessary nor even advisable, and certainly is not acted on in practice.

Much must depend upon the nature both of crop and soil. When the former throws out long and widely penetrating roots, the more uniformly the manure is distributed the better; but if the rootlets are short, it is clearly more advisable that it should be deposited at no great distance from the seed. Practically this is observed in the case of the potato and turnip, which are short rooted, and where the manure is generally deposited close to the seed.

But this course is never adopted with the long rooted cereals, the manure being usually applied to the previous crop, so that the repeated ploughings to which the soil is subjected in the interval may distribute what remains as widely and uniformly as possible. In soils which are either excessively tenacious or light, the accumulation of the manure close to the plants has also the effect of producing an artificial soil in their immediate neighbourhood, containing abundance of plant-food, and having physical properties better fitted for the support of the plant.

SOIL FERTILITY PROGRAMMES

One of the first steps in developing a soil fertility programme is to analyse the intended system including budgeting for nutrient gains and losses. The choice and rates of fertilizer application are then based on the experiences of the farmer and others and make up for losses calculated in the budget and indications from any soil or plant testing are taken into account.

Healthy Soil-Healthy Plant

The aims of a fertilizer programme should go beyond creating the maximum dry matter production or yield of plants. In the organic system it becomes more important than ever to design a soil management and fertilizer programme that improves soil ecology and produces a healthy plant. This entails looking at the types of inputs and the overall balance of elements applied. One of the greatest potential problems in plant quality that can impact feed value and susceptibility to pest and disease is the excess uptake of nitrogen and to some extent phosphorus.

The Perceived Problem with Soluble Salt Fertilizers

One of the early and sustained opposition to chemical fertilizer use from organic proponents has been the fast release nature of soluble salt fertilizers. These are described as releasing major plant nutrients in excess of their requirements causing luxury uptake in excess of the plants requirements and affecting plant health and food quality. The excesses were also considered to contribute to environmental degradation through leaching and surface run-off loss of the nutrients to waterways (in the case of nitrogen and phosphorus).

As it turns out, there can be significant increase in susceptibility to some plant diseases (including rusts and powdery mildews) and plant pests (including aphids and caterpillars) when there are excess levels of nitrate, ammonium, free amino acids or phosphate as can occur following soluble salt fertilizer application. There is also some evidence that these excess levels in the plant cause poor storage qualities and can have anti-nutritive (*e.g.* poor feed utilisation) effects on vertebrate animals.

The amount of environmental damage done will be related to amounts, timing, method of application, slope, soil quality, proximity to waterways and the ability of plant roots to capture the potentially lost nutrients. There is, however, ample evidence that the use of slower releasing mineral fertilizers

(*e.g.* RPR rather than superphosphate) significantly reduces the rate of nutrient loss into waterways.

Excess application of nitrogen and sulphur fertilizers increases the level of nitrate and sulphate (respectively) leaching losses. This is a major cause of soil acidification as each nitrate or sulphate molecule leached out takes a cation (calcium in most cases) – bulk soil pH is largely related to the balance between alkaline cations and acid hydrogen ions.

In livestock systems there are a variety of other potential drawbacks that can be noted for soluble salt fertilizers e.g. :

- Increased dung transfer losses in pastoral systems as animals ingest pasture with excess levels of a nutrient.
- Increased potential for nitrate toxicity e.g. if nitrogen fertilizer is applied at a time of natural nitrogen flush (commonly carried out e.g. after drought breaking rain) causing stock deaths and ill-thrift.
- Nitrogen led growth can be lusher than more balanced growth due to less cell wall thickening creating feed utilisation issues in pasture.
- High nitrate levels in plant tissue are compensated for by extra water uptake, exacerbating a problem with pasture lushness.
- The extra water uptake increases the passive uptake of potassium, the high levels of which (and imbalance with calcium, magnesium, sodium

and boron) can cause metabolic problems in livestock. These problems can also occur with the application of soluble potassic fertilizer shortly before grasing.

- Imbalanced nutritional state of pasture may contribute to larger levels of potentially harmful fungi and mycotoxin production.

Some thought considers that soluble salt fertilizers can cause more direct damage to soil systems including the chemical absorption of water, negative effects on soil biology from residual acidity on the fertilizer particles, and that some cause bad electrochemical conditions. These concerns are less substantiated and of doubtful significance in most situations. Having said that, such fertilizers can contribute to the very real problem of soluble salt build up in intensive (or semi-arid) areas.

The Desire for Rapid Nutrient Availability

A major premise of organic agriculture has been to avoid the excess availability of single nutrients such as nitrogen and phosphorus. This is often achieved by using compost or slowly weathering minerals, which will release a balance of nutrients over a period of time affected by soil conditions of moisture, temperature and biological activity. The organic system also relies on seasonal release of nutrients from the pool of soil organic matter.

The slow and seasonal release of nutrients in organic agriculture can cause a commercial setback. If a crop takes say two weeks longer than conventional crops the premium for early production could disappear before the organic crop is ready (a premium for organic production is often small compared to the premium for early or late production. Slow release of nutrients can also be a problem for some nutrient demanding crops that might require for example high phosphorus levels early in crop development. Many crops have also been shown in conventional growing to have increased from strategically timed nitrogen side dressings *e.g.* sweetcorn, maise, smaller grains, many fruiting vegetables.

Rapid Nutrient Supply in Organic Cropping Systems

One rather artificial way of increasing the rate of growth of plants is to use greenhouses (can be portable to allow market garden flexibility, maximisation of use and minimisation of soil build up of soluble salts). If organic growers adopt tunnel house use, significant commercial gains can be achieved. The risk remains that as other growers adopt the use of tunnel-houses, the commercial gains reduce (this is part of the sometimes vicious market forces which can mean that every advance simply means that commercial growers must work harder on a larger areas to make the same money). It is, however, organic farmers than stand to gain the most from portable tunnelhouses since they offer an improvement in natural soil mineralisation of nutrients and a cleaner crop.

The problem of slow nutrient availability can also be addressed by adjusting the fertilizer programme. One option is to use fast nitrogen releasing materials such as meat, blood or fish meal. The irony is that these materials can cause some of the same imbalance/excess nitrogen issues as soluble salt fertilizers including measurable increases in plant nitrate levels. They are partly for that reason restricted fertilizers in Bio-Gro regulations. It might be argued that they are less of an issue because they are also adding a range of other nutrients at the same time and they are still utilising/feeding the soil life. But such materials are probably best incorporated into a composting or fermentation system to reduce the potential for imbalances. Animal manure should usually be treated in a similar manner or at least be well rotted (thus also avoiding potential harmful organism issues).

A less harsh method of releasing nutrients in sufficient quantity for gross feeding plants is to grow or carry in green manure that is incorporated a short time before crop planting (usually at least two weeks before hand – to avoid affecting the crop with initial nitrogen robbing decomposition or harmful decomposition products). Some green manures such as lupin can be efficient at accessing mineral forms of phosphorus.

Other fertilizer options for fast nutrient release include using a finer particle size of mineral fertilizers. Halving the average diameter of particle doubles the surface area and the rate of weathering increases accordingly. Applying very fine particles can be a problem for machinery in which case it could be mixed with material such as agricultural lime or compost. Applying in suspension is an ideal method though costing money in bringing in an applicator or setting up for application.

Finely ground limestone, limeflour, can have an indirect effect on nutrient availability by stimulating soil biological activity and increasing the rate of mineralisation. This can provide the extra speed of nutrient release required for early crops, nutrient demanding crop establishment and replace the need for nitrogen applications at strategic times in crop growth (probably best to time applications one to two weeks before the stage at which nitrogen fertilizer would be recommended). The use of lime-flour in this manner is only sustainable if there is a programme of organic matter addition or legume led fertility build up (and is limited in how many times it will be effective without such a build up).

Liquid fertilizers are popular for providing strategic boost to plant growth. Any effect will usually be related to either the nitrogen content or a bio-fertilizer aspect to the product (and any limiting trace elements). One advantage of liquid fertilizers is fast response. Attention should be paid to the rest of the fertilizer/input programme to ensure that overall sufficient nutrients are being added to meet losses (liquid fertilizers contain low quantities of the major nutrients). Liquid fertilizers allowable in an organic situation can be even more nutrient limited as they cannot include soluble salt derived nitrogen, or phosphorus. In pasture, the setbacks of slow release are usually less or even non-existent but there may be scope for using finer particle size of minerals especially when converting the pasture into the use of RPR and when relying on elemental sulphur for addressing a sulphur requirement. These fine particles may need to be supplied in suspension if such a service exists in the area or a slurry machine is suitable enough. The main challenge in pasture fertilizer programmes is to grow the legumes well (clovers especially have a poor root system for nutrient uptake in comparison to grass). It should be noted that a purist approach to organics could encourage working within seasonal limitations such that produce is available strictly when in season. The economic feasibility of this is dependent partly on the marketing system and how much the grower needs a high yield or a premium for early produce.

Pastoral Fertilizer Strategies

Although the aim with cropping fertilizer includes producing food of good quality, this is much more of an economic and system performance requirement for pasture fertilizer programmes. Grasing livestock have limited choice of what they eat. The pasture needs to be presented with as balanced a mineral nutrition as possible. This includes no shortages of essential elements, and no excesses of potentially troublesome elements. The level of highly available nitrogen is particularly important as is the balance of the base cations (calcium, magnesium, potassium and sodium).

The aim of a pasture fertilizer then is to grow reasonable quantity of as high a quality of grass as possible. The success of a fertilizer is often simply judged in terms of dry matter of pasture grown but the aim of most livestock systems is livestock performance not just pasture bulk.

Clover and other legumes are a key focus of a good fertilizer strategy – they drive the pasture along with biological nitrogen fixation and clover is a poor competitor with grass. The requirements of clover should be analysed as described in Module 4 Plant Nutrition. If phosphorus, sulphur or trace elements are limiting they are more likely to be affecting the performance of clover and beneficial forbs than grass species. It is these species too that appear to respond better to foliar fertilizers, biofertilizers and improved soil biological activity. A good pasture composition is important for the reasons Livestock Husbandry and because livestock performance on a good amount of clover is far superior to grass alone (probably mostly due to a higher metabolisable energy content).

Improving the quality of feed with fertilizer strategy can include encouraging good pasture composition and simply not applying highly available nutrients. If there is scope for foliar application of nutrients (by spray or suspension) then this can provide an efficient means of applying elements which are not taken up very well by some or all plants from the soil *e.g.* salt to improve palatability and stock health in the face of naturally high potassium levels or ranks grass and trace elements for plant performance and animal health.

Further strategies in a pasture fertilizer programme are to address seasonal feed deficits (these feed pinches are either a hassle/costly in terms of supplementary feed or they lead to depressed growth rates and conditions). In an organic situation, seasonal deficits should ideally be addressed system based strategies such as a variety of pasture and fodder plants supplying year round feed. The options for a fertilizer approach are reasonably limited mostly relying on stimulating soil biological activity at the times of need *e.g.* with lime-flour or bio-fertilizers. The success of this may be variable and is affected by how cold the soil temperatures are.

Integrated Strategies

In mixed cropping, a good strategy is to focus any added mineral fertilizer on the pasture phase to grow good clover levels and general pasture production. The pasture grasses will be particularly efficient at accessing resources such as phosphorus and sulphur from the applied fertilizer and convert them to an organic form (the grass itself) and release some to the surroundings. Many mixed cropping systems will be largely reliant on nitrogen provision from the clover so it makes good sense to support this.

A variation on this strategy is to apply mineral fertilizer onto large areas that are used for providing green manure, pasture clippings or mulching/composting material (including the straw from arable cropping). This material can then be transferred to a smaller area of intensive cropping.

The presence of livestock can be taken advantage of by feeding them with otherwise potentially troublesome wastes *e.g.* orchard or processing wastes for pigs. This provides produce and animal manure. The animal manure can be used in composting operations and the strategic supply of readily available nutrients.

Foliar Fertilizer

A major organic philosophy is to feed the soil to feed the plant. There may, however, be scope for using foliar fertilizers in certain situations. The uptake of certain elements is dramatically higher by foliar means than through the soil. This is the case with most trace elements as well as magnesium and, for many plants, sodium (which may be desired for livestock). Iodine, for example, has practically no measurable effect on plant tissue levels when applied to the soil.

Selenium, however, is taken up efficiently from the soil. The other trace elements fit in between these extremes. Foliar applications are effective regardless of soil pH and other conditions that might prevent uptake from the soil. Foliar fertilizers also give an opportunity to apply biofertilizers to the foliage, which may increase further the efficiency of nutrient uptake and give a strategic increase in growth. As mentioned above, liquid fertilizers contain very small levels of nutrients which although sometimes (though each product need checking) this might be sufficient for trace elements, the requirement for major elements should be addressed to aid sustainability.

OPTIONS FOR IMPROVING SOIL FERTILITY

The use of green manure (GM) legumes as nitrogen-fixing crops has been advocated as one of the most affordable soil-fertilizing technologies for small-scale farmers. In practice, however, resource requirements of GM technology often conflict with the short-term objectives of this target group. Small-scale farmers cannot afford to grow GM legumes simply for the sake of soil fertility unless the seeds of the legumes are edible or in a few cases where GM legumes could be used primarily to combat noxious weeds (*e.g.* Imperata cylindrica).

In order to address the trade-off between soil fertility and food concerns, the development of biomass-rich varieties of local grain legumes. However, the soil fertilizing effect of these varieties is potentially lower than that of GM legumes because of their grain yields, which entail a substantial removal of nutrients from the system. On the whole, the search for niches susceptible to solving the "GM vs. grain legumes dilemma" remains the cornerstone of promoting soil-fertilizing legume options that can be accepted by small-scale farmers.

Study area, Materials and Methods of Soil

The study was conducted from 2000 to 2002 in 4 villages, representative of the major landscapes and land use systems prevailing in southern Bénin. These villages were Agbassakpa (07°04' N, 02°26' E) located on a peneplain built on the Precambrian crystalline basement with ferruginous soils (Luvisols); Azozoundji (07°08' N, 02°03' E) and Zomondji (06°09' N, 01°09' E) located on the plateaux locally called Terres de Barre with ferralitic soils (Nitosols); and Djregbe (06°41' N, 02°61' E) on the costal plain with very poor quartz sand soils (Regosols).

The choice of the villages was based on the need to represent the existing land use intensification gradient : Land fallowing (3 to 5 years) was still practised in Agbassakpa and Azozoundji, but hardly in Zomondji and not at all in Djregbe. Access to mineral fertilizers was restricted in all 4 villages. The length of the growing period ranges from 211-270 days, rainfall is bimodally distributed averaging 1,100 mm a-1. The farming system is rain-fed and maise-based.

In each village, 8 legume options comprising 5 GM species and 3 food grain legume species were introduced to farmers. The introduced GM species were Aeschynomene histrix (accession I.12463), Centrosema molle (syn. C. pubescens) (I.152), Mucuna pruriens and Pueraria phaseoloides (commercial varieties), and Stylosanthes guianensis (I.15557) while the food grain legume comprised Arachis hypogaea (69-101), Glycine max (TGX 1448-2E), and Vigna unguiculata (IT84D-449, Mawuwena).

Researcher-managed demonstration plots were established in each village. These were in addition to individual trials that were freely designed and managed by volunteer farmers. Seeds were distributed free of charge and based on the participants' choices. To monitor farmers' experimentation process and legume diffusion pathways, both quantitative and qualitative methods were used.

The number of adopters was counted seasonally, the areas planted with the introduced legume options were mapped, measured and reasons for adoption, re- or dis-adoption were assessed using periodic workshops, field days and focus group discussions. The rationale behind the utilisation of the species was assessed by eliciting the local soil taxonomy and the value, *i.e.*, yield potentials, attributed to the fields planted with the introduced species. To get a more systematic picture of the comparative advantage of the technology options that were introduced, the ethno-economic values of the legume fields were included in the cost/benefit calculations involved.

This was done using the Partial Budget Analysis (PBA) of legume utilisation. Because of their food advantage, preference of grain legumes over GM legumes was taken as the baseline scenario. Therefore, in-depth analysis (PBA, soil taxonomy etc.) was made only in cases where GM legumes were preferred to grain legumes.

Results and Discussion

In Zomondji, farmers' preferences were clearly for the grain legumes. The soil-fertilizing effect of Mucuna and P. phaseoloides was acknowledged, but the species were not chosen because of land constraints. Also Djregbe's farmers were more in favour of grain legumes, because of their food property.

In contrast, Agbassakpa's and Azozoundji's farmers favoured the GM legume options in addition to the grain legumes. The GM legumes were evaluated according to their "leaf size", "aggressiveness of growth habit" and "soil covering speed". As a result, Mucuna was preferred to every other species, which were qualified either as "second Mucuna" (P. phaseoloides) or simply as "small-leafed" species.

The weed suppressing property of Mucuna was acknowledged, not as a primary advantage but just as a further confirmation of the "strength of Mucuna". The grain legumes varieties were judged according to their grain yield and not for their soil-fertilizing effect, which became the exclusivity of Mucuna.

Table. Intensity of Utilisation of the Introduced Legume Options after 2 years of Experimentation.

Legume	Azozoundji ($N^a=125$)		Agbassakpa ($N^a=90$)	
	No. of users	Average area (m^2) per user (% of afs ^{c1})	No. of users	Average area (m^2) per user (% of afs ^{c2}.)
Aeschynomene. histrix	8	169.2 (3.9)	1	225 (0.8)
Arachis. hypogaea	116	114.7 (2.7)	87	212.4 (0.7)
Centrosema molle	0	0 (0)	0	0 (0)
Glycine max	121	341.9 (8.1)	43	264.5 (0.9)
Mucuna pruriens	108	244.5 (5.8)	43	287.5 (1,0)
Pueraria. phaseoloides	0	0 (0)	0	0 (0)
Stylosanthes guianensis	1	138 (3.2)	16	261 (0.9%)
Vigna unguiculata	112	(n.a.^{b})	17	635.3 (2.2%)

Notes : a : Total number of participants; b : n.a.=non-available; c : afs : Average farm size comprising all types of fields including fallows. In Azozoundji, afs c1= 4229,5 m^2; in Agbassakpa, afs c2=28970,3 m^2.

The maps of the legume fields show that the species were planted along the local soil fertility gradient, respectively on fields classified as Fangle, Kunxo and Sisa.

Table. Local Soil Taxonomy in Agbassakpa and Azozoundji.

Soil Category	Category Subset	Suitability for Crops	Fertility Level	Need of Fertiliser
Fangle	Fangle	Potentially for maize	Potentially fertile	2 seasons of grain legume to get smoother
Kunxo	Fertile	Maize	Fertile	No need
	Middle fertile	Maize/grain legume rotation	Middle fertile	Grain legume rotation
	Poor	Only grain legume	Poor	Mineral fertiliser
Sisa	Sisa	No crop	Exhausted	Not worth of fertilisation

The grain legumes were grown either on Fangle or Kunxo while Mucuna was planted on Sisa soil. How could farmers choose Mucuna while having the option to use cowpea (Vigna unguiculata) — the grain legume that traditionally has been most used for soil fertility in both villages? In the light of the negative rate of return (-42.7 per cent) yielded by a shift from cowpea-maise to Mucuna-maise rotations, farmers choice of Mucuna appeared irrational. However, considering that the Sisa fields allocated to Mucuna cannot sustain cowpea, the opportunity costs of Mucuna is to be equated to zero, at least according to farmers' perceptions. Thus, the Mucuna technology becomes more profitable than that of cowpea : the marginal rate of return with a replacement of cowpea by Mucuna would then be 397.7 per cent (data not shown).

Table. Partial Budget Analysis of a Maise Crop after Mucuna (MM) and Cowpea (CM) in Azozoundji.

Item	Mucuna/Maize (MM)	Cowpea/Maize (CM)
Gross farm benefits		
1. Average grain yield of subsequent maize on Sisa soils for CM and other soils for CM (kg/ha)	1,500	750
2. Price (FCFA[a]/kg)	112	112
3. Gross margin gate benefits (FCFA/ha) (1x2)	168,000	84,000
Variable input costs (FCFA/ha)		
4. Land preparation	25,500	8,625
5. Opportunity costs for lost season	129,716.54	0
6. Total variable input costs (4+5)	155,216.54	8,625
Net benefit		
7. Net benefit (kg/ha) (3-6)	12,783.46	75,375
8. Change in net benefits with a shift from cowpea to Mucuna soil fertilising technology		-62,592
9. Change in total variable input costs with a shift from cowpea to Mucuna technology		146,592
10. Marginal rate of return (%) (100 × 8 ÷ 9)		-42.7

Note: a: Franc de la Communauté Financière Africaine: 656 FCFA =1 Euro

Smallholder farmers can choose GM legumes to improve poor soils that are not suitable for the production of food crops. In the Republic of Bénin, these potential GM niches were found in non-sandy areas, where land is moderately available, *i.e.*, scarce enough scarce to impose a shifting cultivation based on soil fertility taxonomy.

FACILITATED LEARNING IN SOIL FERTILITY MANAGEMENT

Continued natural resource degradation, increasing levels of poverty, unintended negative impacts of globalisation, low level adoption of technologies, poor governance and marketing infrastructure, and increasing population pressure have been observed to be great impediments to increasing food production in sub-Saharan Africa. In recent years various publications have addressed the magnitude of resource degradation and technical solutions to the observed constraints have been proposed.

So far, impacts have been rather limited since many of these technical options require relatively high capital investments, need a well-functioning infrastructure and a conducive policy and market environment, all of which are constraining factors in most parts of SSA. In response to the low success rate, the research community and development organisations are shifting their focus to developing Low External Input Agriculture technologies (LEIA).

However, the impacts of LEIA is still a subject of debate with some authors highlighting success stories while others are voicing their inadequacy to meet increasing food demands. A new line of thought attempts to combine low and high input technologies in an Integrated Nutrient Management (INM) approach which attempts to maximise the use of local resources and optimise application of external inputs. In the search for INM-practices, LEIA and organic farming practices have hardly been examined systematically. Furthermore, methodologies of study have been anecdotal in nature and lacking active participation of farm households, relevant agricultural players and policy makers who have to satisfy multiple goals.

This chapter describes the experiences and results of a project in four districts in Kenya and Uganda during the period 1997-1999 in which farmers, extensionists, *NGOs, researchers and district policy makers joined hands through a learning process to :*

- Diagnose constraints to soil fertility management and opportunities for improving the same;
- Identify, test and evaluate low-external-input technologies; and to
- Formulate enabling policies for soil fertility management with district policy stakeholders.

Methodology

The study was conducted in four research sites in Kenya and Uganda, two with a high agricultural potential (fertile soils, high and reliable rainfall) and two with a medium to low agricultural potential (low soil fertility, low and unreliable rainfall). LEIA and Conventional farm management, with 1418 households each, were studied in each site. LEIA management' was defined as farm households trained in Low-External Input technologies (composting, application of liquid manure etc.) and have applied at least three of those techniques on more than 50 per cent of the cultivated area over a minimum of 3 consecutive years.

Conventional management' was defined as farm households representative of the farming system in the study site, have comparable production resources to that of LEIA group, but not practising any of the defined LEIA techniques. The nutrient monitoring approach (NUTMON) distinguishes a diagnostic phase and an iterative and participative technology and policy development phase.

The diagnosis consisted of the following activities :
- Farm households' assessment of natural resources using soil maps, transect walks and nutrient flow maps,
- Soil sampling and analysis according farmers soil maps,
- Monthly monitoring using NUTMON questionnaire and
- Analysis. Analysis of the data consisted of
- Analysis of the largely qualitative, farmers' assessment of natural resource management,
- Analysis of the quantitative nutrient flows using the NUTMON methodology and soil sample results and
- Integration of the two previous steps and discussing results with participating farmers.

Impact assessment of selected LEIA techniques was done through a Participatory Technology Development (PTD) process in which technologies were jointly selected, experiments designed, monitored, data collected and evaluated, and results shared with farmers using participatory tools.

Based upon the participative diagnosis, the results of the PTD, an inventory of historic developments in the district, and an inventory of the existing and relevant policies in the research sites, draft scenarios for future developments in the research sites were formulated. They were discussed in District policy stakeholders workshops where they were finalised and a prioritised action plan drawn by the policy makers attending the workshop.

Results

Soil and nutrient flow maps enabled farmers to visualise nutrient flows and provided insight into soil nutrient status.

The application of the NUTMON model showed that there were marginal differences in nutrient balances between the conventional and LEIA farm.

	Machakos f(LPA)		Nyeri (HPA)		Pallisa (LPA)		Kabarole (HPA)	
	CONV	LEIA	CONV	LEIA	CONV	LEIA	CONV	LEIA
N-stock (kg/ha)	3900	6400	12200	12300	3100	3000	6800	8300
N-flow (kg ha^{-1},yr^{-1})	-21	-25	-99	-91	-3	-4	-126	-95
N-flow (‰ of stock,yr^{-1})	-5	-4	-8	-7	-1	-1	-18	-11
P-stock (kg/ha)	2000	1700	7900	8000	1000	2500	10300	9000
P-flow (kg ha^{-1}, yr^{-1})	2	1	-23	-27	0	0	-70	-57
P-flow (‰ ofstock,yr^{-1})	1	1	-3	-3	0	0	-7	-6
K-stock (kg/ha)	7800	10200	10400	15300	6100	6300	7800	8400
K-flow (kg ha^{-1},yr^{-1})	-9	2	-23	18	2	1	-55	-7
K-flow ‰ of stock,yr^{-1}	-1	0	-2	1	0	0	-7	-1

LPA – Low-Medium Potential Area

HPA – High Potential Area

CONV - Conventional farm management practices

LEIA – Low-external-input farm management practices

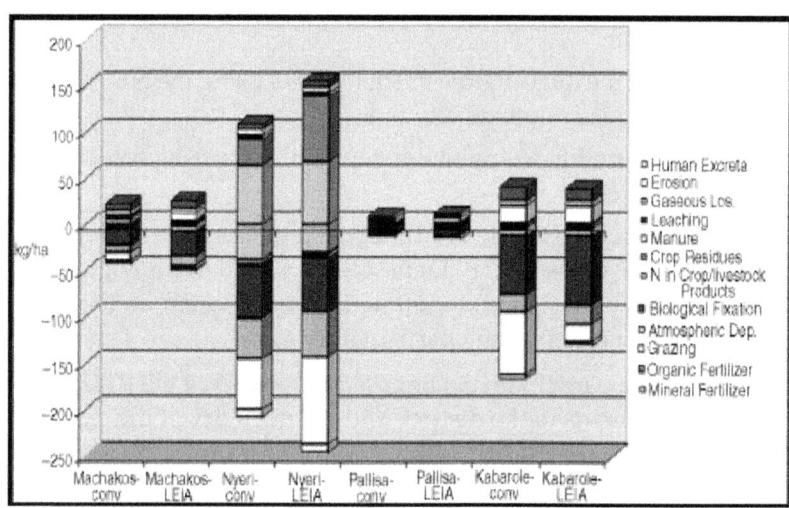

Fig. Average from Nitrogen Flows Per Type, Research Site and Management Type.

The differences between the districts were much more profound. The high potential areas, although different in farming system, both showed a relatively high N, P, K nutrient content of the soil, but also more negative nutrient balances at farm level, especially for N (90–125 kg ha^{-1} year^{-1} representing an annual 7-18 per cent loss of the stock). The latter was mainly due to high erosion, leaching and gaseous losses, despite relative high uses of mineral and organic fertilizers. In Machakos district (Kenya), intensive crop farming on relatively poor soils results in negative nutrient balances and an annual decline in N-stock of 5 per cent at farm level, mainly due to very low levels of external inputs applied. In the low potential areas of Pallisa district (Uganda), keeping of livestock under free-range systems results in nutrients from communal grasing lands being transported into cropping land. The economic performance indicators showed no clear differences between the management systems. However, analysis of labour data showed that LEIA management requires more total farm labour than conventional management. The farms in high potential areas realised on average higher net farm income levels both per farm and per unit area. In Kenya off-farm income is of crucial importance to the total family income, reducing the differences in financial resources between the two areas. When valuing the depleted nutrients against replacement costs, it was observed that huge differences between the districts occur. In Pallisa the replacement cost accounted for 5 per cent of the net farm income while the figure for Nyeri was 11 per cent. In Machakos and Kabarole a considerable proportion of the net farm income was based upon nutrient mining with res pective figures of 25-30 per cent and 60-70 per cent. The economic efficiency of crop activities, expressed in gross margins per acre, tended to be slightly higher for the LEIA farm management systems (although not statistically significant).

Table. Summarised results from tested low-external input technologies.

Nyeri (HPA, Kenya)	• In conventional farms the existing practice of manure and DAP application can be replaced by an application of compost alone without any significant impact on yield and economic returns. • On farms with LEIA management the addition of liquid manure had no significant impact on yields and economic returns. • Residual impacts of the treatments in the following season were limited. • From the economic point of view an application of 8-9 ton of manure/compost per acre is preferred above 17 ton/acre.
Machakos (LPA, Kenya)	• In both management systems a high application of compost with an additional application of liquid manure gave a significant higher yield and is attractive in economic terms. • In both management systems a 50% reduction of the high level compost application level showed similar treatment effects at 50% reduced yield levels compared to the high dose treatments. • On conventional farms the treatment with reduced dose of compost was economically less attractive, while on LEIA farms a slightly higher gross margin was realised but lower returns to labour. • When the economic impact is expressed in returns per labour day however, the addition of liquid manure was less attractive in both farming systems. • On farms with conventional management only a very slight residual effect could be observed in maize yield in the following season, while on LEIA farms a strong residual impact was observed.

Kabarole (HPA, Uganda)	• A high level of manure application had a positive impact on the yield of crops in at least two consecutive cropping seasons. • Although manure had a considerable impact on yields, the economic returns in two seasons were lower than farmers' practice when manure is valued at Ush 14000 per ton.
Pallisa (LPA, Uganda)	• Application of compost, deep tillage and mulch are low-external-input options, which can improve yields in food crops. • Although the selected low-external-input technologies had a considerable impact on yields, the economic impact (over a 2-season period) at farm level is absent or very limited.

LPA – Low-Medium Potential Area

HPA – High Potential Area

The technology identification and experimental design exercise resulted in a research plan for on-farm testing in each research district. The results show that significant increases in yield and economic returns can be realised with relative high application levels of compost, but that availability of material and labour inputs soon become limiting factors. Policy discussion workshops were held in each district with district policy-makers from various ministries, researchers, extension staff, NGO-staff, staff from development projects and others. The conditions required for implementation and a prioritised action plan are presented for Nyeri district as an example.

Chapter 9

COMPONENTS OF SOIL

Soil micro-biology is the study of organisms in soil, their functions, and how they affect soil properties. It is believed that between two to four billion years ago, the first ancient bacteria and micro-organisms came about in Earth's primitive seas. These bacteria could fix nitrogen, in time multiplied and as a result released oxygen into the atmosphere. This release of oxygen led to more advanced micro-organisms. Micro-organisms in soil are important because they affect the structure and fertility of different soils. Soil micro-organisms can be classified as bacteria, actinomycetes, fungi, algae, and protozoa. Each of these groups has different characteristics that define the organisms and different functions in the soil it lives in.

SOIL

It is the outer, loose material of earth's surface which is distinctly different from the underlying bedrock and the region which support plant life. Agriculturally, soil is the region which supports the plant life by providing mechanical support and nutrients required for growth. From the micro-biologist view point, soil is one of the most dynamic sites of biological interactions in the nature. It is the region where most of the physical, biological and bio-chemical reactions related to decomposition of organic weathering of parent rock take place.

Components of Soil

Soil is an admixture of five major components *viz.* organic matter, mineral matter, soil-air, soil water and soil micro-organisms/living organisms.

The amount/ proposition of these components vary with locality and climate :

- *Mineral / Inorganic Matter* : It is derived from parentrocks/bed rocks through decomposition, disintegration and weathering process. Different types of inorganic compounds containing various minerals are present in soil. Amongst them the dominant minerals are Silicon, Aluminum and iron and others like Carbon, Calcium Potassium, Manganese, Sodium,

Sulphur, Phosphorus etc., are in trace amount. The proportion of mineral matterin soil is slightly less than half of the total volume of the soil.

- *Organic matter/components* : Derived from organic residues of plants and animals added in the soil. Organic matter serves not only as a source of food for micro-organisms but also supplies energy for the vital processes of metabolism which are characteristics of all living organisms. Organic matter in the soil is the potential source of N, P and S for plant growth. Microbial decomposition of organic matter releases the unavailable nutrients in available from. The proportion of organic matter in the soil rangesfrom 3-6 per cent of the total volume of soil.

- *Soil Water* : The amount of water present in soil varies considerably. Soil water comes from rain, snow,dew or irrigation. Soil water serves as a solvent and carrier of nutrients for the plant growth. The micro-organisms inhabiting in the soil also require water for their metabolic activities. Soil water thus, indirectly affects plant growth through its effects on soil and micro-organisms. Percentage of soil-water is 25 per cent total volume of soil.

- *Soil air (Soil gases)* : A part of the soil volume which is not occupied by soil particles *i.e.* pore spaces are filled partly with soil water and partly-with soil air. These two components (water & air)together only accounts for approximately half the soil's volume. Compared with atmospheric air, soil is lower in oxygen and higher in carbon dioxide, because CO_2 is continuous recycled by the micro-organisms during the process of decompositionof organic matter. Soil air comes from external atmosphere and contains nitrogen, oxygen CO_2 and water vapour (CO_2 > oxygen). CO_2 in soil air (0.3-1.0 per cent) is more than atmospheric air (0.03 per cent). Soil aeration plays important role in plant growth, microbial population,and microbial activities in the soil.

- *Soil micro-organisms* : Soil is an excellent culture media for the growth and development of various micro-organisms. Soil is not an inert static material but a medium pulsating with life. Soil is now believed to be dynamic or living system. Soil contains several distinct groups of micro-organisms and amongst them bacteria, fungi, actinomycetes, algae, protozoa and *viruses* are the most important. But bacteria are more numerous than any other kinds of micro-organisms. Micro-organisms form a very small fraction of the soil mass and occupy a volume of less than one per cent. In the upper layer of soil (top soil up to 10-30 cm depth *i.e.* Horizon A), the microbial population is very high which decreases with depth of soil. Each organisms or a group of organisms are responsible for a specific change / transformation in the soil. The final effect of various activities of micro-organisms in the soil is to make the soil fit for the growth & development of higher plants. Living organisms present in the soil are grouped into two categories as follows :

- Soil-flora (micro-flora) *e.g.* Bacteria, fungi, Actinomycetes, Algae.

- Soil-fauna (micro fauna) animal like *e.g.* Protozoa, Nematodes, earthworms, moles, ants, rodents.

Relative proportion / percentage of various soil micro-organisms are : Bacteria-aerobic (70 per cent), anaerobic (13 per cent), Actinomycetes (13 per cent), Fungi/molds (03 per cent) and others (Algae Protozoa viruses) 0.2-0.8 per cent. Soil organisms play key role in the nutrient transformations.

Scope and Importance of Soil Microbiology

Living organisms both plant and animal types constitute an important component of soil. Though these organisms form only a fraction (less than one per cent) of the total soil mass, but they play important role in supporting plant communities on the earth surface. While studying the scope and importance of soil micro-biology, soil-plant-animal ecosystem as such must be taken into account.

Therefore, the scope and importance of soil micro-biology, can be understood in better way by studying aspects like :

- *Soil as a living system* : Soil inhabit diverse group of living organisms, both micro-flora (fungi, bacteria, algae and actinomycetes) and micro-fauna (protozoa, nematodes, earthworms, moles, ants). The density of living organisms in soil is very high *i.e.* as much as billions / gm of soil, usually density of organismsis less in cultivated soil than uncultivated / virginland and population decreases with soil acidity. Top-soil, the surface layer contains greater number of micro-organisms because it is well supplied with Oxygenand nutrients. Lower layer / sub-soil is depleted with Oxygen and nutrients hence it contains fewer organisms. Soil ecosystem comprises of organisms which are both, autotrophs (Algae, BOA) and heterotrophs (fungi, bacteria). Autotrophs use inorganic carbon from CO_2 and are "primary producers" of organic matter, whereas heterotrophs use organic carbon and are decomposers/consumers.
- *Soil microbes and plant growth* : Micro-organisms being minute and microscopic, they are universally presentin soil, water and air. Besides supporting the growth of various biological systems, soil and soil microbes serve as a best medium for plant growth. Soil fauna & flora convert complex organic nutrients into simpler inorganic forms which are readily absorbed by the plant for growth. Further, they produce variety of substances like IAA, gibberellins, antibiotics etc., which directly or indirectly promote the plant growth.
- *Soil microbes and soil structure* : Soil structure is dependent on stable aggregates of soil particles—Soil organisms play important role in soil aggregation.Constituents of soil are *viz.* organic matter, polysaccharides, lignins and gums, synthesised by soil microbes plays important role in cementing/binding of soil particles. Further, cells and mycelial strands of fungi and actinomycetes, Vormicasts from earthwormis also found to play important role in soil aggregation. Different soil micro-organisms,

having soil aggregation/soil binding properties are graded in the order as fungi > actinomycetes > gum producing bacteria > yeasts. Examples are : Fungi like Rhizopus, Mucor, Chaetomium, Fusarium, Cladasporium, Rhizoctonia, Aspergillus,Trichoderma and Bacteria like Azofobacler, Rhizobium Bacillus and Xanlhomonas.

- *Soil microbes and organic matter decomposition* : The organic matter serves not only as a source of food for micro-organisms but also supplies energy for the vital processes of metabolism that are characteristics of living beings. Micro-organisms such as fungi, actinomycetes, bacteria, protozoa etc., and macro-organisms such as earthworms, termites, insects etc. plays important role in the process of decomposition of organic matter and release of plant nutrients in soil. Thus, organic matter added to the soil is converted by oxidative decomposition to simpler nutrients / substances for plant growth and the residueis transformed into humus. Organic matter/ substances include cellulose, lignins and proteins (in cell wall of plants), glycogen (animal tissues), proteins and fats (plants, animals). Cellulose is degraded bybacteria, especially those of genus Cytophaga and other genera (Bacillus, Pseudomonas, Cellulomonas, and Vibrio Achromobacter) and fungal genera (Aspergillus, Penicilliun, Trichoderma, Chactomium, Curvularia). Lignins and proteins are partially digested by fungi, protozoa and nematodes. Proteins are degraded to individual amino acids mainly by fungi, actinomycetes and Clostridium. Under unaerobic conditions of water-logged soils, methane are main carbon containing product which isproduced by the bacterial genera (strict anaerobes) Methanococcus, Methanobacterium and Methanosardna.

- *Soil microbes and humus formation* : Humus is the organic residue in the soil resulting from decomposition of plant and animal residues in soil, or it is the highly complex organic residual matter in soil which is not readily degraded by micro-organism, or it is the soft brown/dark coloured amorphous substance composed of residual organic matter along with dead micro-organisms.

- *Soil microbes and cycling of elements* : Life on earthis dependent on cycling of elements from their organic/ elemental state to inorganic compounds, then to organic compounds and back to their elemental states.The biogeochemical process through which organic compounds are broken down to inorganic compounds ortheir constituent elements is known "Mineralisation", or microbial conversion of complex organic compounds into simple inorganic compounds & their constituent elements is known as mineralisation. Soil microbes plays important role in the bio-chemical cycling of elements in the biosphere where the essential elements (C, P, S, N & Iron etc.) undergo chemical transformations. Through the process of mineralisation organic carbon, nitrogen, phosphorus, Sulphur, Iron etc. are made available for reuse by plants.

- *Soil microbes and biological N_2 fixation* : Conversion of atmospheric nitrogen in to ammonia and nitrate by micro-organisms is known as biological nitrogen fixation. Fixation of atmospheric nitrogen is essential because of the reasons :

Fixed nitrogen is lost through the process of nitrogen cycle through denitrification.

Demand for fixed nitrogen by the biosphere always exceed sits availability.

The amount of nitrogen fixed chemically and lightning process is very less (*i.e.* 0.5 per cent) as compared to biologically fixed nitrogen.

Nitrogenous fertilizers contribute only 25 per cent of the total world requirement while biological nitrogen fixation contributes about 60 per cent of the earth's fixed nitrogen

Manufacture of nitrogenous fertilizers by "Haber" process is costly and time consuming. The numbers of soil micro-organisms carry out the process of biological nitrogen fixation at normal atmospheric pressure (1 atmosphere) and temp (around 20 °C). Two groups of micro-organisms are involved in the process of BNF.A. Non-symbiotic (free living)

- *Symbiotic (Associative)Non-symbiotic (free living)* : Depending upon the presence or absence of oxygen, non-symbiotic N_2 fixation prokaryotic organisms may be aerobic heterotrophs (Azotobacter, Pseudomonas, Achromobacter) or aerobic autotrophs (Nostoc, Anabena, Calothrix, BGA) and anaerobic heterotrophs (Clostridium, Kelbsiella, Desulfovibrio) or anaerobic Autotrophs (Chlorobium, Chromnatium, Rhodospirillum, Meihanobacterium etc).

- *Symbiotic (Associative)* : The organisms involved are Rhizobium, Bratfyrhizobium in legumes (aerobic) : Azospirillum (grasses), Actinonycetes frantic (with Casuarinas, Alder).

- *Soil microbes as biocontrol agents* : Several eco friendly bio-formulations of microbial origin are used in agriculture for the effective management of plant

- *Diseases, insect pests, weeds etc., e.g.* : Trichoderma spand Gleocladium sp are used for biological control of seed and soil borne diseases. Fungal general Entomophthora, Beauveria, Metarrhizium and protozoa Maltesiagrandis. Malameba locustiae etc. are used in the management of insect pests. Nuclear polyhydrosis virus (NPV) is used for the control of Heliothis / American boll worm. Bacteria like Bacillusthuringiensis, Pseudomonas are used in cotton against Angular leaf spot and boll worms.

- *Degradation of pesticides in soil by micro-organisms* : Soil receives different toxic chemicals in various forms and causes adverse effects on beneficial soil micro-flora/micro-fauna, plants, animals and human beings. Various microbes present in soil act as the scavengers of these harmful chemicals in soil. The pesticides/chemicals reaching the soil are acted upon by several physical, chemical and biological forces exerted by microbes in the soil and they are degraded into non-toxic substances and there by minimise the damage caused by the pesticides to the ecosystem. For example, bacterial genera like Pseudomonas, Clostridium, Bacillus, Thiobacillus, Achromobacter etc., and fungal genera like Trichoderma, Penicillium, Aspergillus, Rhizopus, and Fusarium are playing important role in the degradation of the toxic chemicals / pesticides in soil.

- *Biodegradation of hydrocarbons* : Natural hydrocarbons in soil like waxes, paraffin's, oils etc., are degradedby fungi, bacteria and actinomycetes. E.g. ethane (C_2H_6) a paraffin hydrocarbon is metabolised and degraded by Mycobacteria, Nocardia, StreptomycesPseudomonas, Flavobacterium and several fungi.

SOIL HUMUS

Humus is the organic residue in the soil resulting from decomposition of plant and animal residues in soil, or it is the highly complex organic residual matter in soil which is not readily degraded by micro-organism, or it is the soft brown/dark coloured amorphous substance composed of residual organic matter along with dead micro-organisms.

Composition of Humus : In most soil, percentage of humus ranges from 2-10 per cent, whereas it is up to 90 per cent in peat bog. On average humus is composed of Carbon (58 per cent), Nitrogen (3-6 per cent, Av. 5 per cent), acids -humic acid, fulvic acid, humin, apocrenic acid, and C : N ratio 10 : 1 to 12 : 1. During the course of their activities, the micro-organisms synthesise number of compounds which plays important role in humus formation.

Functions/Role of Humus :
- It improves physical condition of soil
- Improve water holding capacity of soil
- Serve as store house for essential plant nutrients
- Plays important role in determining fertility level of soil
- It tend to make soils more granular with better aggregation of soil particles
- Prevent leaching losses of water soluble plant-nutrients
- Improve microbial/biological activity in soil and encourage better development of plant-root system in soil
- Act as buffering agent *i.e.* prevent sudden change insoil PH/soil reaction
- Serve as source of energy and food for the developmentof soil organisms
- It supplies both basic and acidic nutrients for the growth and development of higher plants
- Improves aeration and drainage by making the soil more porous
- Types of Micro-organisms in Soil.

Living organisms both plants and animals, constitute an important component of soil. The pioneering investigations of a number of early micro-biologists showed for the first time that the soil was not an inert static material but a medium pulsating with life. The soil is now believed to be a dynamic or rather a living system, containing a dynamic population of organisms/micro-organisms. Cultivated soil has relatively more population of micro-organisms than the fallow land, and the soils rich in organic matter contain much more population than sandy and eroded soils. Microbes in the soil are important to us in maintaining soil fertility / productivity, cycling of nutrient elements in the biosphere and sources of industrial products such as enzymes, antibiotics, vitamins, hormones, organic acids etc. At the same time certain soil microbes are the causal agents of human and plant diseases. The soil organisms are broadly classified in to two groups *viz.* soil-flora and soil-fauna, the detailed classification of which is as follows :

- Soil Organisms
- Soil Flora (Microflora) : a. Bacteria b. Fungi, Molds, Yeast, Mushroom c. Actinomycetes, Stretomyces d. Algae *e.g.* BGA, Yellow Green Algae, Golden Brown Algae.
- Bacteria is again classified as :

Heterotrophic *e.g.* symbiotic & non-symbiotic N_2 fixers, Ammonifier, Cellulose Decomposers, Denitrifiers

Autrotrophic *e.g.* Nitrosomonas, Nitrobacter, Sulphur oxidisers, etc.

- Macroflora : Roots of higher plants
- Soil Fauna
- Micro-fauna : Protozoa, Nematodes
- Macro-fauna : Earthworms, moles, ants & others.

As soil inhabit several diverse groups of micro-organisms, but the most important amongst them are : bacteria, actinomycetes, fungi, algae and protozoa. The characteristics and their functions/role in the soil are described in the next topics.

SOIL MICRO-ORGANISM

Bacteria

Amongst the different micro-organisms inhabiting in the soil, bacteria are the most abundant and predominant organisms. These are primitive, prokaryotic, microscopic and unicellular micro-organisms without chlorophyll.

Morphologically, soil bacteria are divided into three groups viz. :
- Bacilli
- Cocci (round/spherical), (rod-shaped)
- Spirilla/Spirllum (cells with long wavy chains).

Bacilli are most numerous followed by Cocci and Spirilla in soil. The most common method used for isolation of soil bacteria is the "dilution plate count" method which allows the enumeration of only viable/living cells in the soil. The size of soil bacteria varies from 0.5 to 1.0 micron in diameter and 1.0 to 10.0 microns in length. They are motile with locomotory organs flagella. Bacterial population is one-half of the total microbial biomass in the soil ranging from 1, 00000 to several hundred millions per gram of soil, depending upon the physical, chemical and biological conditions of the soil. Winogradsky, on the basis of ecological characteristics classified soil micro-organisms in general and bacteria in particular into two broad categories *i.e.* Autochnotus (Indigenous species) and the Zymogenous (fermentative). Autochnotus bacterial population is uniform and constant in soil, since their nutrition is derived from native soil organic matter (*e.g.* Arthrobacter and Nocardia where as Zymogenous bacterial population in soil is low, as they require an external source of energy, *e.g.* Pseudomonas & Bacillus.

The population of Zymogenous bacteria increases gradually when a specific substrate is added to the soil. To this category belong the cellulose decomposers, nitrogen utilising bacteria and ammonifiers. As per the system proposed in the Bergey's Manual of Systematic Bacteriology, most of the bacteria which are predominantly encountered in soil are taxonomically included in the three orders, Pseudomonadales, Eubacteriales and Actinomycetales of the class Schizomycetes. The most common soil bacteria belong to the genera Pseudomonas, Arthrobacter, Clostridium Achromobacter, Sarcina, Enterobacter etc. The another group of bacteria common in soils is the Myxobacteria belonging to the genera Micrococcus, Chondrococcus, Archangium, Polyangium, Cyptophaga.

Bacteria are also classified on the basis of physiological activity or mode of nutrition, especially the manner in which they obtain their carbon, nitrogen, energy and other nutrient requirements.

They are broadly divided into two groups :
- Autotrophs
- Heterotrophs.

Autotrophs

Autotrophic bacteria are capable synthesising their food from simple inorganic nutrients, while heterotrophic bacteria depend on pre-formed food for nutrition. All autotrophic bacteria utilise CO_2 (from atmosphere) as carbon source and derive energy either from sunlight (photoautotrophs, *e.g.* Chromatrum. Chlorobium). Rhadopseudomonas or from the oxidation of simple inorganic substances present in soil (chemoautotrophs *e.g.* Nitrobacter, Nitrosomonas, Thiaobacillus)

Heterotrophs

Majority of soil bacteria are heterotrophic in nature and derive their carbon and energy from complex organic substances/organic matter, decaying roots and plant residues. They obtain their nitrogen from nitrates and ammonia compounds (proteins) present in soil and other nutrients from soil or from the decomposing organic matter. Certain bacteria also require amino acids, B- Vitamins, and other growth promoting substances also. Functions/Role of Bacteria : Bacteria bring about a number of changes and bio-chemical transformations in the soil and thereby directly or indirectly help in the nutrition of higher plants growing in the soil. The important transformations and processes in which soil bacteria play vital role are : decomposition of cellulose and other carbohydrates, ammonification (proteins ammonia), nitrification (ammonia-nitrites-nitrates), denitrification (release of free elemental nitrogen), biological fixation of atmospheric nitrogen (symbiotic and non-symbiotic) oxidation and reduction of sulphur and iron compounds. All these processes play a significant role in plant nutrition,

Actinomycetes

These are the organisms with characteristics common to both bacteria and fungi but yet possessing distinctive features to delimit them into a distinct category. In the strict taxonomicsense, actinomycetes are clubbed with bacteria the same class of Schizomycetes and confined to the order Actinomycetales. They are unicellular like bacteria, but produce a mycelium which is non-septate (coenocytic) and more slender, tike true bacteria they do not have distinct cell-wall and their cell wall is without chit in and cellulose (commonly found in the cell wall of fungi). On culture media unlike slimy distinct colonies of true bacteria which grow quickly, actinomycetes colonies grow slowly, show powdery consistency and stick firmly to agar surface. They produce hyphae and conidia / sporangia like fungi. Certain actinomycetes whose hyphae undergo segmentation resemble bacteria, both morphologically and physiologically. Actinomycetes are numerous and widely distributed in soil and are next to bacteria in abundance. They are widely distributed in the soil, compost etc. Plate count estimates give values ranging from 10^4 to 10^8 per gram of soil. They are sensitive to acidity/low PH (optimum PH range 6.5 to 8.0) and water logged soil conditions. The population of actinomycetes increases with depth of soil even up to horizon 'C' of a soil profiler. They are heterotrophic, aerobic and mesophilic (2530°C) organisms and some species are commonly present in compost and manures are the rmophilic growing at 55-65°C temperature (*e.g.* Thermoatinomycetes, Streptomyces). Actinomycetes belonging to the order of Actinomycetales are grouped under four families *viz.* Mycobacteriaceae, Actinomycetaceae, Streptomycetaceae and Actinoplanaceae. Actinomycetous genera which are agriculturally and industrially important are present in only two families of Actinomycetaceae and Strepotmycetaceae.

In the order of abundance in soils, the common genera of actinomycetes are Streptomyces (nearly 70 per cent), Nocardia and Micro-monospora although Actinomycetes, Actinoplanes, Micro-monospora and Streptosporangium are also generally encountered.

Functions / Role of actinomycetes :

- Degrade/decompose all sorts of organic substances likecellulose, polysaccharides, protein fats, organic-acids etc.

- Organic residues/substances added soil are first attacked by bacteria and fungi and later by actinomycetes, because they are slow in activity and growth than bacteria and fungi.
- They decompose/degrade the more resistant and indecomposable organic substance/matter and produce a number of dark black to brown pigments which contribute to the dark colour of soil humus.
- They are also responsible for subsequent further decomposition of humus (resistant material) in soil.
- They are responsible for earthy/musty odor/smell of freshly ploughed soils.
- Many genera species and strains (*e.g.* Streptomyces if actinomycetes produce/synthesise number of antibiotics like Streptomycin, Terramycin, Aureomycin etc.
- One of the species of actinomycetes Streptomycesscabies causes disease "Potato scab" in potato.

Fungi

Fungi in soil are present as mycelial bits, rhizomorph or as different spores. Their number varies from a few thousand to a few million per gram of soil. Soil fungi possess filamentous mycelium composed of individual hyphae. The fungal hyphae maybe as eptate/coenocytic (Mastigomycotina and Zygomycotina) or septate (Ascomycotina, Basidiomycotina & Deuteromycotina).

As observed by C.K. Jackson, most commonly encountered genera of fungi in soil are : Alternaria, Aspergillus, Cladosporium, Cephalosporium Botrytis, Chaetomium, Fusarium, Mucor, Penicillium, Verticillium, Trichoderma, Rhizopus, Gliocladium, Monilia, Pythium, etc. Most of these fungal genera belong to the sub-division Deuteromycotina / Fungi imperfeacta which lacks sexual mode of reproduction.

As these soil fungi are aerobic and heterotrophic, they require abundant supply of oxygen and organic matter in soil. Fungi are dominant in acid soils, because acidic environment is not conducive/suitable for the existence of either bacteria oractinomycetes. The optimum PH range for fungi lies-between 4.5 to 6.5. They are also present in neutral and alkaline soils and some can even tolerate pH beyond 9.0.

Functions/Role of Fungi :
- Fungi plays significant role in soils and plant-nutrition.
- They plays important role in the degradation / decomposition of cellulose, hemi cellulose, starch, pectin, lignin in the organic matter added to the soil.
- Lignin which is resistant to decomposition by bacteriais mainly decomposed by fungi.

- They also serve as food for bacteria.
- Certain fungi belonging to sub-division Zygomycotina and Deuteromycotina are predaceous in nature and attackon protozoa & nematodes in soil and thus, maintain biologica lequilibrium in soil.
- They also plays important role in soil aggregation and in the formation of humus.
- Some soil fungi are parasitic and causes number of plant diseases such as wilts, root rots, damping-off and seedling blightseg. Pythium, Phyiophlhora, Fusarium, Verticillium etc.
- Number of soil fungi forms mycorrhizal association with the roots of higher plants (symbiotic association of a fungus with the roots of a higher plant) and helps in mobilisation of soil phosphorus and nitrogen *e.g.* Glomus, Gigaspora, Aculospora, (Endomycorrhiza) and Amanita, Boletus, Entoloma, Lactarius (Ectomycorrhiza).

Algae

Algae are present in most of the soils where moisture and sunlight are available. Their number in soil usually ranges from 100 to 10,000 per gram of soil. They are photoautotrophic, aerobic organisms and obtain CO_2 from atmosphere and energy from sunlight and synthesise their own food. They are unicellular, filamentous or colonial.

Soil algae are divided in to four main classes or phyla as follows :

- Cyanophyta (Blue-green algae)
- Chlorophyta (Grass-green algae)
- Xanthophyta (Yellow-green algae)
- Bacillariophyta (diatoms or golden-brown algae).

Out of these four classes/phyla, blue-green algae and grass-green algae are more abundant in soil. The green-grass algae and diatoms are dominant in the soils of temperate region while blue-green algae predominate in tropical soils. Green-algae prefer acid soils while blue green algae are commonly found in neutral and alkaline soils. The most common genera of green algae found in soil are Chlorella, Chlamydomonas, Chlorococcum, Protosiphon etc., and that of diatoms are Navicula, Pinnularia, Synedra, Frangilaria.

Blue green algae are unicellular, photo-autotrophic prokaryotes containing Phycocyanin pigment in addition to chlorophyll. They do not posses flagella and do not reproduce sexually. They are common in neutral to alkaline soils. The dominant genera of BGA in soil are : Chrococcus, Phormidium, Anabaena, Aphanocapra, Oscillatoria etc. Some BGA posses specialised cells known as "Heterocyst" which is the sites of nitrogen fixation. BGA fixes nitrogen (non-symbiotically) in puddle paddy/water logged paddy fields (20-30 kg/ha/ season). There are certain BGA which possess the character of symbiotic nitrogen fixation in association with other organisms like fungi, mosses, liverworts and aquatic ferns Azolla, *e.g.* Anabaena-Azolla association fix nitrogen symbiotically in rice fields.

Functions/Role of Algae or BGA

- Plays important role in the maintenance of soil fertility especially in tropical soils.
- Add organic matter to soil when die and thus increase the amount of organic carbon in soil.
- Most of soil algae (especially BGA) act as cementing agent in binding soil particles and thereby reduce/ prevent soil erosion.
- Mucilage secreted by the BGA is hygroscopic in natureand thus helps in increasing water retention capacity of soil for longer time/period.
- Soil algae through the process of photo-synthesis liberate large quantity of oxygen in the soil environment and thus facilitate the aeration in submerged soils or oxygenate the soil environment.
- They help in checking the loss of nitrates through leaching and drainage especially in un-cropped soils.
- They help in weathering of rocks and building up of soil structure.

Protozoa

These are unicellular, eukaryotic, colourless, and animal like organisms (Animal kingdom). They are larger than bacteria and size varying from few microns to a few centimeters. Their population in arable soil ranges from 10, 000 to 1, 00,000 per gram of soil and are abundant in surface soil. They can withstand adverse soil conditions as they are characterised by "cyst stage" in their life cycle. Except few genera which reproduce sexually by fusion of cells, rest of them reproduces asexually by fission /binary fission. Most of the soil protozoa are motile by flagella orcilia or pseudopodia as locomotors organs.

Depending upon the type of appendages provided for locomotion, protozoa are :
- Rhizopoda (Sarcondia)
- Mastigophora
- Ciliophora (Ciliata)
- Sporophora (not common Inhabitants of soil).

Class-Rhizopoda

It consists protozoa without appendages usually have naked protoplasm without cell-wall, pseudopodia astemporary locomotory organs are present some times. Important genera are Amoeba, Biomyxa, Euglypha, etc.

Class Mastigophora

It belongs flagellated protozoa, which are predominant in soil. Important genera are : Allention, Bodo, Cercobodo, Cercomonas, Entosiphon Spiromonas, Spongomions, and Testramitus. Many members are saprophytic and some posses chlorophyll and are autotrophic in nature. In this respect, they resemble unicellular algae and hence are known as "Phytoflagellates".

Ciliophora

The soil protozoa belonging to the class ciliate/ciliophora are characterised by the presence of cilia (short hair-like appendages) around their body, which helps in locomotion. The important soil inhabitants of this class are Colpidium, Colpoda, Balantiophorus, Gastrostyla, Halteria, Uroleptus, Vortiicella, Pleurotricha etc.

Sporophora

Protozoa are abundant in the upper layer (15 cm) of soil. Organic manures protozoa. Soil moisture, aeration, temperature and pH are the important factors affecting soil protozoa.

Function/Role of Protozoa :

- Most of protozoans derive their nutrition by feeding or ingesting soil bacteria belonging to the genera Enterobacter, Agrobacterium, Bacillus, Escherichia, Micrococcus, and Pseudomonas and thus, they play important role in maintaining microbial / bacterial equilibrium in the soil.
- Some protozoa have been recently used as biological control agents against phytopathogens.
- Species of the bacterial genera *viz.* Enterobacter and Aerobacter are commonly used as the food basefor isolation and enumeration of soil protozoans.
- Several soil protozoa cause diseases in human beings which are carried through water and other vectors, *e.g.* Amoebic dysentery caused by Entomobea histolytica.

Soil Micro-Organisms in Biodegradation of Pesticides and Herbicides

Pesticides are the chemical substances that kill pests and herbicides are the chemicals that kill weeds. In the context of soil, pests are fungi, bacteria insects, worms, and nematodes etc., that cause damage to field crops. Thus, in broad sense pesticides are insecticides, fungicides, bactericides, herbicides and nematicides that are used to control or inhibit plant diseases and insect pests. Although widescale application of pesticides and herbicides is an essential part of augmenting crop yields; excessive use of these chemicals leads to the microbial imbalance, environmental pollution and health hazards. An ideal pesticide should have the ability to destroy target pest quickly and should be able to degrade non-toxic substances as quickly as possible. The ultimate "sink" of the pesticides applied in agriculture and public health care is soil. Soil being the storehouse of multitudes of microbes, in quantity and quality, receives the chemicals invarious forms and acts as a scavenger of harmful substances. The efficiency and the competence to handle the chemicals vary with the soil and its physical, chemical and biological characteristics.

Effects of Pesticides

Pesticides reaching the soil in significant quantities have direct effect on soil micro-biological aspects, which in turn influence plant growth.

Some of the most important effects caused by pesticides are :

- Alterations hi ecological balance of the soil micro-flora,
- Continued application of large quantities of pesticides may cause ever-lasting changes in the soil micro-flora,
- Adverse effect on soil fertility and crop productivity,
- Inhibition of N_2 fixing soil micro-organisms such as Rhizobium, Azotobacter, Azospirillum etc. and cellulolytic and phosphate solubilising micro-organisms,
- Suppression of nitrifying bacteria, Nitrosomonas and Nitrobacter by soil fumigants ethylene bromide, Telone, and vapam have also been reported,
- Alterations in nitrogen balance of the soil,
- Interference with ammonification in soil,
- Adverse effect on mycorrhizal symbioses in plantsand nodulation in legumes, and
- Alterations in the rhizosphere micro-flora, both quantitatively and qualitatively.

Persistence of Pesticides in Soil

How long an insecticide, fungicide, or herbicide persists in soil is of great importance in relation to pest management and environmental pollution.

Persistence of pesticides in soil for longer period is undesirable because of the reasons :

- Accumulation of the chemicals in soil to highly toxic levels,
- May be assimilated by the plants and get accumulated in edible plant products,
- Accumulation in the edible portions of the root crops,
- To be get eroded with soil particles and may enter into the water streams, and finally leading to the soil, water and air pollutions.

The effective persistence of pesticides in soil varies from a week to several years depending upon structure and properties of the constituents in the pesticide and availability of moisture in soil. For instance, the highly toxic phosphates do not persist for more than three months while chlorinated hydrocarbon insecticides (*e.g.*, DOT, aldrin, chlordane etc.) are known to persist at least for 4-5 years and some times more than 15 years.

From the agricultural point of view, longer persistence of pesticides leading to accumulation of residues in soil may result into the increased absorption of such toxic chemicals by plants to the level at which the consumption of plant products may prove deleterious / hazardous to human beings as well as livestock's. There is a chronic problem of agricultural chemicals, having entered in food chain at highly in admissible levels in India, Pakistan, Bangladesh and several other developing countries in the world. For example, intensive use of DDT to control insect pests and mercurial fungicides to control diseases in agriculture had been known to persist for longer period and thereby got accumulated in the food chain leading to food contamination and health hazards. Therefore, DDT and mercurial fungicides has been, banned to use in agriculture as well as in public health department.

Biodegradation of Pesticides in Soil

Pesticides reaching to the soil are acted upon by several physical, chemical, and biological forces. However, physical and chemical forces are acting upon/ degrading the pesticides to some extent, micro-organism's plays major role in the degradation of pesticides. Many soil micro-organisms have the ability to act upon pesticides and convert them into simpler non-toxic compounds. This process of degradation of pesticides and conversion into non-toxic compounds by micro-organisms is known as "biodegradation". Not all pesticides reaching to the soil are biodegradable and such chemicals that show complete resistance to biodegradation are called "recalcitrant".

The chemical reactions leading to biodegradation of pesticides fall into several broad categories which are discussed in brief in the following paragraphs.

- *Detoxification* : Conversion of the pesticide molecule to anon-toxic compound. Detoxification is not synonymous with degradation. Since a single chance in the side chain of a complex molecule may render the chemical non-toxic.
- *Degradation* : The breaking down / transformation of acomplex substrate into simpler products leadingfinally to mineralisation. Degradation is often considered to be synonymous with mineralisation, *e.g.*

In which an organism make the substrate more complex or combines the pesticide with cell metabolites. Conjugation or the formation of addition product is accomplished by those organisms catalysing the reaction of addition of an amino acid, organic acid or methyl crown to the substrate, for *e.g.*, in the microbial metabolism of sodium dimethly dithiocarbamate, the organism combines the fungicide with an amino acid molecule normally present in the cell and thereby inactivate the pesticides/chemical.

Activation

It is the conversion of non-toxic substrate into atoxic molecule, for *e.g.* Herbicide, 4-butyric acid (2, 4-D B) and the insecticide Phorate are transformed and activated microbiologically in soil to give metabolites that are toxic to weeds and insects.

Changing the Spectrum of Toxicity

Some fungicides/pesticides are designed to control one particular group of organisms / pests, but they are metabolised to yield products inhibitory to entirely dissimilar groups of organisms, for *e.g.* the fungicide PCNB fungicide is converted in soil to chlorinated benzoic acids that kill plants. Biodegradation of pesticides / herbicides is greatly influenced by the soil factors like moisture, temperature, PH and organic matter content, in addition to microbial population and pesticide solubility. Optimum temperature, moisture and organic matter in soil provide congenial environment for the break down or retention of any pesticide added in the soil. Most of the organic pesticides degrade within a short period (3-6 months) under tropical conditions. Metabolic activities of bacteria, fungi and actinomycetes have the significant role in the degradation of pesticides.

Criteria for Bio-remediation/Biodegradation

For successful biodegradation of pesticide in soil, following aspects must be taken into consideration :

- Organisms must have necessary catabolic activity required for degradation of contaminant at fast rate to bring down the concentration of contaminant,
- The target contaminant must be bio-availability,
- Soil conditions must be congenial for microbial / plant growth and enzymatic activity and
- Cost of bio-remediation must be less than other technologies of removal of contaminants.

According to Gales (1952) principal of microbial infallibility, for every naturally occurring organic compound there is a microbe / enzyme system capable its degradation.

Strategies For Bio-remediation

For the successful biodegradation / bio-remediation of a given contaminant following strategies are needed :

- *Passive/intrinsic Bio-remediation* : It is the natural bio-remediation of contaminant by tile indigenous micro-organisms and the rate of degradation is veryslow.
- *Bio-stimulation* : Practice of addition of nitrogen and phosphorus to stimulate indigenous micro-organisms in soil.

- *Bioventing* : Process/way of Bio-stimulation by which gases stimulants like oxygen and methane are added or forced into soil to stimulate microbial activity.
- *Bio-augmentation* : It is the inoculation/introduction of micro-organisms in the contaminated site/soil to facilitate biodegradation.
- *Composting* : Piles of contaminated soils are constructed and treated with aerobic thermophilic micro-organisms to degrade contaminants. Periodic physical mixing and moistening of piles are done to promote microbial activity.
- *Phytoremediation* : Can be achieved directly by planting plants which hyper accumulate heavy metals or indirectly by plants stimulating micro-organisms in the rhizosphere.
- *Bio-remediation* : Process of detoxification of toxic/unwanted chemicals/contaminants in the soil and other environment by using micro-organisms.
- *Mineralisation* : Complete conversion of an organic contaminant to its inorganic constituent by a species or group of micro-organisms.

Chapter 10

MANAGEMENT EFFECTS ON SOIL BIOTA

What happens to the organisms that live in the soil when land is managed for agriculture, forestry and mining? The production of food, fibre, wood products and mineral ores sometimes involves profound disturbances to the habitat of soil organisms.

We clear the land of trees for grasing or cropping, we selectively log trees in forests or clear fell forests, we physically disturb the soil through ploughing or digging, we replace native plants with exotic ones, large herbivores graze the land, we use agro-chemicals for suppressing weeds, invertebrate and microbial pests of plants and the parasites of livestock. How do the activities listed below affect soil biota?

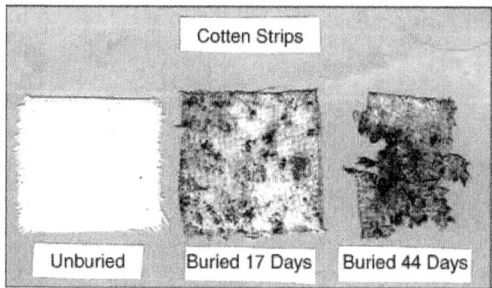

This photo shows cotton strips which have been buried for 17 and 44 days and a fresh, unburied strip. The microbes have been using the cellulose fibres in the cloth as a source of energy and, by 44 days in the soil, have started to rot the strip completely away in spots. You can also see how the colonies of bacteria and fungi colonising the strips have progressively stained the strips over time.

GRASING OF SOIL BIOTA

Effects of Grasing on the Habitat, Micro-climate and Food Supply of Soil Biota

Fig. A Fenceline Photo with Heavily and Lightly Grazed Pastures Compared.

- Heavy grasing results in low plant litter levels. Litter provides both a living-space for soil animals as well as forming organic residues in the soil as
- food for soil biota.
- Litter protects the soil environment against climatic extremes. In pastures at Tamworth, in northern NSW, Australia, summer temperatures at 5 cm depth in overgrazed, bare pastures can reach 50 °C. However, the soil temperature can be <25 °C at the same depth in lightly grazed pastures with a good litter layer.
- Litter layers slow the evaporation of moisture from soil from 10 mm/day to <2 mm/day in summer in the same pastures.
- Trampling by high numbers of large herbivores such as sheep and cattle can compact the soil, squeesing the pores and making them smaller. It is then difficult for small soil animals to access the soil as they cannot form their own tunnels. In addition, compacted soil is hard for the tunneling animals such as earthworms to move through.
- Soil nutrients and organic matter (in the form of dung) are concentrated into stock camps on higherground or in corners of paddocks, where stock

congregate to rest. This aggregation occurs at the expense of nutrients and organic matter over the rest of the paddock.

Overgrasing affects the soil biota in following ways :

- Reduces the numbers and biomass of soil mesofauna and macro-fauna
- Soil microbial activity is higher in the stock campsthan over the general paddock
- Diversity of some species of soil biota declines (*e.g.* springtails).

CULTIVATION

Effects of Cultivation on the Habitat

Fig. Freshly ploughed soil.

- Burial of the protective layer of dead plants on the soil surface. This affects the living space for litter-dwellers and the soil micro-climate (increased soil moisture loss and extremes of temperature)
- Physical disruption of soil habitat *e.g.* pore channel continuity, pore size – this limits the mobility of those animals which do not tunnel
- Physical injury to animals
- Reduction of organic matter levels in soil
- Inversion of soil brings animals to surface where they desiccate and are eaten by birds
- Inversion of soil traps smaller soil biota

Effects of cultivation on soil organisms :

- Numbers of large animals (earthworms, millipedes) are lowered and they can take up to two years to recover to pre-cultivation levels because they have long life cycles. Cropping soils that are repeatedly cultivated have fewer earthworms.
- Micro-arthropod numbers (springtails, mites) can also decline by 50 per cent after cultivation. They can recover their original numbers within 6 months as they have short life cycles.

- Soil microbial biomass declines after cultivation probably as a result of the reduced soil organic matter levels, their energy source. Microbial biomass can also be reduced, in part, by the physical disruption of extensive fungal hyphal networks.

FERTILIZERS

Fig. Native unfertilized Pasture,

Fig. Fertilized Pasture.

Fertilizers are used to increase agricultural productivity and to replace nutrients that are either removed in agricultural products (meat, wool, crops) or eroded by water and wind. Agricultural fertilizers fall into several groups. Inorganic chemical fertilizers include phosphate, nitrogen and sulphur fertilizers while organic fertilizers include manures and compost. Lime is used mainly as a soil ameliorant to reduce acidity.

INORGANIC FERTILIZERS

Mostly, inorganic fertilizers increase soil biological activity because they increase plant production leaving organic residues of high quality as food for soil biota. There may be some short-term adverse effects, possibly due to pH change but, in the long-term, effects of inorganic fertilizers on overall biological activity are mostly stimulatory. However, species types and diversity of soil biota may change with fertilizer use.

Fig. Micro-arthropods (springtails and mites) from native unfertilized pasture (left) and fertilized pastures (right). Note larger numbers collected in same volume of fertilized pasture. Note greater diversity of types of animals in the native unfertilized pasture.

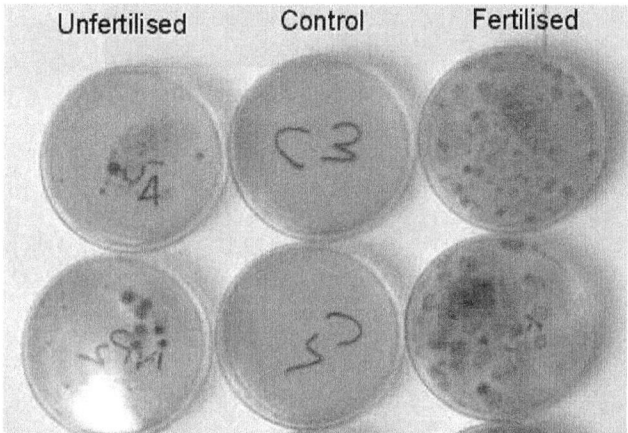

Fig. Agar Plates Growing Microbes from Native Unfertilized Pasture (left) and Fertilized Pasture (right). Note Larger Numbers of Microbes on Plates from Fertilized Pastures in same Volume of Soil.

P AND S FERTILIZERS

Superphosphate (a source of P and S) is used widely on both pastures and on cropping soils and is not considered to be toxic to soil biota in the long term. The increased secondary production that follows from their use, along with the increased quality of organic residues that form on those farming systems, increases biological activity.

Effects of superphosphate on soil biota include :

- Increase in earthworm numbers by between 2 to 5 times.
- Increase in springtail, mite and nematode numbers by up to 3-4 times.

- Increase of microbial abundance in soil by 2 times.
- Species diversity of some groups decline (*e.g.* springtails).

Effects of elemental S fertilizers on soil biota include :
- Decrease in some types of Protozoa.
- Decrease in fungal biomass.
- Increase in abundance of sulphur-oxidising bacteria.

N FERTILIZERS

Nitrogen content of soil can be increased with the use of legumes in pastures or green manure rotations in cropping soils and this can increase soil biota abundance. Nitrogen can also be applied as inorganic fertilizers. There are different forms of nitrogen-containing fertilizer which are commonly used and they have variable effects on soil biota. As a general rule, if soil pH declines after application of the fertilizer, a reduction occurs in abundance of some soil biota. Urea is generally considered to be less harmful than the anhydrous ammonia as the pH change is less. These deleterious effects may be only short-term. Long-term effects of nitrogen fertilization can be stimulatory to soil biota with increased plant production providing better quality residues. However, soil microbes decline if soil becomes acidic with prolonged use of anhydrous ammonia.

Effects of urea fertilizers on soil biota :
- Increase in free-living nematode abundance.
- Species diversity of nematodes can change : root and bacterial feeders can increase with N fertilizer use while fungal feeders and omnivores decrease.

Effects of anhydrous ammonia on soil biota :
- Decrease in soil microbes in immediate vicinity of fertilizer (short term) or after prolonged use of this fertilizer.
- Decrease in earthworm abundance in immediate vicinity of fertilizer.

LIME

Effects of lime on soil biota include :
- Change in earthworm species from acid to alkaline tolerant. Overall numbers may remain the same.
- Increase in bacterial to fungal ratio. Fungi more acid tolerant than bacteria.

ORGANIC FERTILIZERS

Organic amendments in the form of composts or manures, generally increase biological activity in soil through an increased supply of energy (food for soil biota) and nutrients to the detrital food web.

Fig. Manure Spreading on Cropping Soil.

Effects of organic fertilizers on soil biota :
- Increase in earthworm numbers.
- Increase in nematode, springtail and mite abundance.
- Increase in microbial biomass.

Pesticides

Chemicals used to control pests and diseases in plants and animals can have undesirable toxic side-effects on the non-target soil biota. Pesticides have variable effects on soil biota. The same pesticide may affect different species of soil biota in different ways. Also, effects of pesticides depend on the nature of the chemical, its dose, the method of application, temperature and moisture conditions in soil, crop residue management, the rate of decay of the chemical and the extent of leaching from the site. For example, some herbicides are more persistent if sprayed onto a mulch than if sprayed onto bare soil as they may be quickly inactivated by adsorption onto soil particles.

However, some generalisations of effects of pesticides on soil biota can be made. A general rule-of-thumb is that if the toxic effect of a pesticide on a non-target organism lasts longer than 60 days, then the chemical can be regarded as persistent and its toxic effects only slowly reversible.

Another general rule is that toxicity of pesticides from the least to the most toxic to soil biota, follows the order : herbicides < insecticides < fungicides. The following points can be made about the eco-toxic effects of pesticides.

Fig. Spraying Herbicides.

Herbicides

- Newer herbicides affect enzyme pathways in plants which are not found in soil biota
- Older herbicides can have some toxic effects on soil biota *e.g.* springtails, mites
- Many fungi are tolerant to herbicides
- Earthworms can increase in numbers following herbicide application as their food supply, in the form of dead plants, is increased.

Insecticides

- Organochlorines are more active in soil than organophosphates or carbamates
- Earthworms are not very susceptible to insecticides except carbamates.

Fungicides

- Benomyl is toxic to earthworms.
- Some fungicides contain copper which is toxic to earthworms and microbes – see the inhibition of the decay of cellulose by soil microbes in cotton strips buried in soil containing copper. Staining of the strips are areas where microbes have colonised.

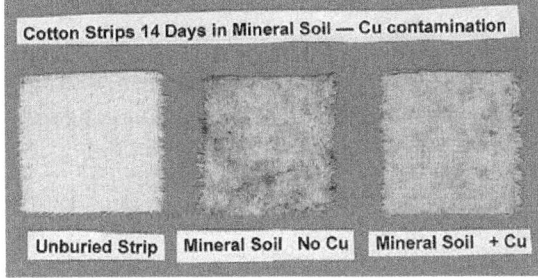

Fig. Toxic Effect of a Copper-based Pesticide (+Cu Strip) on Soil Microbes Decomposing a Cotton Strip.

PLANT RESIDUE RETENTION

Fig. On Left a Cultivated Field with no mulch Retained Compared with a Field with Zero Tillage with a Thick Mulch Layer.

In zero tillage cropping systems, a layer of mulch is retained on the soil surface and is not cultivated into the soil. Under these systems soil biota generally increase in abundance. *Effects of retention of organic surface residues on soil biota include* :
- Increase in earthworm numbers (x6 times).
- Increase in springtails and mites.
- Increase in microbial biomass.

Crop Rotations

Crop rotations occour when different crops are planted in succession, on the same area of land. Benefits of this practice are that they provide a disease break as diseases specific to one crop are denied a host for several years.

Crop rotations that include a legume phase (*e.g.* lucerne) increase the nitrogen content of organic residues and soil. The quality of the diet for decomposer organisms is increased.

Fig. Cropping soil with Lucerne Growing in Rotation.

Effects of legume in crop rotations on soil biota include :
- Increase in abundance of some small soil animals.
- Increase in earthworm abundance.
- Increase in soil microbial biomass.

Irrigation

The increased soil moisture which follows irrigation favours many soil biota. However, water-logged soil conditions decreases the oxygen content of soil and there are few soil macrofauna (*e.g.* earthworms) present under these conditions.

Effects of irrigation on soil biota :
- Irrigation allows earthworms to remain active in summer.
- Increase in springtail and mite abundance.
- Increase in Protozoa abundance.
- Increase in soil microbial biomass.

Fig. Irrigation Keeps Soil Moistur.

Conjugation (Complex Formation or Addition Reaction)

Burning of vegetation is used to remove stubble from cropping soils or in grasslands to improve the quality of rough pastures such as savannah grasslands where it eliminates woody plants and inedible dry vegetation. In addition, many wildfires occour throughout Australia in the summer months which burn whole forests. The adverse effects of fire are generally restricted to litter dwellers and true soil species are little affected. However, repeated burning of crop residues or forests depletes the soil of organic matter and biological activity falls as the food supply to soil biota is reduced.

Fig. Regrowth after Fire in Semi-arid Australia.

Effects of fire on soil biota :
- Litter dwelling organisms die.
- True soil dwellers are little affected in uncultivated soils.
- Some large invertebrates can survive fires bysheltering under rocks.
- Soil microbes increase after fire due to "limingeffects" of wood ash.

Fallow and Top-soil Storage

Top-soil is often scraped off and stored in heaps. Soil can be stored for months on building sites and then subsequently spread out again once building is finished.

Soil is stored for longer periods (*e.g.* years) during surface mining and the soil re-spread and rehabilitated with plants once mining operations have finished. However, during storage, mycorrhizal fungal spores germinate and die in this stored soil as there are few plants for them to grow into and they are dependent on plants for survival. Vegetation planted into this top-soil once it has been spread, often fail to thrive. One explanation is that the soil is depleted of the spores of the mycorrhizal fungi which help in plant nutrition. A similar thing happens during long fallow where the soil surface is bare of plants for many months.

Tree Clearing

Fig. Forests Cleared to Make way for Cow Pastures.

Effects of tree clearing on soil biota have not been studied extensively.

Some effects that have been noted are :

- Micro-arthropod and microbial biomass abundance higherunder eucalypt trees than in grassy interspaces between trees.
- Increase in earthworm numbers in temperate grassland after clearing deciduous European forests.
- Decline in earthworm numbers in tropical soils after clearing forests.

Where do Soil Biota Live?

Organisms are generally considered to be "soil" organisms if they live in soil and the layers of dead organic matter that overlie the soil. Thus, soil biota live in the soil, in the surface layers of dead plant leaves (or litter), in dead vertebrate and invertebrate animals (carrion), in dead trees and logs, dead roots and in dung pellets and pats.

Organisms are generally considered to be "soil" organisms if they live in soil and the layers of dead organic matter that overlie the soil. Thus, soil biota live in the soil, in the surface layers of dead plant leaves (or litter), in dead vertebrate and invertebrate animals (carrion), in dead trees and logs, dead roots and in dung pellets and pats.

There are some true soil dwelling biota that rarely see the light of day but many "soil" animals migrate up and down the soil/litter/dung/carrion profile as these layers are moistened and then dry out again. For example, springtails will migrate into the litter from the soil when the litter is moistened with rain or dew but retreat into the moister regions of the soil once the day warms up and dries the litter out again.

Earthworms also migrate from the soil below dung pats up into the pats and drag bits of dung down into their burrows. Migrations allow soil biota to exploit the rich food sources in the surface organic residues.

Even in the soil, biota are not uniformly distributed throughout. They aggregate at areas which are rich in organic matter such as around a small piece of dead leaf which has fallen into the soil, or in the faecal pellet of a springtail or around a dead root, or in the rhizosphere. The rhizosphere is that region which extends a few millimeters out from a root where root exudates and sloughed cells from roots provide an aggregation of organic materials which provide a rich food source for soil animals and microbes.

Ninety per cent of soil organisms live in the top 10 cm soil. This is also the zone where most plant roots live, where most organic matter in soil occurs, and most nutrients are also found. It's no wonder that soil organisms congregate in this region of highest nutrition.

Chapter 11

ABIOTIC SOIL COMPONENTS

CHEMICAL INPUT ON ABIOTIC SOIL CHARACTERISTICS

Nitrogen

Recent concerns over nitrogen deficiencies have led to others concerning excess nitrogen availability and the potential for forest decline and surface water pollution. High input of N can lead to N-saturation with serious environmental impacts on soil chemistry and water quality and on fluxes of radioactively active (or "greenhouse") gases. In Table the characteristics of N-saturated forest soil are listed. The values for the characteristics are endpoints. To get information about the deposition level upon which these characteristics start to change, the "critical load" concept has been introduced. The definition is "the maximum deposition of elements that will not cause chemical changes leading to long-term harmful effects on ecosystem structure and function". A regional assessment of critical loads is very important to formulate optimal policies for emission reductions. The generic approach to map critical loads is presented in Figure. In Table, an illustration is given of some critical chemical amounts for forest soil (water).

Table. Characteristics of N-saturated forest soils.

Characteristic	Value	Method
N cycled	25-50% NO_3^-	Anderson and Ingram, 1989;
	50-75% NH_4^+	Faber and Verhoef, 1991
DOC concentration	Low	Complete or partial oxidation
Z/N ratio	Low	Kirstei, 1979
Ca, Mg concentration	Low	Flame at omit absorbance spectrophotometry
S-, Ali concentration	High	pH based on free protons, on the exchangeable fraction extracted with KCl, or on the fraction titrated with a base; and ICP
N_2O production	High	Lloyd, 1985; Hairison et. al,1990
CH_4 production	Low	Lloyd, 1985; Hairison et. al,1990

Sulphur

Sulphur is transformed in soils by processes similar to those occurring in the nitrogen cycle. Like nitrogen, sulphur can be oxidised, reduced, assimilated, or mineralised from organic matter.

The major differences between the two cycles are :
- 1. No process is equivalent to N-fixation and
- 2. Losses of gaseous sulphur from soils are not equivalent to N-losses due to denitrification.

Recent interest in soil sulphur transformations results from an increased awareness of the fertilizer value of the element and the recognition of the importance of the sulphate ion, which reaches the soil in acid rain, as the major counter-anion involved in cation leaching from soils.

Sulphate adsorption by soils is an important property affecting the availability of sulphate to plants and the leaching of sulphate and associated cations. Sulphate adsorption is particularly important in soils subjected to acid precipitation, since it determines the impact of acid rain on cation mobility and leaching. Soil temperature and moisture influence sulphate adsorption, whereas desorption on waterlogging may also be an important reaction in soils exposed to atmospheric pollution.

Flowchart to Map Critical Loads and Areas where They have been Exceeded :

Select receptor type (*e.g.*, soil ecosystems)

↓

Determine critical chemical values

↓

Select computation method (*i.e.*, model)

↓

Quantify receptor distribution

↓

Collect input data

↓

Conduct critical load calculations

↓

Draw maps according to procedures.

Many agents have been used to extract sulphate and other sulphur ions from the soil. Important is 0.01 M $Ca(H_2PO_4)_2$ which appears to remove sulphate from the same pool of soil sulphur that is available for plants. Sulphate can be measured by methods including gravimetrically, turbidimetrically with barium chloride, spectrophotometric ally using methylene blue, by titrimetric methods, adsorption chromatography, ion exchange chromatography, ion-selective electrodes, and thin-layer or gas chromatography.

Recent studies on the damage to ecosystems caused by oxides of sulphur and acid rain involve the use of lysimeters, and collectors to measure through-fall, stem flow, and litter deposition. Unfortunately, most of these studies have omitted the microbial transformations of the element.

Recent laboratory studies have been concerned with the microbial cycling of sulphur in soils exposed to heavy atmospheric pollution from point sources. The first approach was to remove soils from the field at intervals throughout the season, and to have them analysed in the laboratory for S-ions and sulphur oxidising micro- organisms. In the latter approach, soils were sampled from sites exposed to point-source pollution and from relatively unpolluted sites that had essentially the same soil type, vegetation, and climate as the polluted sites.

Table. Critical Levels of Chemicals for Forest Soils.

Criteria	Unit	Soil
[Al]	Mol per m^3	0.2
Al/Ca	Mol per mol	1
pH	–	4.0^a
[Alk]^c	Mol per m^3	-0.3^a
NO$_3$	Mol per m^3	0.1^b
NH$_4$/K	Mol per mol	5

Note: ^{a}For forest top soils, pH of 3.7 and alkalinity of -0.4 are suggested.
^{b}Related to vegetation changes.
^{c}[HCO$_3$] + [RCOO] - -[AL]

In this way, an assessment of effects of pollution on relatively unpolluted soil could be determined as well as the time taken for heavily polluted soils to regain characteristics more typical of relatively unpolluted soil. Exposure duration was about 1.5–2 years. Similar exposure periods were found in a reciprocal transplant experiment with soil cores over a gradient of N and S input over Europe.

Phosphorus

Both forests and grasslands are frequently phosphorus deficient to a variable degree, and this deficiency limits their productivity. Fertilizer application is, therefore, a principal means of increasing timber and grass production.

In forestry, tree needle analysis has been used for many years as the main guide in the assessment of phosphorus fertilizer requirements. Recent publications, however, indicate that this type of analysis is unrealistic as a predictor of fertilizer responses in commercial forest trees. Analysis of the forest or grassland soils has been proposed as an alternative.

Extraction methods for P_i rely on three different principles :

- Anion exchange resin acts as a sink for solution Pi and thereby offsets the equilibrium between dissolved and soluble Pi. "Exchangeable" Pi as well as some of the more soluble precipitated p forms will enter the solution, bind to the resin, and can then be measured.
- Changes in pH cause changes in the solubility of Pi. Acid will extract calcium Pi. Alkaline solutions will solubilise Al and Fe bound Pi. Different Pi compounds have different solubilities at various pH values, and this can be used to characterise soil Pi composition or to evaluate labile Pi.
- Specific anions can bring Pi into solution by competing for adsorption sites and/or lowering the solubilities of cations that bind Pi. Fluoride has for instance been used under conditions of controlled pH, to release P from Al-bound forms, by forming insoluble aluminium fluoride. Organic anions have also been used to bind or chelate cations and release Pi into solution.

Methods to extract P_0 have employed alkaline solutions (*e.g.*, $NaHCO_3$ or NaOH) or various organic solvents (*e.g.*, acetylacetone which dissolves organic matter). Little progress has been made towards characterising P_0 extracts in terms of the mechanisms for bringing P_0 into solution, binding modes in the soil and its availability to plants. A new, physiologically based root bioassay has been developed, that appears to be sensitive in assessing P-deficiency in plants. The bioassay relies on the negative relationship between the rate of metabolic uptake of ^{32}p-labelled phosphorus by roots from a standardised solution in the laboratory and the amount of phosphorus supply in the original rooting environment. This method has been successfully applied to forest stands and grasslands.

Carbon

The CO_2 concentration in the atmosphere has increased by 25 per cent over the past 100 years, and a consensus exists that a doubling of the concentration may occur by the middle of the next century. The largest terrestrial carbon sources and sinks influencing CO_2 fluxes are the forests, which account for approximately two-thirds of the photo-synthesis. The effects of a doubling of the CO_2 concentration on the growth and development of trees is known for a few species. A general finding is an increase of the tissue density of the leaves, a change in leaf structure, and an increase in the C/N ratio of the tissues.

This changed C/N ratio may reduce the decomposition rates of plant material and modify the nutrient availability (Coûteaux *et. al.*, 1991). On the other hand, increased atmospheric concentrations of CO_2, together with trace gases such as methane (CH_4), nitrous oxide (N_2O), and chlorofluorohydrocarbons (CFCs), are effecting changes in the global heat balance, resulting in significant changes in climate over the next century. Twice as much carbon is found in the top metre of soil compared to the amount in the atmosphere, and CO_2 emissions from soils will increase as organic decomposition is enhanced at higher temperatures. CO_2

emissions are particularly sensitive to temperatures between 0°C and 5°C. Based on a recent model predicting global emissions from soil organic matter, a world temperature rise of 0.3°C per decade has been estimated to result in an additional release of CO_2 from soil organic matter over the next 60 years, equivalent to about 19 per cent of that released by combustion of fossil fuels if present use of fuel were to continue unabated. These calculations suggest that increased decomposition of soil organic carbon could make an important contribution to the greenhouse effect.

Methods to measure C/N ratios have already been given. Measurements of CO_2 evolution under field conditions are described and standardised.

ORGANIC CHEMICALS AND METALS

Effects of organic chemicals on the abiotic soil properties are rarely mentioned in the literature. Some chemicals, such as the herbicide paraquat, are incorporated into clay particles, but the extent to which this may influence the swelling and shrinking behaviour of the clay is unknown. Other chemicals or their degradation products are incorporated into the soil organic matter. This is a physical or a biological process, which may result, in case of chlorinated organics, in chlorination of the soil organic matter. The extent to which this process occurs and the consequences for the soil characteristics are unknown. Presently, no methods exist to determine the potential impact of organic chemicals on soil abiotic properties.

Metals generally occur in the soil solution as positively charged cations, competing for negatively charged adsorption places on the soil particles. An overload of metals will affect the ionic balance of the soil; it may also lead to a release of other, less strongly bound, metals or cations from the soil. Often this process is slow, not affecting soil abiotic properties to a great extent. Only in the case of flooding a soil with salt water, containing an excess of cations, was the swelling and shrinking properties of clays shown to be severely affected. No methods can be given to measure the impact of metals on soil abiotic properties.

MICROCOSM TESTS INCLUDING THOSE ON SOIL MICRO-FLORA

Microcosm Tests

Single-species tests are carried out under rather artificial conditions, and disregard ecological interactions between different species. To evaluate effects of chemicals under more natural conditions, model ecosystems, microcosms or micro-ecosystems have been designed that simulate certain aspects of real ecosystems, and are yet simple enough for experimental use. Decomposing invertebrates have been considered for such systems, because their activities can be assessed conveniently in terms of system functions such as leaf litter fragmentation and nutrient conversions.

Several terrestrial model ecosystems have been described, without attempt to arrive at standardisation. The system may either be closed or open to the ambient air, and contains intact core samples from a natural habitat or a more or less standardised soil (*e.g.,* Bond *et. al.,* 1976). For ecotoxicological tests, the use of standardised soils seems to be most appropriate, since it allows the chemical to be mixed homogeneously through the soil, and it minimises experimental variation between replicate units. The effects of various pretreatments, such as drying, sterilising, inoculation, litter type, age of the litter, however, have a significant impact on the behaviour of the system and need to be investigated thoroughly.

Natural rainfall may be simulated, and leachate can be collected. Various chemical analyses of the leachate solution may indicate aspects of decomposer activity : dissolved organic carbon, NH_4, NO_3, pH, and Ca. The advantage of this procedure is that repeated sampling in time from the same soil column is possible.

A disadvantage of the leaching procedure is that the humidity of soil and litter is unstable, and is difficult to standardise. Moreover, toxicants added to the system may be displaced through the column or leached out. Microbial respiration can be estimated by measuring CO_2 production in the microcosms. For that purpose, CO_2-free air (20.8 per cent O_2; 79.2 per cent N_2) is guided through the soil column, and the resulting CO_2 is subsequently measured by infrared gas analysis. Verhoef and Dorel, Verhoef *et. al.* and Verhoef and Meintser have used these types of microcosms, filled with pine litter, to study the effects of gaseous (NH_3) and wet (($NH_4)_2SO_4$) atmospheric deposition. N-deposition eliminates the stimulation of mineral leaching by the collembolan *Tomocerus minor*. Neither survival nor growth of the animals are affected. Reproduction, however, is negatively influenced. In pine litter, which has been confronted with high N input for several decades, *T. minor* slows down mineral leaching by stimulating microbial growth.

Van Wensem and Van Wensem *et. al.* added chemicals to poplar leaf litter, which is incubated for 4 weeks, after which some replicates are terminated to determine DOC, NH_4, NO_3, and pH. The remaining replicates are incubated for another four weeks, with eight isopods *(Porcellio scaber)* added to each system. Survival and growth of the isopods can be assessed, as well as particle size distribution and concentrations of minerals of the remaining litter. The organotin fungicide triphenyltin hydroxide increased the concentration of soluble ammonium in the litter, due partly to excretion by isopods and partly to stimulating effects on the micro-flora.

In systems with isopods, the organotin decreased ammoniation in treatment levels higher than 10 µg per g, but in systems without isopods the organotin had no significant effect. The addition of isopods in this case, therefore, made the system quite sensitive, which was unexpected (triphenyltin is a fungicide), and would not have been noticed in a single-species test using isopods.

Mothes-Wagner *et. al.* described a more complex microcosm system, consisting of 25 litres of natural or standardised soil that are inoculated with nematodes *(Pelodera strongyloides)* and enchytraeides *(Enchytraeus coronatus)* and sown with bush beans *(Phaseolus vulgaris)*. After emergence of the beans, spider mites *(Tetranychus urticae)* are introduced. After a pre-incubation period of about six months in the laboratory, in a greenhouse, or in the field, the systems can be treated with the test chemical. Test parameters are survival, reproduction, and population growth of the introduced organisms. Further more, measurement of several histological and enzymatic parameters in these organisms is recommended. As no substantial test results are available, the predictive value and sensitivity of this system cannot be evaluated.

Tests on Soil Microbial Processes

Tests on single species of isolated micro-organisms in artificial substrates are not regarded as representative for the soil ecosystem, and will, therefore, not be considered here. Based on a series of workshops held during the 1970s and 1980s, Somerville and Greaves formulated several recommended tests to assess the side effects of pesticides on the soil micro-flora. For all microbial tests in soil, the use of freshly sampled soil containing an active micro-flora was considered essential. Prolonged storage and drying of the soil should be avoided.

For a proper assessment of the effect of chemicals, at least two different soil types should be used. A short description is given here of several tests on microbial processes related to the conversion of nutrients in soil. Unless stated otherwise, all tests are carried out in the dark at a temperature of 20±2°C, and the test chemicals are mixed homogeneously through the soil. Generally, soils are tested at a moisture content corresponding to field capacity or to 40-60 per cent of the water holding capacity.

Test for Soil Respiration and Mineralisation of Substrates

In these tests the production of CO_2 from small soil samples (≤ 100 g) treated with the test chemical is measured continuously or semi-continuously. The tests should run for a minimum of 30 days. Tests may be performed in either unamended soil or in soil amended with a substrate. For this purpose mostly 0.5 per cent (w/w) lucerne or horn meal is used. The disadvantage of this soil respiration test is that the activity of the total soil micro-flora is determined. When certain species are affected by the test chemical, this will often not be noticed, as other (less sensitive) species may take over the activity of the sensitive ones.

During the past decade some new test methods have been developed which aim to determine chemical effects on more specific groups of soil micro-organisms. One is the addition of a readily degradable substrate and the determination of the short-term respiration rate. Such a test was described by Haanstra and Doelman using glutamic acid as a substrate. Soils are amended with glutamic acid and the CO_2-production is measured.

Glucose may be used as a substrate. The duration of the test is no longer than 100-120 hours. The test appeared to be quite sensitive to heavy metals. These short-term respiration tests may be combined with a biomass determination, and seem to be more sensitive than the traditional respiration tests. Another alternative may be found in the addition of more persistent substrates such as lignin or cellulose.

Only a few soil micro-organisms are capable of degrading these substrates, and, when they are affected by the test chemical, no others can take over their activity.

The disadvantage of the previously described soil respiration or substrate degradation methods is that less sensitive species of micro-organisms may grow on the substrate during the test.

This results in a shift among the micro-flora towards more resistant species, masking the possible elimination of sensitive species. For this reason, Van Beelen *et. al.*developed test methods using the mineralisation of low concentrations of ^{14}C-acetate, ^{14}C-chloroform or other labelled substrates. The amount of substrate applied is very low (1 µg per L) to ensure that no growth of the micro-flora will occur.

This amount of substrate is added to a slurry of the test soil, prepared by mixing the homogenised soil with an equal weight of ground water. The test chemicals are added to the slurry in the desired concentration levels. Samples are incubated at 10°C. The test duration depends on the capacity of the micro-flora in the soil sample to mineralise the test substrate, and is chosen depending on the half-life of the acetate mineralisation. Acetate mineralisation is measured by determining the amount of $^{14}CO_2$ released from the sample and by determining the amount of ^{14}C remaining in the suspension at the end of the test.

Test for Ammoniation and Nitrification

In ammoniation tests, the release of inorganic nitrogen from soil organic matter or a substrate (*e.g.*, plant material or horn meal) is studied in a way comparable to soil respiration tests. The influence of nitrification, *i.e.*, the conversion of ammonia into nitrate, may also be studied in these tests. Ammoniation is performed by a wide variety of soil micro-organisms, and is, therefore, relatively insensitive to perturbation.

The advantages of nitrification are :
- That fewer species of micro-organisms are involved in this process and,
- That the process is considered to be of ecologicaland agricultural importance. Therefore, either combining these parameters in one test or running a separate test on nitrification is recommended. Nitrification tests can be performed in soil amended with either $(NH_4)_2SO_4$ or with organic substrates suchas lucerne or horn meal. For this purpose, substrate equivalent to approximately 100 mg N per kg soil is added, and the disappearance of NH_4 and the appearance of NO_3 is monitored. In case the rate of NO_3 formation does not follow the disappearance rateof the NH_4^+, the soil should also be checked for the formation of NO_2. To check whether the test soil is capable of nitrification and whether the organic matter-amendment is suitable for ammoniation and nitrification, studies are also recommended.

Test for Nitrogen Fixation

Tests on both symbiotic and asymbiotic nitrogen fixation can be identified. Tests on symbiotic nitrogen fixation in fact consider the unique relationship between the host plant and *Rhizobium*, and, therefore, include in one test effects on both the plant and the bacteria. These experiments are conducted in a soil suitable for growth of the plant. The plant, seeds, or soil can be inoculated with *Rhizobium* if no suitable bacteria are present in the soil. Effects on plant growth and the degree of nodulation should be included. In tests on asymbiotic nitrogen fixation, the degree of acetylene reduction (or formation of ethylene from acetylene) by soil samples is determined in relation to the addition of the test chemical.

Test for Denitrification

Denitrification is the conversion of nitrate to atmospheric nitrogen, and will especially take place under anaerobic conditions. This process may be relevant in soil, as microsites may become anaerobic. In this test, soils are generally flooded with a layer of water, and nitrate is supplied as a substrate. Additionally, an organic substrate such as glucose is added to the soil. Since the process cannot be quantified, the formation of nitrogen gas, the disappearance of nitrate, and the formation of nitrite are measured as test parameters.

Tests on Enzyme Activity in Soil

As in the tests on microbial processes, chemicals are also mixed homogeneously through the soil in tests on enzyme activity. Moisture content is adjusted to field capacity or to 40-60 per cent of water holding capacity, and all incubations are done in the dark at 20°C. Several soil enzymes, relevant to microbial processes in soil, can be used as test parameters, as noted below.

Soil Enzyme : Urease

At several intervals, small soil samples (6-7g) are taken, and incubated with 5 ml of demineralised water and 1.0 ml of a solution containing 60 mM urea. Incubation is at 35°C for 5 hours on a shaking water bath. A phenylm ercury acetate solution in 2 M KCl is added to the soil samples to stop the urease reaction. After 10 minutes of shaking, the soil suspensions are filtered. The filtrates are analysed photometrically at 525 nm for urea concentrations.

Soil Enzyme : Dehydrogenase

Soil samples (5-10g) are incubated with a solution of TTC (2,3,5 triphenyl tetrazolium chloride) in 0.1 M tris buffer solution (pH 7.6), and incubated for 24 hours at 30 or 37°C. The reduced triphenyl formazan formed is extracted with methanol and quantified by measuring the absorbance at 485 nm. Dehydrogenase reflects a broad range of microbial oxidative activities, and does not consistently correlate to microbial numbers, CO_2 evolution or O_2-consumption.

Additionally, dehydrogenase activity may depend upon the nature and concentration of amended C-substrates and alternative electron acceptors. Rossell and Tarradellas concluded that short-term (substrate-induced) dehydrogenase activity may reflect the impact of chemicals on the physiologically active biomass of the soil micro-flora.

Soil Enzyme : Phosphatase

At several intervals, 0.5 g soil samples are taken, and incubated with 5 mM p-nitrophenylphosphate (p–NPP) for 1 hour in a shaker at 20°C. Phosphatase activity is measured as the amount of p-nitrophenol formed using a spectrophotometer. Phosphatase is said to bear little relation to total phosphate availability in soils. Its relevance for microbial activity in soil may, therefore, be questionable.

Somerville and Greaves stated that soil enzyme activities would be of little value to monitor side effects of pesticides on micro-flora.

The main reasons for this were :
- The total enzymatic activity of the soil is made upof various fractions, and quantifying the contribution of each to the catalysis of a particular substrate is extremely difficult; furthermore, many enzymes are formed extracellularly, and will still be active when the micro-organisms responsible for their productionhave been eliminated.
- There is no universally agreed methodology, and almostany result can be achieved by varying assay conditions(temperature, pH, substrate). Although tests on enzymeactivity have been described by many authors, few data are available to judge the reproducibility of these methods. Also soil animals, such as collembola and isopods, significantly influence the activity of several enzymes, such as urease, dehyd rogenase, and cellulase. Therefore, discriminating between direct and indirect effects of the tested chemicals on micro-organisms is difficult. Other enzymes that are more or less frequently used as test parameters for microbial activity in soil are : arylsulphatase, bglucosidase, b-acetylgluco-saminidase, saccharase, galactosidase, protease, and phosphodiesterase.

Field Tests

The reliability of microcosm studies in the laboratory to interpret field conditions is much debated. Microcosms differ from the field situation concerning the influence of temperature and moisture dynamics, the influence of root presence, and the composition of the soil biota community. A recent study compared microcosm studies in the laboratory with mesocosm studies and direct field measurements concerning microbial respiration, enzyme activities, and availability of macro-nutrients in interaction with soil animals; these soil process variables appeared to be of the same order of magnitude.

The tests described in the preceding sections provide only a rough estimate of the possible hazard imposed by a chemical in the environment. In many cases, this degree of precision is sufficient. Usually the laboratory test is considered to be a "worst case" situation, since test animals are exposed to a constant concentration that is relatively available, because the test substrate is prepared freshly. Under field conditions, exposure may be lower since the chemical is not distributed unifonnly over the habitat, and bio-availability will often be lower due to various sorption processes. By contrast, the laboratory test considers the test organism under optimal conditions, without secondary stresses, such as those of food shortage, drought, and cold. The uncertainties attached to the laboratory-to-field extrapolation can be avoided by conducting experiments under semi-field or field conditions.

Various organisations have recommended test protocols for field investigations. In several cases, guidelines for field tests are part of the national registration procedures for pesticides. Furthennore, considerable scientific research has been done in which side-effects of pesticides have been published. Some attempts have also been made to develop standardised procedures for field tests to assess the effects of pesticides on earthworms.

Cage Tests Using Selected Arthropod Species

Some of the arthropods used as laboratory test species can also be exposed to chemicals under semi-field conditions, while exposed in cages. Hassan *et. al.* lists protocols developed for *Trichogramma cacoeciae, Phygadeuon trichops, Coccygomimus turionellae, Phytoseiulus persimilis, Aleochara bilineata, Chrysoperla camea,* and *Drino inconspicua*. The usual procedure is to treat a group of plants with a spray of the chemical. A cage is then put over the plants after which test animals, with hosts or food, are introduced. The cage is installed either in a greenhouse or outdoors under a cover to provide shelter from rain and excessive sunshine. After an adequate duration of exposure, the performance of beneficials is compared with water-treated controls. Cage tests have been described in detail for testing with pollinators. Bees *(Apis mellifera)* from small colonies are made to forage on a flowering crop in cages measuring minimally 2 × 2 × 3 m, with a 3 mm mesh netting. The product is applied to the plants, and not to the cage walls, by spraying. The EPPO-guideline does not require replication of the treatment. Effects are recorded at several intervals, preferably 0, 1, 2, 4, and 7 days after treatment. Observations are made on the number of dead bees, on foraging activity, and on behaviour. The results are compared with a blank control (usually water-sprayed) and a positive control (a reference product known to be hazardous to bees, *e.g.*, parathion).

Honey Bee Field Test

OEPP/EPPO also provides a guideline for field tests on honey bees. A chemical to be tested is applied to a plot of at least 1500 m^2, with the crop for which the chemical is intended, or another crop attractive to bees (rape, *Phacelia),* in full flower. Per treatment, three colonies of honey bees are placed in or on the edge of the plot. Test plots should be separated by at least 500 to 1000 m² to avoid bees foraging on the wrong plot. Replication of the treatment is considered desirable, but is not required in the EPPO guideline. A blank control (untreated, or treated with a reference product known to present a low hazard to bees), as well as a positive control (*e.g.*, parathion, dimethoate) are applied to separate test plots.

After treatment, observations are made at several intervals, preferably after 0, 1, 2,4,7, and 14 days. Meteorological data are recorded during the entire period of the trial. Several parameters are estimated such as the number of foraging bees in the crop, behaviour of bees on the crop and around hives, mortality of bees (using dead bee traps), pollen collection (using pollen traps), pollen in collected honey, number of bees on frames, brood status in frames, and residues in dead bees, pollen wax, and honey. For the test to be valid, mortality in the negative control should not exceed 15 per cent, while mortality in the positive control should be statistically significant.

Arthropod Fauna in Arable Crops

Hassan *et. al.* summarise the recommendations made by the IOBC Working Group Pesticides and Beneficial Organisms for full-scale field tests. The methods are suitable for a variety of crops, but have been applied mostly to winter wheat.

The trial is laid out in a replicated block design : three large fields with similar agronomic history are each divided into three treatment areas, where each treatment area covers at least 3 ha. The treatments are a spray with the product to be tested, a blank treatment (water spray), and a positive control (*e.g.*, dimethoate).

Sampling is planned on seven occasions : 10 and 5 days before treatment, 2, 5, 10, and 20 days after treatment, and just before harvest. Sampling activities are concentrated in the central parts of each treatment area. Crop foliage fauna is collected with a suction net sampler (*e.g.*, the Dietrich vacuum sampler). Soil surface fauna are sampled using pitfall traps, left in the field for 5 days. Visual inspection, water traps, and sticky traps can provide additional information on those arthropods sampled inefficiently by pitfalls or suction samplers.

The fauna collected are identified to at least the family level. Some groups where species can be recognised easily can be further subdivided (*e.g.*, carabid beetles).

In addition to the large-scale experiments suggested for arthropods in winter wheat, smaller set-ups have been suggested by Edwards and Thompson and Eijsackers and Van de Bund. Treatment plots of 3 × 3 m are recommended for microarthropods (Collembola, mites), while 10 × 10 m plots are suitable for studies on beetles and spiders. In all cases, however, the plots should be fenced, preferably using a polythene sheet, protruding 15 cm below ground and 40 cm above. The barrier should limit the immigration from neighbouring plots by surface-active arthropods such as beetles.

The statistical treatment of data from field experiments is not harmonised. Yet, such harmonisation may be important since the probability of finding effects of the treatment will depend on the power of the statistical analysis. Stewart-Oaten et. al. (1986) suggested the use of "pseudoreplication in time," to allow for a detailed evaluation of the effects of a treatment in relation to a control. In this design, also called BACI (Before After Control Impact comparison), the correlation between the observations from control plots and treatment plots before treatment is used to assess the effects in the treatment plots as deviations from the expectations made on the basis of the control plots.

Arthropod Fauna in Orchards

Hassan et. al. summarise the standardised methods developed by the IOBC Working Group "Integrated Protection in Orchards." The methods consist of catching the fauna in collectors placed under the trees to which a chemical has just been applied. The collectors can be trays, canvas sheets, or funnels ("Steiner funnels"), with at least 0.5 m^2 of collecting area. In each trial, both the control (water spray) and the test treatment are followed by a "cleaning" treatment, 48 hours after the initial treatment. The "cleaning" treatment consists of dichlorvos at double the recommended dose, which will remove all beneficials present. The effectiveness of the treatment can thus be expressed in relation to the total population of beneficials present in the treated tree. The use of a reference chemical with each treatment, e.g., phosalone, is also recommended.

The design of the trial is a complete randomised block design, or a balanced incomplete block design. Each replicate is represented by one tree with one or more fauna collectors; six to eight replicates per treatment are recommended. The trees should be separated by at least one untreated tree. The fauna are gathered from the collectors 24 hours and 48 hours after the treatment, as well as 24 hours after the "cleaning" treatment. Only the arthropods, of which there is at least an average of ten individuals per collector, are considered. The fauna are identified to the species or the family level.

Earthworm Field Tests

Although no standardised guidelines exist yet for the study of pesticide side-effects on earthworms, recomm endations for the performance of such a test have been formulated by Kula.

Field studies with earthworms can be performed on arable land or on permanent grass; in both cases, a minimum number (100 individuals per m^2 of earthworms is required, and the relevant species *(Aporrectodea caliginosa* and *Lumbricus terrestris)* must be present. Minimum plot size should be 100 m^2, and at least four replicate plots should be used per treatment. A study should include a control, the highest recommended dose, and a toxic standard (benomyl). In many cases also a manyfold (*e.g.*, five-fold) of the recommended dose should be studied. At each sampling time, at least two samples (sampling area 0.25 m^2 should be taken from each replicate plot. Preferably treatment should take place in the spring, and samples should be taken 1, 4 to 6, and 12 months after application. For the sampling of earthworms, the formaldehyde method and electrical sampling methods seem to be the most useful.

CHEMICAL COMPOSITION OF SOIL

The industrialisation of our society has led to an increased production and emission of both xenobiotic and natural chemical substances. Many of these chemicals will end up in the soil. Various soil constituents have a great capacity to retain chemicals, especially those with apolar molecules or positively charged divalent and trivalent ions. Consequently, the soil is a net sink for all kinds of chemicals, and concentrations are often considerably higher than in any other environmental compartment. This situation may lead to smaller or larger impacts on the functioning of soil ecosystems. Important ecological functions of the soil are those associated with organic matter decomposition, mineralisation of nutrients, and synthesis of humic substances. For that reason, an increasing need exists for methods to assess the side effects of these chemicals on soil ecosystems.

An overload of chemicals will affect both abiotic soil properties and, directly and indirectly, soil biota. This paper will, therefore, start with a description of the possible sources and consequences of chemical pollution for abiotic soil properties. Subsequently, methods are described for the determination of the effects of chemicals on soil organisms.

When considering methods to assess the effects of chemicals on soil biota, two types of tests can be distinguished. The first contributes to the prediction of the potential effects of single chemicals on soil ecosystems. For that purpose, mainly single-species laboratory tests are conducted; at times, more complex micro- ecosystem, mesocosm, or field studies are carried out. This type of testing, which aims at establishing dose-response relationships and the estimation of LC_{50}, EC_{50}, or NOEC values, may be called "prognosis." The second type of method is aimed at assessing the potential ecological risk of a certain case of soil pollution. In such a situation, several chemicals may be involved. To determine whether a specified case of soil pollution poses a real hazard for soil biota, both laboratory and field studies may be performed. This type of testing may be called "diagnosis."

Quantification of Input of Chemicals in the Soil

Before methods to assess the effects of chemicals on soil components can be presented, information must be provided about the various ways chemicals enter the soil and how these inputs can be quantified.

Major Elements

Focusing on major elements, Table summarises the most important processes contributing to the input of chemicals to a soil ecosystem. The relative importance of the input differs greatly among elements as well as among areas of low versus high pollution. The precision in the measurement of input quantities is mainly a question of equipment, methods, and experimental design.

Table. Major Inputs of Elements to Soil Ecosystems.

	N	S	P	K	Na	Ca	Mg	Cl
Wet deposition	X	X		X	X	X	X	X
Dry deposition								
Gaseous input	X						X	
Particle input	X	X	?	?	X	?	X	X
N_2-fixation						X		
Mineral weathering		X	X	X	X		X	X

Wet Deposition

This is the major process for the input of nitrogen, sulphur, and chloride, and is significant for other elements. Excluding problems such as the definition of input by mist and fog, the measurement of wet deposition is essentially the estimation of rainfall and its elemental content.

Type and design of rain gauges can be found in *Tropical Soil Biology and Fertility : A Handbook of Methods* and their position relative to the ground surface is usually the reason for an under-estimation in rainfall, depending on wind and evaporation losses in exposed areas. To avoid chemical changes in the sample, frequent sampling and rapid analysis are preferred to preservation.

Dry Deposition

This is defined as the direct transfer of gases and particles to different ecosystem surfaces (= receptors). Two methods are noted here :

- Micro-meteorological methods are indirect, and based on the assumption that transmission of chemical compounds is a process similar to transmission of heat and momentum. The method has been proven toproduce useful results for SO_2 deposition to grasscovered areas, but does not fully satisfy to measure deposition to forests.
- Receptor-(or ecosystem-) oriented mass balance methods are more direct methods, but with problems and limitations. For chloride, sodium, and sulphur at high deposition, throughfall measurements have been useful to estimate dry deposition, when combined withwet deposition measurements.

Nitrogen Fixation

This method is well known as the acetylene-reduction (AR) method for measuring nitrogenase activity. Isotope techniques, such as incubation in $^{15}N_2$-

containing atmosphere are attractive; but for long-running experiments, practical problems arise. Further methods are the ^{15}N-isotope dilution technique and the classic total nitrogen difference method, based on a comparison of total N-yield in a N-fixing crop and that of a non-N fixing reference crop.

Mineral Weathering

Methods to estimate current weathering rates include mass balance of watersheds, radiometric methods, and mineral bag technique. Mass balance of watersheds or lysimeters is widely used and gives the best estimates of current weathering (*e.g.* Likens *et. al.,* 1977). The approach is indirect, leaving weathering as the residue in the mass balance equation :

$E(rw) = [E(efflux) - E(influx)] + E(ps)$ where E(rw) is element release by weathering, E (efflux) is elemental losses mainly from leaching, E (influx) is input of elements mainly by dry and wet deposition, and E(ps) is the change in elemental storage in plants and soil.

The results of mass balance studies depend on the accuracy of estimation of all possible sources and sinks for elements in the soil. Main sinks are elemental storage in biomass and humus, but also accumulation in the soil by microbial activity, redox reactions, surface exchange processes, and formation of secondary minerals. These sinks may easily turn into sources if biological or chemical conditions are altered. For instance, present-day acid deposition depletes base cations from the exchange sites in the soil profile.

Radiometric methods using ^{87}Sr/ ^{86}Sr ratios are used to estimate calcium weathering.

For the mineral bag technique, selected soil or mineral fraction is put into a non-biodegradable mesh bag, and placed in the field for a specific period of time. The bag is then returned to the laboratory for analysis.

Organic Chemicals and Metals

Major sources for the input of organic chemicals and heavy metals in soils are agriculture, industries, and traffic. Both diffuse and point sources can be identified, and some chemicals (*e.g.*, pesticides) are applied in quite a controlled way enabling a proper prediction of the input in the soil. To assess the deposition of pesticides that are generally applied under controlled conditions, sheets of aluminium foil or other inert substances can be placed on the soil and analysed after spraying. For other chemicals that may be released in a less controlled manner, chemical analysis of soil samples is needed to quantify the input. In all cases, rather specific analytical techniques are required to determine chemical concentrations in soils. Before analysis, complicated extraction and purification steps are often required. Generally, extraction with an organic solvent (*e.g.*, hexane, acetonitrile, toluene or acetone) is applied, followed by analysis by HPLC, GC, or GC-MS.

To determine the soil content of heavy metals, digestion of soil samples with strong acids (e.g., $HNO_3/HClO_4$) is required. After that, the destruate can be analysed by atomic absorption spectrophotometry.

RISK OF CHEMICALS FOR SOIL ORGANISMS (PROGNOSIS)

A brief description is presented of single-species laboratory tests, microcosm tests, and field tests. For an extended overview of these tests, the reader is referred to Van Straalen and Van Gestel.

Single-Species Laboratory Toxicity Tests

Among soil invertebrates, only earthworms have seriously been considered as test organisms during the past decade, and some standardised test methods are available. For micro-fauna and meso-fauna, only few tests are available, although these animals are among the most numerous and species-rich groups of soil animals. Many species, however, are promising test animals, because they are easy to culture and their size allows for small-scale experimental set-ups with many replications. Besides these soil animals, higher plants have also been considered for testing, and standardised tests with some plant species are available.

In several tests, artificial substrates (nutrient solution, agar, silica gel, filter paper) are used, the composition of which greatly affects toxicity. However, extrapolation of these test results to the field remains problematic. If concentrations in test solutions can be equated with pore water concentrations, sorption data may be used to express the toxicity per unit of soil. The validity of this extrapolation, however, still has to be investigated. The same extrapolation problems may arise for tests in which the main route of exposure is via the food. For such tests, a conversion of food concentrations to soil concentrations may be needed.

Higher Plants

A test with higher plants has been described in an international test guideline (OECD, 1984b), while some others are under discussion. The OECD guideline 208 on higher plant toxicity testing uses several plant species, representing different agricultural crops and both monocotylodoneous and dycotylodoneous species.

Protozoans and Nematodes

Protozoans and nematodes live in the soil pore water, and the best way to test them is to use methods similar to those used in aquatic toxicology. Among the protozoans the ciliates *Tetrahymena pyriformis, Colpoda cucullus,* and *Paramecium aurelia* have been considered for test animals, as have the nematode species *Caenorhabditis elegans, Panagrellus silusiae,* and *Plectus parietinus* however, an accepted test procedure is unavailable.

Isopods and Millipedes

Isopods are an interesting group of animals in heavy metal research, because of their unique ability to concentrate extreme amounts of metals in their bodies. Their use as a test animal for soil toxicity studies, however, is restricted to a few cases, and no attempts have yet been made to arrive at standardisation. *Porcellio scaber, Oniscus asellus,* and *Trichoniscus pusillus* are three species frequently investigated. Among these, *T. pusillus* seems to be the most suitable as a test species, as it has a somewhat shorter life-cycle compared to *P. scaber* and *O. asellus*. All three species are very easy to culture, and do not require special conditions.

Usually isopods are kept on a plaster substrate, and are fed with partly decomposed leaves, either intact or ground, to which chemicals can be added. Increase in growth over several weeks is observed, but is rather variable, even for one individual. Reproduction is difficult to assess, because, after mating, females may retain the sperm for a long period before producing eggs, which are carried in a brood pouch. Tests require a minimum period of four weeks.

Millipedes (Diplopoda) are another important group of saprotrophic soil invertebrates, but they have never been considered seriously as test animals. The most widely investigated species is *Glomeris marginata*. The species *Cylindroiulus britannica* is also well suited as a test animal. Test conditions for millipedes are similar to those for isopods.

Oribatid Mites

A reproduction toxicity test using the partheno genetic oribatid mite *Platynothrus peltifer* has been described by Denneman and Van Straalen. This seems to be the only oribatid used so far in soil toxicity experiments, although oribatids comprise hundreds of species, and are usually the most numerous group of arthropods in forest soils.

In the test with *P. peltifer,* the animals are exposed to contaminated algae, and the number of eggs are counted. The test is very laborious, as the animals hide their eggs in small crevices; it is also a rather lengthy test (9 to 12 weeks) because of the low rate of egg production in this species and its long life-cycle (1 year), which are remarkable features for such a small animal ($\pm$ 1 mm). *P. peltifer* appeared to be rather resistant to cadmium, copper, and lead in terms of lethality, but very susceptible in terms of egg production. Due to their peculiar habits, species such as *P. peltifer* tend to be forgotten in the development of toxicity tests. It is, however, the most sensitive soil invertebrate tested so far for cadmium, while it is more sensitive than springtails for copper and lead.

Collembola

Collembola are a relatively well investigated group of soil animals. Several species have been used frequently in toxicity experiments : *Onychiurus* spp. The first three species are parthenogenetic (thelytokous); *O. cincta* is sexual, sperm being transferred indirectly through spermatophores deposited on the substrate by the male.

Three different exposure systems have been described :
- Through feeding on fungi grown on contaminated agar,
- Through feeding on directly contaminated food, and,
- Residual exposure (treated substrate, *e.g.* sand, leaves, soil).

When testing Collembola with contaminated fungi, the animals are kept on a plaster of Paris substrate in a Petri dish, and fed on a piece of agar, overgrown with hyphae (*e.g., Verticillium bulbillosum*). Egg production, growth and survival are recorded regularly throughout a period of several weeks. The advantage of this system is that substances are offered in a natural way, *i.e.*, after being taken up and possibly transformed to naturally occurring complexes. Concentration levels in the fungus, however, are difficult to maintain or to set to specific values. When testing Collembola with directly contaminated food, chemicals are added in water or acetone solution to the food (algae, yeast, ground leaf material). Food can be offered as droplets on filter paper discs, while the animals are kept on a plaster or sand substrate. In this manner, concentrations can be manipulated easily, while growth, egg production, and survival are monitored over a period of several weeks.

The third system of testing Collembola is to use the artificial soil medium developed for earthworm toxicity tests. Juvenile Collembola *(Folsomia candida)* are placed in artificial soil with dry yeast provided for food. After 28 days, the number of remaining animals and their offspring are counted after flotation extraction. The *Folsomia* test is very easy to carry out; it requires little attention during the test; and it gives reproducible results. Another advantage is the use of artificial soil similar to the earthworm test; thus, experimental results can be compared between earthworms and springtails. The only disadvantage of the test is that reproduction cannot be observed directly, and cannot be separated from juvenile mortality and hatching success. The *Folsomia* test is now undergoing the process of international standardisation.

Enchytraeids

Enchytraeids can be cultured easily on agar and on (artificial) soil substrates, when fed rolled oats. The toxicity tests described in the literature all use species of the genus *Enchytraeus*. The well-known species *Cognettia sphagnetorum* can also be bred easily in the laboratory, but its tendency to fragmentate upon handling makes it less suitable for toxicity tests. Westheide *et. al.* described a test in which the test chemical is incorporated in 1.5 per cent nutrient agar. Two species are used, *Enchytraeus* cf. *globuliferus* and *E. minutus*. Reproduction, measured as the number of cocoons and juveniles produced, is the endpoint studied in this test. This method seems to provide an easy and reproducible test, but, because of the use of an agar substrate, results cannot be translated to real soil.

Römbke (1989) described a test with *Enchytraeus albidus*, using the OECD artificial soil prescribed for earthworm toxicity tests. Adult *E. albidus* are exposed for 28 days to different concentrations of the test chemical, mixed homogeneously through the artificial soil. Survival of adult worms and the number of juveniles produced are the endpoints studied. In this way, both acute and sublethal effects are combined in one test. The reproducibility of the method cannot be judged and, because only one chemical was tested, no conclusions can be drawn with respect to the sensitivity of the sublethal endpoint. An important positive aspect of this test is that the substrate used is the same artificial soil used in the internationally accepted earthworm toxicity tests.

Lumbricids

In the guidelines of OECD (1984a) and EEC (1985), *Eisenia fetida* and its sibling species *E. andrei* are recommended. Both species are commonly found in compost and dung heaps, and can be cultured easily in the laboratory on a substrate of horse dung or cow dung. According to the existing guidelines on acute toxicity testing with earthworms, other real soil dwelling species may also be used. Such species are, however, hard to culture in the laboratory, because they have long generation times and need large volumes of soil. So, for practical reasons, the use of the two *Eisenia* species is recommended. Three acute toxicity tests exist. In the filter paper contact test (OECD, 1984a), adult earthworms of the species *Eisenia* spp. are exposed to filter paper wetted with a solution of the test substance. Mortality is assessed, and the 48-hour LC_{50} value is expressed as µg per cm^2. The method has been shown to be easy, fast, and highly reproducible. Several authors including Heimbach have demonstrated, however, that this test has no predictive value for the effect of chemicals on earthworms in the soil; it can only be used to rank chemicals.

In the artificial soil test, adult earthworms of the species *Eisenia* spp. are exposed to the test chemical, which is mixed through an artificial soil substrate for 14 days. This artificial soil is made up by mixing (dry weight) 10 per cent sphagnum peat, 20 per cent kaolin clay, 70 per cent quartz sand, while some $CaCO_3$ is added to adjust the pH to 6.0±0.5. The moisture content of the substrate is adjusted to about 55 per cent (w/w) or to 40-60 per cent of the water holding capacity. Mortality is the only test parameter, and LC_{50} values are expressed as mg per kg dry soil. Van Gestel and Ma (1990) have demonstrated that results obtained in this artificial soil can easily be translated to natural soils by using sorption data. For this reason, the use of the artificial soil is acceptable, and the test can be concluded to have enough predictive value with respect to effects that occur in the field. The test has been shown to be reproducible.

In the Artisol test adult earthworms of the species *Eisenia* spp. are exposed to chemicals mixed through a substrate of amorphous silica gel (Artisol) for fourteen days. Survival is the only test parameter, and results are expressed in terms of LC_{50} values. The silica gel substrate does not bear any resemblance to natural soil;

thus for reasons of ecological realism and extrapolation towards natural soil, this test cannot be recommended.

Recently, two sublethal toxicity tests have been described. In both tests, the OECD artificial soil and the earthworm species *Eisenia* spp. are used. In the first test (Van Gestel *et. al.*, 1989), chemicals are mixed homogeneously through the artificial soil, and after three weeks of exposure effects on the growth and cocoon production by adult earthworms are determined. The worms are fed by supplying a small amount of (untreated) cow dung in a small hole in the middle of the soil. By incubating cocoons produced for five weeks in untreated artificial soil, effects on hatchability (per cent fertile cocoons, number of juveniles per cocoon) and the total number of offspring per adult worm can be determined.

The second method was developed to determine the sublethal effects of pesticides on earthworms. The pesticide is sprayed onto the soil surface, and earthworms are fed by applying about 0.5 g cow dung per animal to the soil surface once a week. Pesticides are applied in two treatment levels, corresponding with the recommended dose and a five fold dose. After six weeks incubation, adult worms are removed from the substrate and weighed. The test substrate containing cocoons and juveniles is incubated for another four weeks. Food is given when required. After ten weeks, the juveniles are extracted from the substrate by hand-sorting or heat extraction and counted. Effects on earthworm growth and on the total number of offspring produced per tray are determined. The method has been subjected to a (German) ring test, and will be revised on the basis of the results of it. The method is only applicable for pesticides, and the recovery of all juveniles from the artificial soil by hand-sorting is difficult. This hinders comparison of this method with that of Van Gestel *et. al.*. Furthermore, the method of pesticide application is not standardised, which may be the reason for the variability observed in the first ring test.

Molluscs

In the limited number of toxicity tests using terrestrial molluscs, exposure was via the food. Russel *et. al.* (1981) described a method for toxicity experiments with the garden snail *Helix aspersa*. The snails were kept in polyethylene boxes, filled with a substrate of moist quartz sand covered by a piece of woven glass towel. The snails are fed a diet of ground Purina Lab-Chow (formulation for rats, mice, and hamsters) supplemented with $CaCO_3$. Parameters affected include survival, reproductive behaviour, dormant state, new shell growth, and food consumption. Similar test methods using the snail *Helix pomatia* or the slug *Arion ater* have been described by other authors.

Beneficial Arthropods

Arthropods that may improve the production of agricultural products are designated as "beneficials," and commercial interest exists in designing and applying pesticides in such a way that beneficials are least affected. The working group on "Pesticides and Beneficial Orga nisms" of the International Organisation for Biological and Integrated Control of Noxious Animals and Plants (IOBC) has contributed significantly to designing ecotoxicological test methods and decision schemes to evaluate the hazard of pesticides.

The hymenopteran groups Ichneumonidae, Braconidae, and Chalcidoidea contain a large number of parasitoid species. The female insect deposits an egg in or on a host (usually an insect egg or larva), which is then gradually eaten as the offspring develop. The host selection process and the life-cycle of the parasitoid are finely tuned to the host, and many species will attack only a single or a few host species. Furthermore, other hymenopteran species used in toxicity tests include *Diaeretiella rapae,* an internal parasite of aphids such as *Myzus persicae, Phygadeuon trichops,* a parasite of *Delia* species (bulb flies), *Coccygomimus (=Pimpla) turionellae,* a polyphagous parasite of Lepidoptera (Tortricidae, Geometridae, Noctuidae), and *Opius* sp., a parasite of leaf mining insects. The methods used for these species are similar to those described for *Trichogramma* and *Encarsia*.

Within the order of the Coleoptera, the families Carabidae (ground beetles), Staphylinidae (rove beetles), and Coccinellidae (lady birds) contain representatives that are common in agricultural fields and are recognised for their predation of pests.

Among the various arthropod groups, spiders seem to be particularly sensitive. This phenomenon often appears in field tests with pesticides, where catches of surface active spiders are reduced in a manner similar to that of predatory mites following pesticide application. The families Erigonidae and Linyphiidae (money spiders) are important groups with a great species richness. The recommendations made by the International Commission for Plant Bee Relations (ICPBR) have been included in a guideline of the European and Mediter ranean Plant Protection Organisation (EPPO) to evaluate the hazards of pesticides to the honey bee *Apis mellifera*. Several countries have slightly different national guidelines to test pesticides on honey bees.

Chapter 12

AEROBIC MICROBIAL OF SOILS

Aerobic microbial transformation of solid materials or "Solid Substrate Fermentation" (SSF) can be defined in terms of a solid porous matrix which can absorb water with a relatively high water activity. The solid/gas interface should be a good habitat for the fast development of specific cultures of moulds, yeasts or bacteria, either by isolated or mixtures of species. The mechanical properties of the solid matrix should stand compression or gentle stirring as required for a given fermentation process. This requires small granular or fibrous particles, which do not tend to break or stick to each other. The solid matrix should not be contaminated by inhibitors of microbial activities and should be able to absorb or contain available microbial foodstuffs such as carbohydrates (cellulose, starch, sugars) nitrogen sources (ammonia, urea, peptides) and mineral salts.

Traditional fermentations are typical examples of SSF :
- Lated with solid strains of the mould Aspergillus oryzae.
- Indonesian tempeh or Indian ragi which use steamed and cracked legume seeds as solid substrate and a variety of non- toxic moulds as microbial seed.
- French "blue cheese" which uses perforated fresh cheeseas substrate and selected moulds, such as Penicillium roquefortii as inoculum.
- In addition to traditional fermentations new versions
- of SSF have been invented. For example, it is estimated that nearly a third of industrial enzymeproduction in Japan which is made by SSF process andkoji fermentation has been modernised for large scaleproduction of citric and itaconic acids.
- Composting which was produced for small-scale production of mushrooms has been modernised and scaled up in Europe and United States. Also, various firms in Europe and USA produce mushroom spawn by cultivating aseptically Agaricus, Pleurotus or Shii-Take on sterile grains in

static conditions. Generally, most of the recent research activity on SSF is being done in developing nations as a possible alternative for conventional submerged cultures which are the main process for pharmaceutical and food industries in industrialised nations. SSF seems to have theoretical advantages over LSF. Nevertheless, SSF has several important limitations. Most of the processes are commercialised in South-east Asian, African, and Latin American countries. Nevertheless, a resurgence of interest has occurred in Western and European countries over last 10 years.

- Potentially many high value products as enzymes,metabolites, antibiotics, could be produced in SSF. But improvements in engineering and socio-economic aspects are required because processes must use cheapsubstrate locally available, low technology applicablein rural region, and processes must be simplified.
- The greatest socio-economical potential of SSF is the raising of living standards through the production of protein rich foods for human consumption. Protein deficiency is a major cause of malnutrition and the problem will become worse with further increases in the world population. Two ways can be explored for that :
- Production of protein-enriched fermented foods for direct human consumption. This alternative involves starchy substrates for its initial nutritional calorific value. Successful production of such food will require demonstration of economical feasibility, safety, significant nutritional improvement, and cultural acceptability.
- The second alternative consists to produce fermented products for animal feeding. Starchy fermented substrates with protein enrichment could be fed to monogastric animals or poultry. Fermented lignocellulosic substrates by increasing in the fibre digestibility could be fed to ruminants. In this case, the economical feasibility should be decisive in comparison to the common model using protein of soybean cake, a by-product of soybean oil. Since 15 years, the Orstom group investigated on solid fermentation process for improving protein content of cassava and other tropical starchysubstrates using fungi (especially from Aspergillusgroup) in order to transform starch and mineral saltsinto fungal proteins.
- Protein enrichment of Cassava and starchy substrates
- Production of organic acids or ethanol by SSF from starchy substrate and Cassava
- Digestibility of fibres and lignocellulosic materials for animal feeding
- Degradation of caffeine in coffee pulp and ensilingfor conservation and detoxification
- Enzymes and fungal metabolites production by SSF using sugarcane bagasse

MICRO-ORGANISMS

Bacteria, yeasts and fungi can grow on solid substrates, and find application in SSF processes. Filamentous fungi are the best adapted for SSF and dominate in research works.

Bacteria are mainly involved in composting, ensiling and some food processes. Yeasts can be used for ethanol and food or feed production. But filamentous fungi are the most important group of micro-organisms used in SSF process owing to their physiological, enzymological and biochemical properties. The hyphal mode of fungal growth and their good tolerance for low Aw and high osmotic pressure conditions make fungi efficient and competitive in natural micro-flora for bioconversion of solid substrates.

Koji and *Tempeh* are the two most important applications of SSF with filamentous fungi. *Aspergillus oryzae* is grown on wheat bran and soybean for *Koji* production, which is the first step of soy sauce or citric acid fermentation. *Koji* is a concentrated hydrolytic enzymes required in further steps of the fermentation process. Tempeh is an Indonesian fermented food produced by the growth of *Rhizopus oligosporus* on soybeans. The fermented product is consumed by people after cooking or toasting. The fungal fermentation allows better nutritive quality and degrades some antinutritional compounds contained in the crude soybean.

The hyphal mode of growth gives also the filamentous fungi the power to enter into the solid substrates. The cell wall structure attached to the tip and the branching of the mycelium ensure firm and solid structure. The hydrolytic enzymes are excreted at the hyphal tip, without large dilution like in the case of LSF, that makes very efficient the action of hydrolytic enzymes and allows penetration into most solid substrates. Penetration increases the accessibility of all available nutrients within particles.

SUBSTRATES

In general, substrates for SSF are composite and heterogeneous products from agriculture or by-products of agro-industry. This basic macromolecular structure (*e.g.* cellulose, starch, pectin, lignocellulose, fibres etc.) confers the properties of a solid to the substrate. The structural macromolecule may simply provide an inert matrix within which the carbon and energy source (sugars, lipids, organic acids) are adsorbed (sugarcane bagasse, inert fibres, resins). But generally the macro-molecular matrix represents the substrate and provides also the carbon and energy source.

The most significant problem of SSF is the high heterogeneity, which makes difficult to focus one category of hydrolytic processes, and leads to poor trials of model ling. Lignocellulose occurs within plant cell walls, which consists of cellulose micro-fibrils embedded in lignin, hemicellulose and pectin. Each category of plant material contains variable proportion of each chemical compound.

Pectins are polymers of galacturonic acid with different ratio of methylation and branching. Exo-and endo pectinases and demethylases hydrolyse pectin in galacturonic acid and methanol. Hemicellulases are divided in major three groups: xylans, mannans and galactans. Most of hemicellulases are heteropolymers containing two to four different types of sugar residue. Lignin represents between 26 to 29 per cent of lignocellulose, and is strongly bounded to cellulose and hemicellulose, hiding them and protecting them from the hydrolase attack. So the lignocellulose hydrolysis is a very complex process. Effective cellulose hydrolysis requires the synergetic action of several cellulases, hemicellulases and lignin peroxydases. But lignocellulose is a very abundant and cheap, natural, renewable material, so a lot of works were dedicated to micro organisms breakdown, especially fungal species. Starch is another very important and abundant natural solid substrate. Many micro-organisms are capable to hydrolyse starch, but generally the efficient hydrolysis requires previous gelatinisation. Some recent works concern the raw (crude or native) starch like it occurs naturally.

Within the plant, cell starch is stored in the form of granules. During the process of gelatinisation, starch granules swell when heated in the presence of water, which involves the breaking of hydrogen bonds, especially in the crystalline regions. Many micro-organisms can hydrolyse starch, especially fungi, which are suitable for SSF application involving starchy substrates.

Glucoamylase, a-amylase, b-amylase, pullulanase and isoamylase are involved in the processes of starch degradation. Mainly a-amylase and glucoamylase are of importance for SSF.

Micro-organisms generally prefer gelatinised starch. But large quantity of energy is required for gelatinisation, and it would be attractive to use organisms growing well on raw (ungelatinised) starch. Different works are dedicate to isolate fungi producing enzymes able to degrade raw starch, as has been done by Soccol *et. al* (1991), Bergmann *et. al.* (1988) and Abe *et. al.* (1988).

In our lab we developed many studies concerning SSF of cassava, a very common tropical starchy crop, in the view of upgrading protein content, both for animal feeding using *Aspergillus sp*. Initial protein content (1-4 per cent) could be increased until 18-20 per cent Dry Matter basis. Recently Soccol using selected strains of *Rhizopus* bio-transformed cassava in starchy fermented flours containing 10-12 per cent of good protein, comparable to cereals. Such bio-transformed Cassava flour can be used as cereal substitute for breadmaking until 20 per cent without sensible change for the consumer.

PH CONTROL AND RISKS OF CONTAMINATION

The pH of a culture may change in response to microbial metabolic activities. The most obvious reason is the secretion of organic acids such as citric, acetic or lactic acids, which will cause the pH to decrease, in the same way than ammonium salts consumption. On the other hand, the assimilation of organic acids which may be present in certain media will lead to an increase in pH, and urea hydrolysis result in an alcalinisation.

Oxygen Uptake

Aeration fulfils four main functions in solid state processes, namely :
- To maintain aerobic conditions,
- For carbon dioxide desorption,
- To regulate the substrate temperature and,
- To regulate the moisture level. Solid state processallows free access of atmospheric oxygen to the substrate; aeration may be easier than in submerged cultivations because of the rapid rate of oxygen diffusion.

SSF is a well-adapted process for cultivation of fungi on natural vegetal materials, which are breakdown by excreted hydrolytic enzymes. In contrast with LSF, in SSF processes, water related to the water activity is a limiting factor, both parameters no involved in LSF where water is in large excess. On the other hand, oxygen is a limiting factor in LSF but not in SSF where aeration is facilitated by the porous and particular structure and high surface contact area which facilitate transfers between gas and liquid phases.

SSF are aerobic processes where respiration is a predominant processes for energy supply to the mycelium; but it can cause severe limitation of the growth when heat transfer is not efficient enough causing rapid elevation of the temperature. Is the reason why it is so important to study and control respirometry in SSF? We developed a laboratory technique to measure CO_2 and O_2 *on line* in SSF. A special lecture will be dedicated to the theory, modelling and basic concept of respirometry. Also it will be organise training cessions at the lab, to practice respirometric measurement and kinetics analysis.

BIOMASS MEASUREMENT OF SOIL

Biomass is a fundamental parameter in the characterisation of microbial growth. Its measurement is essential for kinetic studies on SSF. Direct determination of biomass in SSF is very difficult due to problems of separation of the microbial biomass from the substrate. This is especially true for SSF processes involving fungi, because the fungal hyphae penetrate into and bind the mycelium tightly to the substrate. On the other hand, for the calculation of growth rates and yields it is the absolute amount of biomass, which is important. Methods that have been used for biomass estimation in SSF belong to one of the following categories. Direct measurement of exact biomass in SSF is a very difficult question. For that we can consider the global stoechiometric equation of the microbial growth.

Respiratory Metabolism

Oxygen consumption and carbon dioxide release result from the respiration, the metabolic process by which aerobic micro-organisms derive most of their energy for growth. As carbon compounds within the substrate are metabolised, they are converted into biomass and carbon dioxide. Production of carbon dioxide causes the weight of fermenting substrate to decrease during growth, and the amount of weight lost can be correlated to the amount of growth that has occurred. The measurement of either carbon dioxide evolution or oxygen consumption is most powerful when coupled with the use of a correlation model. If both the monitoring and computational equipment is available then these correlation models provide a powerful means of biomass estimation since continuous *on-line* measurements can be made. Other advantages of monitoring effluent gas concentrations with paramagnetic and infrared analysers include the ability to monitor the respiratory quotient to ensure optimal substrate oxidation, the ability to incorporate automated feedback control over the aeration rate, and the non-destructive nature of the measurement procedure.

Production of Extracelluiar Enzymes or Primary Metabolites

Associated extra-cellular enzymes are another metabolic activities, which may be produced. It is observed frequently a good correlation between mycelial growth and organic acid production, which can be measured by the pH measurement or *a posteriori* correlated by HPLC analysis on extracts. In the case of *Rhizopus*, Soccol (1992) demonstrated a close correlation between fungal protein (Biomass) and organic acids (citric, fumaric, lactic or acetic).

Protein Content

The most readily measured biomass component is protein. The Folin method is more sensitive and allowed a greater dilution of the sample which avoided interference from the starch in the substrate.

Glucosamine

A useful method for the estimation of fungal biomass in SSF is the glucosamine method. This method takes advantage of the presence of chitin in the cell walls of many fungi. Chitin is a poly-Nacetylglucosamine. Interference with this method may occur with growth on complex agricultural substrates containing glucosamine in glucoproteins.

Ergosterol

Ergosterol is the predominant sterol in fungi. Glucosamine estimation was therefore compared with the estimation of ergosterol for determination of the growth of Agaricus bisporus.

Physical Measurement of Biomass

Peñaloza (1991) used another physical parameter to evaluate mycelial growth, based on the difference in the electric conductivity between biomass *versus* the substrate. Good correlation with biomass was obtained and a model was proposed. Recently Auria *et. al.* (1990) monitored the pressure drop in a packed bed during SSF of Aspergillus niger on a model solid substrate consisting of ion exchange resin beads. Pressure drop was closely correlated with protein production. Pressure drop is a parameter, which is simple to measure and can be measured on-line. Further studies are required to determine whether the use of pressure drop in monitoring growth in forcefully aerated SSF bio-reactors is generally applicable. An interesting point of this physical technique resides in the fact that it is sensible to the conidiation : early conidiophore stage makes the pressure drop drastically and a breaking point can be easily observed.

On the other hand, in the production of protein enriched feeds, the protein content itself is of greater importance than the actual biomass concentration, and the variation in biomass protein content during growth becomes less relevant.

Overall, oxygen uptake and carbon dioxide evolution methods are probably the most promising techniques for biomass estimation in aerobic SSF as they provide on-line information. The monitoring and computing equipment is relatively expensive and will not be suitable for low technology or rural applications. None method is ideally suited to all situations so the method most appropriate to the particular SSF application must be chosen on the basis of simplicity, cost and accuracy. The best choice could be to cross two or three, or more, techniques for measurement of various parameters, and the total balance could be highly correlated to the actual biomass.

Environmental Factors

Environmental factors such as temperature, pH, water activity, oxygen levels and concentrations of nutrients and products significantly affect microbial growth and product formation. In sub-merged stirred cultures environmental control is relatively simple because of the homogeneity of the suspension of microbial cells and of the solution of nutrients and products in the liquid phase.

Moisture Content and Water Activity (AW)

SSF process can be defined as microbial growth on solid particles without presence of free water. The water present in SSF systems exists in a complexed form within the solid matrix or as a thin layer either absorbed to the surface of the particles or less tightly bound within the capillary regions of the solid. The optimum Aw for growth of a limited number of fungi used in SSF processes was at least 0.96 whereas the minimum growth Aw was generally greater than 0.9. This suggests that fungi used in SSF processes are not especially xerophilic. The optimum Aw values for sporulation by Trichoderma viride and Penicillium roqueforti were lower than those for growth (Gervais *et. al.* 1988). Maintenance of the Aw at the growth optimum would allow fungal biomass to be produced without sporulation.

Temperature and Heat Transfer

Stoechiometric global equation of respiration is highly exothermic and heat generation by high levels of fungal activity within the solids lead to thermal gradients because of the limited heat transfer capacity of solid substrates. In aerobic processes, heat generation may be approximated from the rate or CO_2 evolution or O_2 consumption. Each mole of CO_2 produced during the oxidation of carbohydrates released 673 Kcal. That is why it is of high interest to measure CO_2 evolution during a SSF process, because it is directly relied to the risk of elevation of temperature. Heat removal is probably the most crucial factor in large scale SSF processes, and conventional convection or conductive cooling devices are inadequate for dissipating metabolic heat due to the poor thermal conductivity of most solid substrates and result in non-acceptable temperature gradients. Only evaporative cooling devices provide sufficient heat elimination. Although the primary function of aeration during aerobic solid state cultivations was to supply oxygen for cell growth and to flush out the produced carbon dioxide, it also serves a critical function in heat and moisture transfer between the solids and the gas phase. The most efficient processes for temperature control consist in evaporating water, what needs in return to complete the loss to avoid desiccation.

Chapter 13

ORGANIC PASTURE SYSTEMS OF SOIL

PASTURES

Compared with crop production systems, there is very little information available on organic pastures for dairy production. This is primarily due to the intensive, housed nature of animal production systems in Europe, where the majority of organic research is undertaken. There is, however, increased interest in the use of pasture as a feed source, particularly in the United Kingdom.

Topics identified in the literature under the heading of organic pastures included composition, herbal leys, nutrient management, grasing management, fertilization, legume content and grassland management before conversion.

New Zealand has a wealth of information on conventional pasture management and many of the theories behind this are applicable to organic systems.

What is lacking in New Zealand is specific research on organic pasture systems and the effect that organic practices will have on pasture production, quality and composition and interactions with other components of the systems (*e.g.*, animals, soils).

Permanent Pasture Composition

There has been no published research carried out in New Zealand on the composition of pastures under organic management. Some research from overseas are applicable in New Zealand, however. A key factor highlighted in this literature is the use of legumes in pastures as a source of nitrogen. A difference between New Zealand and many European systems is that in Europe ryegrass/white clover leys are considered a fertility-building crop rather than permanent pastures. Le Gall *et. al.* described typical pastures in France under organic dairy/beef grasing and how to maintain desired sward composition.

The pastures described in this paper are very similar to New Zealand pastures, and the information is therefore transferable. Frieben highlighted the impact that pre-conversion fertilization has on pasture composition under the organic system. Studying the stand composition and species diversity on permanent pastures on 7 organic farms in Germany and comparing these with conventionally managed grasslands, she found the number of species in organic grassland was greater than in conventional grassland. Cook *et. al.* described soil biodiversity and its interaction with grassland management, emphasising the importance of balancing the whole system. In key research, van Elsen emphasised the importance of species diversity, in an organic farming system.

Plant species diversity, genetic diversity within species, insect diversity and animal diversity are all important for an organic system to function efficiently. Padel *et. al.* reported on forage field measurements undertaken during conversion of dairy farms to organic production in the U.K. Herbage growth fell by 15 per cent during the first year of conversion, but by year 3 had recovered to 93 per cent of pre-conversion values. The white clover content of the herbage increased substantially throughout transition, from less than 5 per cent to over 30 per cent by year 3. Colmenares *et. al.*, (1999) observed the effects of the application of bio-dynamic preparations over a 3.5-year period on permanent grassland in Spain. The results indicated that the bio-dynamic preparations enhanced production and dry matter content.

Leys

In international literature there is much discussion on the use of leys, which are more often than not ryegrass/white clover swards incorporated into a crop rotation. The main purpose of these leys is to increase soil fertility. Petterson *et. al.* collected samples of grass and clover swards from leys during one season on different Swedish farms converting to organic farming, to obtain a survey of general practical differences in the botanical and chemical composition of leys during such a conversion. The main difference of practical importance was a significantly higher percentage of clover in organically grown leys, and differences in botanical composition and with harvest times.

There is less discussion, however, about herbal leys, which are typically made up of a diverse range of species. Foster published a key paper on this topic. It described research carried out on herbal pastures from 1850 to 1984. While this paper does not specifically discuss organic systems, much of the research described was carried out pre-synthetic fertilizer, and can therefore be applied to the organic system. Foster's paper described species and their specific attributes, different mixes commonly used, management, dry matter production and animal production from herbal pastures.

The main body of work carried out in New Zealand on herbal leys was not under organic management as such, but under conditions of nil synthetic fertilizer and herbicide/pesticide application. Ruiz-Jerez *et. al.*, (1991) described an experiment comparing production from a herbal ley sward with a low N and a high N ryegrass/white clover sward. This is a key paper on/for herbal ley production in New Zealand. Two subsequent papers described the nitrate leaching and denitrification, respectively, from the experiment discussed in Ruiz-Jerez *et. al.*

Pasture and Grasing Management

Most of the conventional research carried out in New Zealand over the last 50 years on grasing management in particular would be applicable to organic pastoral systems. The key factor in a pastoral system is to utilise as fully as possible all pasture grown so that it can be converted into milk, regardless of whether a system is under conventional or organic management. Most of the international research on organic pasture/grasing management has been undertaken in the United Kingdom, with the Institute of Grassland and Environmental Research (IGER) being particularly active in this area. Stickland covered the topic of grassland management in an organic system and described species common in New Zealand. Watson and Philips looked at environmentally responsible management of grassland to avoid such problems as N leaching and erosion. Eriksen *et. al.*, (1999) looked at N leaching that can be caused when permanent pastures are ploughed up, and Conacher and Conacher (1998) identified management practices that can prevent some of the negatives of organic farming, such as mining of soil nutrients and elements.

PEST AND DISEASE MANAGEMENT

While there is a vast amount of international research describing the specific treatment of a pest/disease, particularly in cropping systems, there was not much research (and no New Zealand research) dealing with pastures. A few generic papers have been identified that describe the importance of functional diversity in preventing pest and disease infestation of plant communities (including pastures and fodder crops). Finckh *et. al.*, discussed why pests and disease are likely to occur and the management of their prevention. La Torre and Donnarumma, emphasised the importance of prevention in organic systems, and Hopkins and Feber, (1999) described the technique of maintaining a high pest predator population in field margins, again emphasising the importance of bio-diversity.

Weed Management

An abundance of international literature was identified dealing with weed management in organic cropping systems. There was very little (and again no New Zealand specific research) particularly targeted at pastoral systems. A number of generic organic weed control papers have been identified that include methods such as tillage, timing of operations, sowing rates, cultivar selection, mulching, livestock grasing, composting, flame weeding and allelopathy for control of weeds.

Animal Management and Health

There is a huge range of international literature in the area of animal management and health; however, it is not targeted specifically at pastoral systems but rather all organic dairy production systems. Again, no New Zealand-specific research was identified. The focus of much of the research was the use of management practices to maintain good animal health, and most of these good management techniques would be directly applicable to New Zealand pastoral dairy systems. Topics included dairy cow fertility, mastitis, animal nutrition, rearing of young stock, parasite control and general papers covering animal health and husbandry. The paper by Sundrum was a key paper as it critically reviewed literature on animal health, welfare and organic product quality. Various publications confirmed that animal health, represented by expenditure on veterinary costs or incidences of diseases, is generally significantly better on organic farms than on conventional farms. Offerhaus *et. al.* reported on better fertility on organically-managed dairy farms.

Farm Management

Most of the papers described in earlier sections contain information on management of organic farm systems that may be useful to those converted or converting to organics. Austria Bundesanstalt fur alpenlandische Landwirtschaft Gumpenstein discusses managing high yielding cows and the effect of grassland management on milk production. MacNaeidhe and Younie *et. al.* looked at management problems when converting to organics, Anon focused on whole-farm system management, and Disenhaus *et. al.* looked at the role of extension in assisting managers. Again, there was no published research on this topic specific to New Zealand.

Milk Production Potential

There was no published research undertaken in New Zealand on organic dairy production potentials, and it was very difficult to find any data relating to organic production potentials that could be directly transferred into the New Zealand situation. Organic farming systems in Europe tend to have much larger animals that are fed high energy concentrate diets and therefore have greater yield than New Zealand animals.

Milk production levels from organic systems compared with conventional were varied. In Pabst, and Weber *et. al.* milk yield was lower on organic farms but in Jonson was slightly higher. These contrasting results emphasise the overriding importance of good farm management on organic farm production potential.

Wider Environmental Issues

With farm management, it is important to consider the impact of any agricultural system on the environment and this is particularly so under the organic philosophy of farming. Several research articles were identified dealing with the issue of organic production and its relationship with the environment.

Eriksen et. al., Ruiz-Jerez et. al., Ruiz-Jerez et. al. and Watson et. al. looked at the denitrification and nitrogen leaching that can occur from permanent pastures and herbal leys. Watson and Philips described environmentally responsible grassland management techniques, and Cederberg described key environmental differences between an organic and conventional milk production system.

Recommendations

The following questions, in relation to the New Zealand system of farming, are not answered by the current literature. Research in these areas needs to be undertaken for pastoral organic dairy farming to be economically, environmentally and socially sustainable in New Zealand.

- *Soil quality under organic pasture management* : Does organic management positively influence physical soil characteristics long term, *e.g.*, the water-holding capacity of the soil in times of drought, or reduction of the damage occurring through animal treading on pasture in wet conditions?
- *Pasture composition* : Will pasture quality change under organic management? How? What composition dynamicsare likely to occur during the conversion phase when conventional fertilizers are removed?
- *Herbal leys* : What is their value in New Zealand systems as an alternative forage (during drought,etc.)? What is their potential from an animal health perspective?
- *Pasture/Soil Interactions* : What are the key relationships? How is a healthy soil developed and maintained under intensive grasing pressure? How do soil/plant and animal interact? What is the role of trace elements in pasture production and animal health?
- *Pasture systems* : How much is actually grown in an organic system? How do organic dairy pastures need to be managed? Are grasing management techniques thesame as on a conventional farm? How can a pasture system be managed so that it has minimal impact onthe environment? How are pests, diseases and weeds specific to New Zealand controlled?
- *Animal health* : How do organic farms need to be managed to prevent diseases common in a pasture-based dairy production systems?
- *Effluent* : What are the effects of organic farm management on the wider environment? Does organic management with no input of water-soluble fertilizer and decreased stocking rates reduce the amount of detrimental nutrient leaching into surface and groundwater?
- Evaluating the literature on research undertaken in Europe on the different treatments of slurry and themethods and timing of its application onto pasture are some of these research findings applicable to New Zealand conditions?

These are a few of the questions that need to be answered, first to give farmers the confidence to convert to organic dairy production, and second to enable them to farm successfully and in a sustainable manner.

Principles

The principles of Bio-dynamic Agriculture were presented by Dr. R. Steiner, an Austrian scientist and philosopher in 1924 at the request of farmers who were looking for alternatives to conventional agriculture. They were concerned about the decrease of soil fertility, and the decline in both animal health and in seed and food quality that was becoming evident at that time. A series of eight lectures by Steiner, known as the Agricultural Course, became the foundation for the bio-dynamic method of agriculture.

In New Zealand, bio-dynamic methods were first used in 1928 in Havelock North. The Bio Dynamic Association in N.Z. Inc. currently has over 800 members and about 40 Demeter-certified farms. Demeter is the sole internationally recognised and used standard for certification of bio-dynamic farms and produce. The aims of bio-dynamic agriculture are to restore and maintain the vitality and living fertility of soils, and in doing so, produce foods of the highest nutritional quality.

At the centre of the bio-dynamic method and therefore of research on bio-dynamic agriculture are the concept of the farm as a unique individuality, the two field-spray preparations (horn-manure, also known as preparation 500 and horn-silica, also known as preparation 501) and the six compost preparations (known as preparations 502-507). For details of these preparations see Koepf *et. al.*

Early bio-dynamic research in Europe was focused on finding proof that bio-dynamic preparations have a measurable effect. By the end of the 1980s, the focus of research had shifted to examining how the preparations worked in relation to varying farm ecosystems. Since then research has become more focused on understanding the farm individuality.

A further uniquely bio-dynamic method, and one that has been a source of contention, is the ashing of weeds and pests. As this method requires a highly integrated level of knowledge of astronomy and biology it is beyond the scope of this report. Interested readers are referred to the Bio-dynamic Research Institute, Darmstadt branch, Bad Vilbel, Germany.

Soon after his Agricultural Course, Steiner (June 1924) emphasised the importance of making a direct link between research and practice, so that the empirical knowledge of the farmer and the scientific knowledge of a team of disciplinary scientists could assist each other. He also discussed the importance of supporting the farmer by placing scientific knowledge in the context of farm ecosystems. This approach is consistent with participatory research methods now widely used in developing countries and finding increasing application in Europe, North America, and New Zealand.

Research Approaches

In bio-dynamic agriculture one is confronted with concepts of terrestrial and cosmic origin, polarities, and questions of vitality and nutritional quality. The pioneering bio-dynamic researchers, in particular Pfeiffer and Kolisko, found conventional research methods were not always applicable and asked Steiner for advice, which led to the development of new and complementary research methods. Some of the most important bio-dynamic research methods, that have evolved from this early work, are outlined below.

Pictorial Imaging Methods

Amongst other things, Pfeiffer and Kolisko tested measurable substances as well as highly diluted and or potentised substances and their effects on plant and animal with newly developed pictorial imaging methods. This work has been developed and refined by other researchers.

The starting point for this method is that formative forces of a living organism or system are recognisable in each part of the whole.

Pictorial imaging methods make these formative forces visible in a picture and can be applied in the following ways :

- Sensitive copper crystallisation picture, for example, from a drop of blood or plant extract;
- Radial chromatogram from soil moisture;
- Ascending chromatogram from plant extract;
- Degradation/decomposition picture from raw foodproducts;
- Drop picture from water.

These methods are nowadays widely applied in combination with other complementary research methods and conventional methods, in bio-dynamic research (especially in the improvement of product and food quality) and areas such as medical diagnostics. Pictorial imaging techniques require training and experience before a picture can be read with accuracy. Dr. Ursula Balzer-Graf from the Swiss institute Forschungs institut for Vitalqualitat has given this method a scientific base.

In a recent study, Anderson looked at the following methodological and experimental aspects of the crystallisation method : crystallisation chamber techniques, morphological features applicable for visual evaluation of biocrystallograms, computerised image analysis, application of the method in connection with different crop samples, and application of the method in connection with freesing processes of crop samples.

Goethean Phenomenology

The idea of a phenomenological approach was introduced by Goethe and extended by Steiner. Through regular observation of the living plant, animal or other organism in all stages of growth, inner and outer pictures of its processes of movement and changes of form are developed. Bockemuhl developed this into a rational method to research the development of plants and their nutritional and medical properties.

This method is used extensively in research work on the application of bio-dynamic preparations, for example in the work of Bisterbosch on lettuce. Biographical sequencing of plant growth, for example in leaf series of fruit trees, can help orchardists and scientists evaluate and improve cultivation measures. The method is beneficial for forming hypotheses, and the synthesis of different parameters such as the relationships between soil, crop and animal. These relationships can then be considered for the farm or eco-system as a whole. The phenomenological approach can also be applied directly at the farm landscape level.

LONG-TERM FIELD TRIALS

The focus of bio-dynamic research on life processes requires long-term field trials of a minimum of 4 years to be held in different regional locations. A recent example is the Long Term Fertilization trial, supported by the European Commission, over a period of 7 years in Germany, Switzerland and Sweden, in which seven institutes participated.

Gardner discusses the ideas of Steiner for the setting up of test plots on the farm. Statistical models have been designed to exclude unpredictable environmental influences from the investigated variables. The Goethean phenomenological method plays an important role in the synthesis of different parameters in field trials.

Food Quality

Early organic and bio-dynamic research looked at differences in quality of products from conventional and organic agriculture.

Research overseas has since entered a new phase aimed at :

- Improvement of production methods based on the three criteria of organic production, *i.e.*, semi-closed production cycles, natural self regulation, and agro-biodiversity;
- Improvement of the vital quality of seed varieties, food products and food processing;
- Investigation of the links between product qualityand consumer health, in particular the influence of vital quality on certain life processes in human beings.
- In traditional science it is assumed that the nutritional or health value of food results from a measurable composition of substance only. Limiting ones point of view to these material aspects misses the fact that food from plants and animals are alsothe result of the integral organising activity of growing living things (Balzer-Graf, 1999).

Huber defines quality by :
- The product's form;
- The quality of life processes;
- The analysis of its nutrients and other substances;
- The state of integration of its vegetative and generative processes.
- The vital quality of a product, eventually determined by a complex of life processes, plays an important role in the consumer's perception of the value of food in relation to personal health.

New research aimed at adequate and precise assessment of the vital quality of products is under development. The Louis Bolk Institute in the Netherlands developed the concept of vital quality as determined by a complex of life processes at three levels : (a) growth, (b) differentiation or ripening, (c) integration. In their multi-disciplinary research project "Elstar", these three levels were examined in trials and tests on the growth, structure and coherence of apples. The phenomenological method, the pictorial imaging methods and the lesser known biophoton measuring play a central role in research on food quality.

Data resulting from the work with these methods are complemented by a series of other tests that include the following parameters :
- *Bio-chemical* : Respiration, enzyme activities, aroma patterns;
- *Chemical* : Free amino acids, Vitamin C, sugar, nitrate content, pollutants;
- *Physical* : Tissue strength;
- *Microbiological/biochemical* : Storage and intensity of microbial attack, dry matter loss, CO_2 developmentand darkening in degradation tests.

The advantages and disadvantages of the degradation tests have been discussed by Raupp (1998). Parameters have also been set for consumer panel tests on taste, colour and smell, for shelf-life trials with retailers, and for trials on food preparation and food handling by chefs and bakers. Comparative trials of bio-dynamic and other food fed to animals have been studied by Gutknecht. A new International Network for Food, Quality and Health has recently been set up. A training course on pictorial imaging methods for scientists is offered by the Forschungsinstitut fur Vitalqualitat.

Farm Participatory Research Model

The requirement for the bio-dynamic farm to form a semi-closed system with a high degree of self-sufficiency and low dependency on input from outside the farm system has led to the development of farmer participatory research methods. The questions and experiences of the farmer form the basis of the research, allowing a symbiosis of scientific knowledge and empirical knowledge to take place. Results from all trials and tests are judged in the context of the farm as a living organism and as an individuality with its own characteristics of soil, climate, location, history, management style of the farmer, and economic possibilities.

RESEARCH FINDINGS RELATING TO THE BIO-DYNAMIC PREPARATIONS

This section provides an overview of research on the field-spray preparations from the pioneering phase of bio-dynamic agriculture through to a more mature phase. The latter has focused on gaining an increased understanding of the preparations through application of a combination of analytical and new research methods in the context of the farm ecosystem. Findings from recent research on the compost preparations are also presented.

Hornmanure and Hornsilica Preparations

This review is based mainly on the publication of Lammerts van Bueren and Beekman-de Jonge, (1995) on the experiences, research and vision development on the use of the hornmanure preparation (field-spray preparation 500) and the hornsilica preparation (field-spray preparation 501) over 70-year timeframe. There is no translation of this publication from the Dutch. See also Goldstein. Lammerts van Bueren and de Jonge (1995) identify three development periods within this timeframe.

In the first period, from 1924 till the end of the '50s, the field-spray preparations were reported to give positive results when used in conjunction with the compost preparations. A series of publications suggest that there is a 20-30 per cent increase in yield from use of the field-sprays. Pioneering research in the first half of this period was driven by enthusiasm and confidence in the preparations. Towards the '50s research developed more scientific rigour. Pictorial imaging methods were developed as a complement to conventional analytical approaches. Pfeiffer (1948) undertook chemical analysis, bacteria counts, and spectographic analyses of the preparations before and after fermentation in the earth. These tests provide rather amasing analytical data on the preparations and their bacteria-affecting properties. These tests have been repeated in the microbiological research of Dewes in the period 1983-1990.

In the second period, from the '50s until the end of the '80s, the main aim was to establish the effects of the preparations by scientific investigation. The research focused on various scientifically recognised chemical and analytical parameters as measurements for qualitative and quantitative effects. The results obtained could not always be repeated in subsequent experiments in different conditions of soil, climate or landscape.

In the third period, from the '80s onwards, emphases shifted to the restricting and stimulating effects of the two field-spray preparations on the development of crop processes in relation to varying agro-ecosystems.

De Vries reported a balanced development of grass in spring and a stimulation of grass growth in autumn after applying the field-sprays 500 and 501. Von Mackensen studied the effects of the preparations on strawberries and reported a rich setting of fruit, good aroma and fungus free growth, 30 per cent higher yield, and 8-10 days earlier harvest when the hornsilica preparation was sprayed after harvest of the last crop and not in spring. He discussed the polar effects of this preparation as related to environmental factors of soil, light intensity and moisture. By stimulating one phase in the plants growth one can help the polar opposite phase to reach its full potential. Bloksma found that young apple trees in pots in a nursery showed more balanced growth and developed less side branches if the preparations 500 and 501 were applied.

Research and practice show that the field-spray preparations positively influence soil processes such as levelling the pH, and optimising mineralisation, humification, germination and rootgrowth. Work with the hornsilica preparation indicates that the preparation particularly stimulates those plant processes related to warmth and light, such as assimilation, ripening, shelf life and aroma. This is expressed particularly in the lowering of nitrate content and the increasing dry matter and sugar. Both preparations appear to regulate crops in such a way that plants are less susceptible to diseases and pests and have a longer shelf-life.

Application of both preparations may lead to a more integrated development of the crop plant to reach its full potential and provide corresponding quality improvements. The crop actively utilised the environment for its own development rather than submitting to the environment. Application of the preparations creates conditions for a sound arrangement of life processes in an ecosystem. Koenig summarised the three principal effects of the preparations as normalisating, compensating and stimulating.

Field-Spray Preparations and the Compost Preparations

Kotchi found that field-spray preparations worked best if bio-dynamic compost had been applied. Wistinghausen confirmed this. This research shows the importance of an integrated approach to bio-dynamics; application of only some bio-dynamic preparations may give an unbalanced effect.

Use of Horns

Brinton concluded that the bigger the weight :volume ratio of the horn used in making the field-spray preparations, the better the quality of the preparation. Big horns of bulls (which do not have rings) give a low quality of hornmanure. Dewes found that hornmanure from a good horn contained micro-flora similar to those found in worm-castings, but these were not evident in hornmanure made in an imitation horn. Brinton distinguished good quality hornmanure from low quality hornmanure by less rotting, an agreeable smell and less nitrogen loss. Goldstein and Koepf (1982) worked on quality control for preparations. They recommended their results needed further verification through field trials.

Frequency of Applications

Where field-spray preparations are applied with increasing frequency, there is often an increasing effect which is not always positive. Effects may be negative;

apparent in the dehydrogenase activity of soil micro-organisms. Bisterbosch found in her research on lettuce, which included extensive phenomenological observations and food tests, that application of preparations 500 and 501 more than once during the growth season negatively affected product quality. She concluded that a healthy ordering of live processes was disturbed.

Stirring

Pfeiffer mentions a 75 per cent increase of oxygen in the water after 1 hour of manually stirring the bio-dynamic preparations into the water. Schwenk and Filler supported the hypothesis that water transfers information. Schiff described the scientific relevance of the theory that water keeps the memory of dissolved substances. In New Zealand, flowforms are used for stirring. Schikkor suggested that flowforms were better than machine stirring but that hand-stirring gave the best results.

Compost Preparations

A number of scientists have looked at the effects of the bio-dynamic compost preparations on the decomposition of manure and compost in comparative trials. From visual observation, the preparations appear to increase decomposition of manure. Wistinghausen, (1986) found an increase in the cation exchange capacity of manure piles and differences in their ammonium and nitrate content, indicating that the bio-dynamically treated manure was further decomposed. Ahrens saw greater decomposition in bio-dynamically treated straw, shown by a higher ash and lower carbon content and a higher carbon/nitrogen ratio. Abele, (1987) found significantly higher organic matter, higher nitrogen dehydrogenase and cellulolytic activity, more humic acid, greater humification of organic matter and higher levels of Azotobacter and nitrogen fixation by free-living nitrogen fixers.

In a long-term trial with mineralised, organic and bio-dynamically fertilized vegetables, Abele found the lowest nitrate content in bio-dynamically treated samples. Under optimal storage conditions the durability of products shows only small differences. However, under stress conditions (temperature, humidity, chopping up), clear differences occurred in terms of microbial attack and degradation, in favour of bio-dynamically treated samples. More recently Bachinger *et. al.* found the highest organic carbon content in topsoil following the application of composted manure and bio-dynamically treated manure. It took several years for the new organic carbon levels to be established, depending on the type of fertilization applied. Greater enzyme activity and larger microbial biomass with dehydrogenase activity was found in manure fertilizers, especially with bio-dynamic preparations. Positive effects of applying organic fertilization treated with bio-dynamic preparations had also been observed in the subsoil at a depth of 25-55 cm.

In a study on the metabolic activity of the soil micro-organism population in a clover/grass sown pasture, Hoffman *et. al.* observed that bio-dynamically treated grassland showed similar micro-xssbiological characteristics to soil under permanent pasture. Scheller and Raupp, (1997) observed that regular application of farmyard manure led to increases of both amino acid and humus contents of the soil, these could be increased further by the use of the bio-dynamic preparations.

Chapter 14

ANALYSIS METHODS FOR THE DETERMINATION OF ANTHROPOGENIC ADDITIONS OF P TO AGRICULTURAL SOILS

Richard L. Haney[1], Virginia L. Jin[2], Mari-Vaughn V. Johnson[3], Elizabeth B. Haney[4], R. Daren Harmel[1], Jeffrey G. Arnold[1], Michael J. White[1]

[1] United States Department of Agriculture-Agriculture Research Service, Grassland, Soil and Water Research Laboratory, Temple, TX, USA

[2] United States Department of Agriculture-Agriculture Research Service, Agroecosystem Management Research Laboratory, University of Nebraska, East Campus, Lincoln, NE, USA

[3] United States Department of Agriculture-Natural Resource Conservation Service, Temple, TX, USA

[4] Texas A&M University, Texas AgriLife Research & Extension Center, Temple, TX, USA

Email: rick.haney@ars.usda.gov

ABSTRACT

Phosphorus loading and measurement is of concern on lands where biosolids have been applied. Traditional soil testing for plant-available P may be inadequate for the accurate assessment of P loadings in a regulatory environment as the reported levels may not correlate well with environmental risk. In order to accurately assess potential P runoff and leaching, as well as plant uptake, we must be able to measure organic P mineralized by the biotic community in the soil. Soils with varying rates of biosolid application were evaluated for mineralized organic P during a 112-day incubation using the difference between P measured using a rapid-flow analyzer (RFA) and an axial flow Varian ICP-OES. An increase in the P mineralized from the treated soils was observed from analysis with the Varian ICP-OES, but not with the RFA. These results confirm that even though organic P concentrations have increased due to increasing biosolid application, traditional soil testing using an RFA for detection, would not accurately portray P concentration and potential P loading from treated soils.

Keywords

Phosphorus, Anthropogenic Additions, Biosolids, Rapid-Flow Analyzer (RFA), Inductively Coupled Plasma (ICP), Texas Commission of Environmental Quality (TCEQ), Soil Organic C (SOC), Total N (TN), Water-Soluble Organic C (WSOC), Water Soluble Organic N (WSON).

1. INTRODUCTION

Phosphorus (P) is an essential nutrient for plant growth and is applied to agricultural and urban lands in inorganic (fertilizer) and organic forms (manure, biosolids). While the application of inorganic P is relatively uniform on urban and most agricultural lands, intensive P application may occur with the application of biosolids from municipal waste processing centers on agricultural fields. The land application of biosolids is a means of waste management that may provide many benefits to soils including soil conditioning [1] [2] and increased aggregate stability [3] [4] soil water content [5] [6], soil organic C (SOC) [7]-[9], soil nitrogen (N) and phosphorus (P) [6] [10] and potential C mineralization and net N mineralization and immobilization [11]. When biosolids are applied at normal agronomic rates, agricultural productivity may be increased; however, the benefits are limited when crop requirements are exceeded [12]-[14].

Phosphorus concentration in surface water can be tied directly to water quality degradation and can accelerate freshwater eutrophication [15]. Eutrophication is the accumulation of high concentration of inorganic chemicals such as N or P in bodies of water, which is often associated with algal blooms. The negative effects of eutrophication range from decreased species diversity to decreased aesthetics [16]. When large populations of algae die, they decompose on the bottom of the lake or reservoir, resulting in a depletion of oxygen in the ecosystem, which can cause fish and aquatic plant to kill. The algal blooms are not only visually unappealing, but also potentially hazardous to human and animal health.

Total P concentrations of 50 $\mu g \cdot L^{-1}$ are sufficient to sustain chlorophyll concentrations of 10 $\mu g \cdot L^{-1}$, the lower boundary for a trophic state in temperate streams [17] and eutrophy in lakes [18]. Using the Redfield ratio [19], P concentrations exceeding 65 $\mu g \cdot L^{-1}$ are sufficient to support algal growth in many water bodies when nitrate-N concentrations exceed 10 $mg \cdot L^{-1}$. It has been shown that these concentrations of total P can be from ground water in intensive agricultural regions [20]. These non-point sources of P, including both surface runoff and groundwater, can contribute to eutrophication of streams, lakes and rivers. USEPA national lakes assessment survey indicates that based chlorophyll-a concentrations, 20% of lakes are hypereutrophic [21].

Because of the negative environmental effects of the overloading of P in surface waters, P regulations for agricultural lands are being considered, as well as for the land application of biosolids. Regulations for agricultural lands may be centered on a soil-test P value [15]. Soil-test P is most commonly related to plant-available P and may be assessed using many different extractions and analysis methods depending upon which part of the country the soil is analyzed in and the laboratory performing the analysis. As a result, the soil-test P numbers obtained may vary widely.

The most common laboratory analysis for solution P is by colorimetric analyzers (rapid flow analyzer (RFA), which use chemistries to develop color intensity) or inductively coupled plasma (ICP) which uses heat generated by producing plasma to excite atoms to a higher state of energy and measure the light given off as they fall back to a normal state. Differences between P values obtained from colorimetric and ICP analyses have been noted by some studies, while others indicate there is no significant difference between the analysis methods [22]. The difference between the two instruments is often attributed to the fact that the colorimetric method measures H_2PO_4 and HPO_4 (inorganic P) whereas ICP measures orthophosphate P plus other forms of organic and inorganic P [23]. In other words, the organic P may not be available to form the phosphomolybdate complex needed for color formation and detection in colorimetric analysis [22] [24]. The instrumentation used to analyze P can have a significant influence on P values, even though values obtained using the RFA and ICP are generally correlated [25]. Therefore, depending on instrumentation and soil inputs, phosphate values in soils treated with biosolids, where a significant portion of P may be attributed to the organic additions, may be underestimated if using colorimetric analyses versus ICP.

In addition, soil-test P recommendations may not be relevant to the environmental risk associated with soil P levels [15] because they do not reflect organic P concentrations associated with soil organic matter or P mineralized over time due to natural processes in the soil. Flushes of P are associated with the natural drying and re-wetting of soil with natural rainfall events or irrigation and are partially due to the death of the microbes and decomposition of their cells [26]. Even though P associated with organic matter that is converted to dissolved inorganic P contributes to P loading, mineralized P is not routinely tested for regulatory or fertility assessment.

It is important that the methods of analysis for P be as sound and comprehensive as possible in order to accurately assess the P loading associated with runoff from agricultural lands with biosolid application. The objective of this study is to: 1) determine if there is a significant difference between analyses of P from a RFA as compared to an ICP in soils treated with varying levels of biosolids; 2) determine the fraction of organic P mineralized during 112-day incubation.

2. MATERIALS AND METHODS

The study site where samples were obtained is a municipally operated 485-ha, zero-discharge facility in Travis County, TX. Approximately 220 ha of the facility are used in a year round, continuous field rotation for hay production using coastal Bermuda grass (*Cynodon dactylon* L.). Currently, one-third of the total biosolids processed by the facility is land-applied as anaerobically digested, Class B biosolids. Following anaerobic digestion, biosolids are belt-pressed to reduce water content then loaded into manure-spreaders for surface application to a perennial no-tillage forage crop system. Annual mean characteristics of dry biosolids based on monthly analysis (November 2006-October 2007) were pH 8.46 (SM 4500-H B); 17.5% total solids (SM 2540 G); 10,745 mg NH_3-N kg^{-1} (SM 4500-NH_3 D); 14.7 mg NO_2-N + NO_3-N kg^{-1} (EPA 353.2); 52,127 mg TKN kg^{-1} (EOA 351.2). Regulated biosolids elemental constituents for dry biosolids at all samplings fall well below the EPA 40 CRF Part 503 limits [27] [28].

A subset of fields within the 220-ha area was sampled for soils used in the present study. Since 1985, a 14-ha field has been treated with 22 dry Mg biosolids $ha^{-1} \cdot y^{-1}$ (25-yr treatment), an agronomic rate below the maxi- mum annual metal loading rate regulated by the Texas Commission of Environmental Quality (TCEQ); 30 TAC §312.43 (b) (4); TAC (2005). Annual applications in this field, however, were changed to a single application every 2 years (*i.e.* bi-annual applications) starting in 2006 to comply with nutrient management plan recommendations to address high soil N levels. In 2001, 105 ha of adjacent row-crop land was acquired and approved by the former Texas Natural Resource Conservation Commission (now TCEQ) as an experimental exemption area to evaluate the potential soil impacts of applying higher rates of biosolids amendments. The exemption area was split into 20 ha, 36 ha, and 49 ha fields for experimental applications of 22, 45, and 67 Mg $ha^{-1} \cdot y^{-1}$, respectively, starting in 2002 (8-yr treatment). This newer area was converted to coastal Bermuda grass hay production in 2003. A 2-ha unamended control field is located next to the lowest application rate field. The control field has not received applications of biosolids or commercial fertilizer of any kind since 2000. Soils across the site are Bergstrom silt-loams (fine-silty, mixed, superactive, and thermic Cumulic Haplustolls) and are very deep, well-drained, and moderately permeable, with slopes <1%.

Surface soils (0 - 10 cm; 20 cm × 20 cm sampling area) were collected in January 2009 from each field (un-amended control; 22 dry Mg biosolids $ha^{-1} \cdot y^{-1}$ for 25 years; 22, 45, and 67 Mg $ha^{-1} \cdot y^{-1}$ for 8 years) using a hand spade. No biosolids had been applied to any of the treatment fields for at least 3 months prior to soil sampling. Visible biosolids and plant material were removed from the soil surface prior to sampling. Soils were sampled from three random locations per treatment area and composited by treatment. Soils were sieved through a 2 mm mesh, large roots removed, and air-dried prior to laboratory incubations.

Initial total soil organic C (SOC) and total N (TN) in soils were determined by dry combustion at 680°C (Vario Max CHN, Elementar, Hanau, Germany). Soils were extracted with water (10:1 deionized water: dry soil ratio) to simulate available soil nutrients following a natural rain event, as would occur in a field setting. Water-extracts were measured for water-soluble organic C (WSOC) and organic N (WSON) to evaluate the potential in field availability of labile C and N for microbial consumption (Apollo 9000, Teledyne Tekmar, Mason, OH). Inorganic N concentrations in water extracts were measured using continuous flow colorimetry (Flow Solution IV, OI Analytical, College Station, TX). Soil C and N concentrations in the 0-10 cm soil layer were converted to an area basis using soil bulk densities (Mg ha^{-1} for SOC, TN; kg·ha^{-1} for water-soluble C and N). Net N mineralization (mg N kg^{-1}·d^{-1}) over the incubation period was determined as the difference between water-extractable soil inorganic N concentrations (NH^+_4-N + NO^-_2-N + NO^-_3-N) between two consecutive destructive sampling days. Initial P concentrations were colorimetrically determined using a water extract and H3A extract [29]. CO_2-C mineralized after 1-day was determined for each of the surface soil samples using the Haney-Brinton Method [30].

Triplicate soil samples from each treatment were moistened to 50% water-filled pore space and incubated at 25°C ± 1°C in the dark, with destructive sampling at days 0, 7, 14, 28, 56, and 112. Soil samples (40 g each) were incubated in 1-quart canning jars containing vials with 10 mL water to maintain humidity. Soils that were destructively harvested were measured for net P mineralization over time. On the appropriate sampling day, each sample was dried to 40°C, ground and passed through a 2 mm sieve. These sub-samples were weighed into 50-ml plastic centrifuge tubes and received 40 ml of the soil extractant H3A [29], shaken for 10 minutes and centrifuged for 5 min at 3500 rpm. The extracts were filtered through Whatman 2V filter paper.

Orthophosphate P concentrations were measured using an OI Analytical segmented rapid flow analyzer (RFA) (Flow Solution IV, OI Analytical, College Station, Texas) and elemental P concentrations were measured with an axial flow Varian ICP-OES from the same extractant. Net P mineralization (mg P kg^{-1} soil) over the incubation period was determined as the difference between soil extractable P concentrations (PO_4-P and elemental P) from the initial P concentrations (day 0) and consecutive destructive sampling days (days 7, 14, 28, 56, and 112). P mineralization values were compared between the results obtained from the rapid flow analyzer and the axial flow Varian ICP-OES.

Data were tested for normality using the Shapiro-Wilk statistic. One-way ANOVA was used to test the effects of biosolids treatment on net P mineralization rates over time using pair-wise multiple comparison procedures (Holm-Sidak method). Linear regressions were used to examine correlations between net P mineralization from both instruments. All statistical tests were performed using SigmaPlot ver. 11 (Systat, Inc.).

3. RESULTS AND DISCUSSION

Surface soils ranged in pH from 7.2 to 7.7, with the control (unamended soil) having the highest pH value. The amended soils had significantly higher total soil organic carbon (SOC) than the unamended soil (Table 1). The amended soils had correspondingly increased C mineralization after 1 day, except the 22 Mg dry biosolids ha^{-1} 8 yr treatment. Total N, total C:N, WSON, soluble organic C:N, NH_4^+-N, NO_3^--N, inorganic N, total soluble N, and total soluble C:N were greatest in the soil amended with the two highest application rates. There was a statistically significant difference in H3A extractable P among each treatment ($P < 0.001$), with the 22 Mg dry biosolids ha^{-1} 25-yr treatment having the greatest H3A extractable P. The 22 Mg dry biosolids ha^{-1} 25-yr treatment also had significantly more water soluble P than the other treatments ($P < 0.001$). All treatments had significantly greater water soluble P than the control ($P < 0.001$).

Table 1. Mean initial soil pH, bulk densities, and total and water-soluble C and N in surface soils (0 - 10 cm) collected from control and biosolids-applied fields.

Constituent	Unit	Biosolids application rate (Mg dry biosolids ha^{-1})				
		0	22	22	45	67
		Control	25-year	8-year	8-year	8-year
pH (1:10)		7.7	7.4	7.4	7.4	7.2
Bulk density	g·cm^{-3}	1.5	1.5	1.5	1.5	1.5
Total SOC	mg·kg^{-1}	31.4	45.7	41.8	48.1	59.8
H3A extractable P	mg·kg^{-1}	5.7	73.3	30.4	39.0	46.3
Water soluble P	mg·kg^{-1}	2.0	12.1	7.7	6.7	6.2
Total N	mg·kg^{-1}	3.0	4.6	3.8	5.6	9.1
Total C:N		10.4	9.9	11.6	8.7	6.5
Water soluble OC	kg·ha^{-1}	545.0	521.0	534.0	483.0	555.0
Water soluble ON	kg·ha^{-1}	37.7	45.9	55.0	143.5	165.4
Soluble organic C:N		14.5	11.6	9.8	3.4	4.8
NH_4^+-N	kg·ha^{-1}	7.5	7.9	84.0	10.0	12.6
NO_3^--N	kg·ha^{-1}	11.1	55.9	71.4	374.7	573.6
Inorganic N	kg·ha^{-1}	18.6	63.8	79.8	384.7	586.2
Total soluble N	kg·ha^{-1}	56.3	109.7	134.8	528.2	751.7
Total soluble C:N		9.7	4.7	4.1	0.9	0.7
1-day CO_2-C production	mg CO_2-C kg^{-1}	47.7	64.5	47.1	53.4	61.5

P mineralization at 7, 28, 56, and 112 days was highly correlated between the ICP and RFA, while the 14-day mineralization numbers were not as strongly correlated, but still related (Figure 1). When P mineralization values obtained from the RFA and ICP were compared for each treatment, we found that values for the control were not correlated (Figure 2). These results were to be expected since very little P was mineralized from the control when measured both by the ICP and the RFA. P mineralization values for the remaining treatments were highly correlated. Furthermore, H3A and water extractable P, as well as P mineralization values, follow the increased additions of biosolids in the treatments. We speculate that as organic P from the biosolids is mineralized and converted to inorganic P, the inorganic P accumulates in the soil.

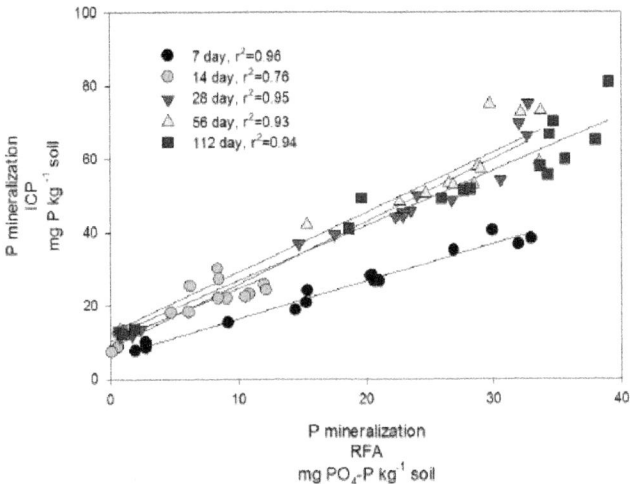

Figure 1. Regression analysis among P mineralized from rapid flow analyzer *vs.* inductively coupled plasma analyzer across days of mineralization.

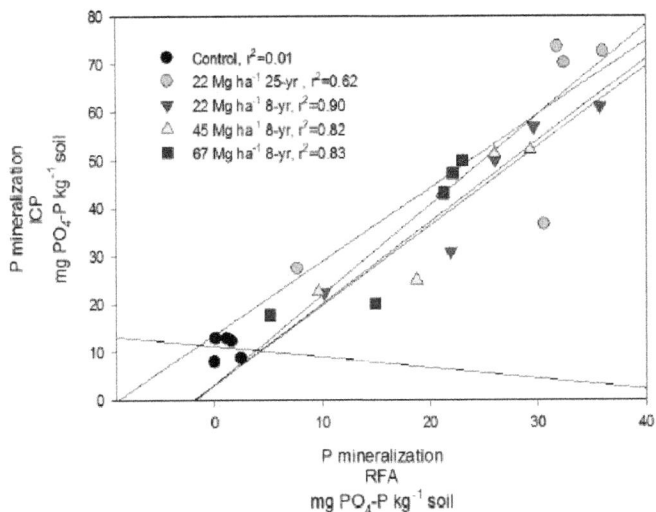

Figure 2. Regression analysis among P mineralized from rapid flow analyzer *vs.* inductively coupled plasma analyzer across treatments.

Even though the relationship among the amounts of P mineralized as measured by the ICP were highly significant ($P < 0.001$), the values measured by the ICP were greater than those measured by the RFA for all treatments (Figure 3). The difference between the ICP and RFA results are clearly depicted in Figure 4, which illustrates the P mineralized from both methods at 56 days (the point at which organic P mineralization peaked).

There was not a significant difference in P mineralization among the biosolid treatments when measured using the RFA, excepting the 22 Mg ha^{-1} 25-yr treatment at 28 days incubation (Figure 5). When measured using the ICP, a significant difference was seen in P mineralization between the 22 Mg ha^{-1} 25-yr treatment and the remaining 8 year treatments after 14 days incubation. If the RFA method was used exclusively, the observer would not see a significant difference among biosolid treatments, which may lead them to believe that there would be no difference in the amount of P mineralized in each soil.

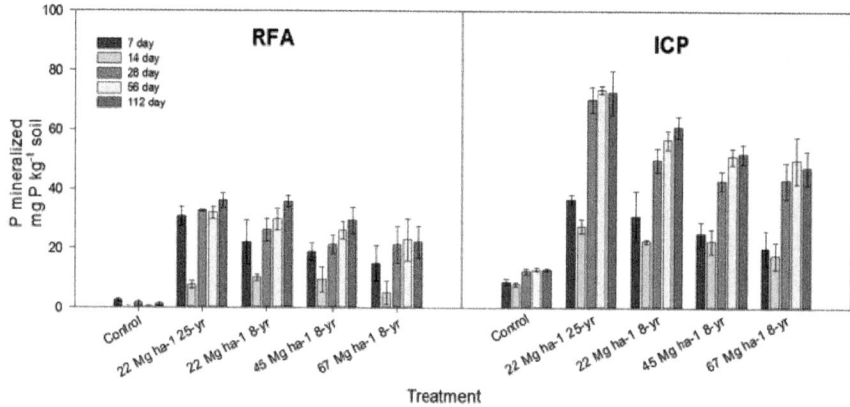

Figure 3. P mineralized from rapid flow analyzer and inductively coupled plasma analyzer across treatments. Error bars represent one standard deviation.

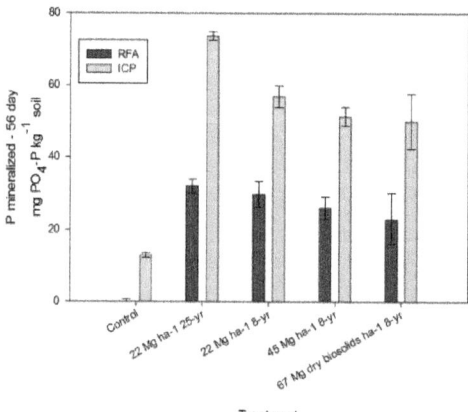

Figure 4. P mineralized from rapid flow analyzer and inductively coupled plasma analyzer across treatments after 56 days. At 56 days organic P mineralized had peaked. Error bars represent one standard deviation.

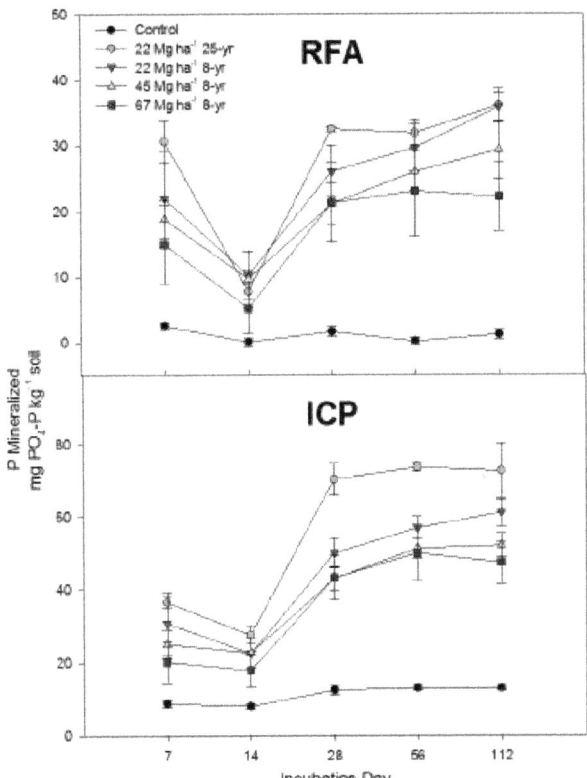

Figure 5. P mineralization over time as determined using the rapid flow analyzer and inductively coupled plasma analyzer methods.

As shown in Figure 6, differences in the mean organic P mineralization values between the control and all treatments and the 22 Mg dry biosolids ha^{-1} 25-yr treatment and all of the 8-yr treatments are greater than would be expected by chance; meaning there is a statistically significant difference (P = <0.001). The 22 Mg dry biosolids ha^{-1} 25-yr treatment had the greatest amount of organic P mineralization, followed by the three 8-yr treatments, then the control. As expected, as organic matter increases organic P mineralization increases, which was easily discernable using the difference in P measured from the ICP and RFA.

There was no correlation between the P mineralized obtained using the RFA method and 1-day CO_2 production (r^2 = 0.19, Figure 7). The correlation increased, but was still weak, when using the ICP method, with 35% of the variability in P mineralized being related to 1-day CO_2 production. One-day CO_2 results are a direct reflection of the activity of the soil microbial biomass. Fierer and Schimel [31] demonstrated that the flush of CO_2 after drying/rewetting soil largely originated from the microbial biomass and Franzluebbers et al. [32] and Ha- ney et al. [33] observed a strong correlation observed between short-term C mineralization and microbial bio- mass C. Release of potentially mineralizable nutrients and decomposition are generally functions of the soil microbial biomass and its activity. The flush of CO_2 after 1 day following rewetting is closely related to potential C and N mineralization [32]. It follows that this flush should also be correlated to P mineralization; however, 1-day CO_2 is only correlated with difference between the P measured using the ICP and the colorimetric method, or the organic P mineralized. When considering only the

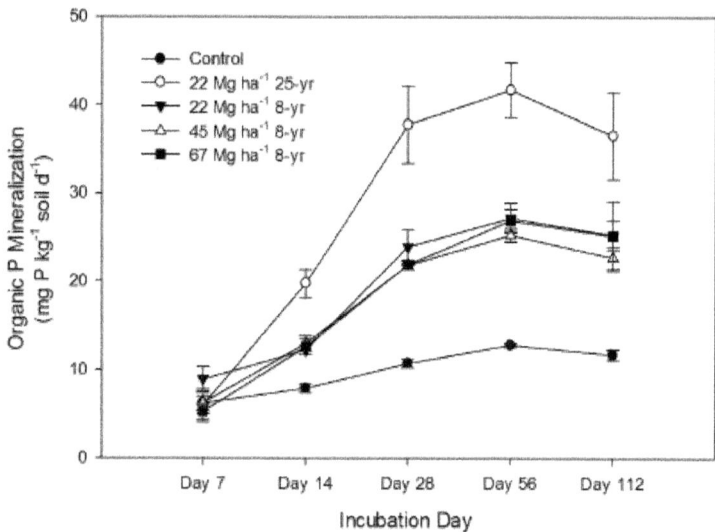

Figure 6. The difference between P mineralized from inductively coupled plasma analyzer minus rapid flow analyzer from biosolids application over time.

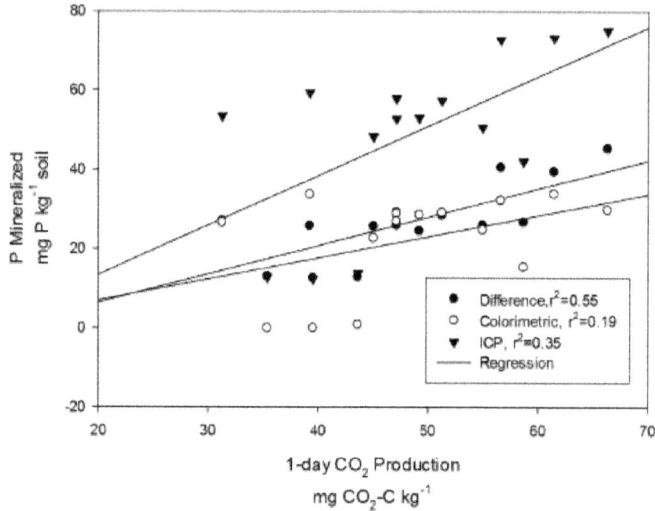

Figure 7. Correlation between P mineralized as measured by the colorimetric method, inductively coupled plasma analyzer, and organic P mineralized and 1-day CO_2 production.

difference in P mineralized by the ICP and the RFA, the correlation further increased as 55% of the variability in P mineralized can be explained by 1-day CO_2 values. The additional P that is detectable by the ICP only may be from the organic P mineralized from the organic fraction (biosolids) of the soil. Little is actually known about the factors that drive microbial P mineralization, but researchers have indicated that microbes mineralize P as a byproduct of C acquisition, without incorporating the P mineralized [34].

The results of this study indicate that as organic P concentrations increase due to increasing biosolid application, traditional soil testing using an RFA for detection would not accurately portray P concentration and potential P loading from treated soils. We recommend using the ICP to measure P mineralized from soils treated with organic P amendments.

REFERENCES

[1] Hargreaves, J.C., Adl, M.S. and Warman, P.R. (2008) A Review of the Use of Composted Municipal Solid Waste in Agriculture. *Agriculture, Ecosystems & Environment*, 123, 1-14. http://dx.doi.org/10.1016/j.agee.2007.07.004

[2] Singh, R.P. and Agrawal, M. (2008) Potential Benefits and Risks of Land Application of Sewage Sludge. *Waste Manage*, 28, 347-358. http://dx.doi.org/10.1016/j.wasman.2006.12.010

[3] Ojeda, G., Alcañiz, J.M. and Ortiz, O. (2003) Runoff and Loses by Erosion in Soils Amended with Sewage Sludge. *Land Degradation & Development*, 14, 563-573. http://dx.doi.org/10.1002/ldr.580

[4] Tejada, M. and Gonzalez, J.L. (2007) Application of Different Organic Wastes on Soil Properties and Wheat Yield. *Agronomy Journal*, 99, 1597-1606. http://dx.doi.org/10.2134/agronj2007.0019

[5] Nielsen, G.H., Hogue, E.J., Forge, T. and Nielsen, D. (2003) Surface Application of Mulches and Biosolids Affect Orchard Soil Properties After 7 Years. *Canadian Journal of Soil Science*, 83, 131-137. http://dx.doi.org/10.4141/S02-034

[6] Kelly, J.J., Favila, E., Hundal, L.S. and Marlin, J.C. (2007) Assessment of Soil Microbial Communities in Surface Applied Mixtures of Illinois River Sediments and Biosolids. *Applied Soil Ecology*, 36, 176-183. http://dx.doi.org/10.1016/j.apsoil.2007.01.006

[7] Klavidko, E.J. and Nelson, D.W. (1979) Changes in Soil Properties from Application of Anaerobic Sludge. *Journal (Water Pollution Control Federation)*, 51, 325-332.

[8] Crecchio, C., Curci, M., Mininni, R., Ricciuti, P. and Ruggiero, P. (2001) Short-Term Effects of Municipal Solid Waste Compost Amendments on Soil Carbon and Nitrogen Content, Some Enzyme Activities and Genetic Diversity. *Biology and Fertility of Soils*, 34, 311-318. http://dx.doi.org/10.1007/s003740100413

[9] Parat, C., Chaussod, R., Lévêque, J. and Andreux, F. (2005) Long-Term Effects of Metal Containing Farmyard Manure and sewage Sludge on Soil Organic Matter in a Fluvisol. *Soil Biology and Biochemistry*, 37, 673-679. http://dx.doi.org/10.1016/j.soilbio.2004.08.025

[10] Bayley, R.M., Ippolito, J.A., Stromberger, M.E., Barbarick, K.A. and Paschke, M.W. (2008) Water Treatment Residuals and Biosolids Coapplications Affect Semiarid

Rangeland Phosphorus Cycling. *Soil Science Society of America Journal*, 72, 711-719. http://dx.doi.org/10.2136/sssaj2007.0109

[11] Jin, V.L., Johnson, M.V.V., Haney, R.L. and Arnold, J.G. (2011) Potential Carbon and Nitrogen Mineralization in Soils from a Perennial Forage Production System Amended with Class B Biosolids. *Agriculture, Ecosystems Environment*, 141, 3-4. http://dx.doi.org/10.1016/j.agee.2011.03.016

[12] Nielson, G.H., Hogue, E.J., Nielson, D. and Zebarth, B.J. (1998) Evaluation of Organic Wastes as Soil Amendments for Cultivation of Carrot and Chard on Irrigated Sandy Soils. *Canadian Journal of Soil Science*, 78, 217-225. http://dx.doi.org/10.4141/S97-037

[13] Speir, T.W., Horswell, J., van Schaik, A.P., McLaren, R.G. and Fietje, G. (2004) Composted Biosolids Enhance Fertility of a Sandy Loam Soil under Dairy Pasture. *Biology and Fertility of Soils*, 40, 349-358. http://dx.doi.org/10.1007/s00374-004-0787-6

[14] Sigua, G.C. and Adjei, M.B. (2005) Cumulative and Residual Effects of Repeated Sewage Sludge Applications: Forage Productivity and Soil Quality Implications in South Florida, USA. *Environmental Science and Pollution Research*, 12, 80-88. http://dx.doi.org/10.1065/espr2004.10.220

[15] Sharpley, A.N., McDowell, R.W. and Kleinman, P.J.A. (2001) Phosphorus Loss from Land to Water: Integrating Agricultural and Environmental Management. *Plant and Soil*, 237, 287-307.http://dx.doi.org/10.1023/A:1013335814593

[16] Steffen, W., Crutzen, P.J. and McNeill, J.R. (2007) The Anthropocene: Are Humans Now Overwhelming the Great Forces of Nature. *AMBIO: A Journal of the Human Environment*, 36, 614-621. http://dx.doi.org/10.1579/0044-7447(2007)36[614:TAAHNO]2.0.CO;2

[17] Dodds, W.K., Jones, J.R. and Welch, E.B. (1998) Suggested Classification of Stream Trophic State: Distributions of Temperate Stream Types by Chlorophyll, Total Nitrogen, and Phosphorus. *Water Research*, 32, 1455-1462. http://dx.doi.org/10.1016/S0043-1354(97)00370-9

[18] Smith, V.H. (2003) Eutrophication of Freshwater and Coastal Marine Ecosystems a Global Problem. *Environmental Science and Pollution Research*, 10, 126-139. http://dx.doi.org/10.1065/espr2002.12.142

[19] Redfield, A.C. (1958) The Biological Control of Chemical Factors in the Environment. *American Scientist*, 46, 205-221.

[20] Burkart, M.R., Simpkins, W.W., Morrow, A.J. and Gannon, J.M. (2004) Occurrence of Total Dissolved Phosphorus in Unconsolidated Aquifers and Aquitards in Iowa. *Journal of the American Water Resources Association*, 40, 827-834.

[21] USEPA (2009) National Lakes Assessment: A Collaborative Survey of the Nation's Lakes. US Environmental Protection Agency, Office of Water and Office of Research and Development, Washington DC, EPA 841-R-09-001.

[22] Pittman, J.J., Zhang, H. and Schroder, J.L. (2005) Differences of Phosphorus in Mehlich-3 Extracts Determined by Colorimetric and Spectroscopic Methods. *Communications in Soil Science and Plant Analysis*, 36, 1641-1659. http://dx.doi.org/10.1081/CSS-200059112

[23] Kuo, S. (1996) Phosphorus. In: Sparks, D.L., Ed., *Methods of Soil Analysis: Part 3*, SSSA Book Series No. 5, SSSA and ASA, Madison, 869-919.

[24] Sobeck, S.A. and Ebeling, D.D. (2007) Mass Spectrometric Analysis for Phosphate in Soil Extracts; Comparison of Mass Spectrometry, Colorimetry, and Inductively Coupled Plasma. *The Journal of Analytical Sciences Digital Library*.

[25] Wolf, A.M., Kleinman, P.J.A., Sharpley, A.N. and Beegle, D.B. (2005) Development of a Water-Extractable Phosphorus Test for Manure: An Interlaboratory Study. *Soil Science Society of America Journal*, 69, 695-700. http://dx.doi.org/10.2136/sssaj2004.0096

[26] Oehl, F., Frossard, E., Fliessbach, A., Dubois, D. and Oberson, A. (2004) Basal Organic Phosphorus Mineralization in Soils under Different Farming Systems. *Soil Biology and Biochemistry*, 36, 667-675. http://dx.doi.org/10.1016/j.soilbio.2003.12.010

[27] USEPA (1992) Technical Support Document for Land Application of Sewage Sludge, Volume I. Office of Water, United States Environmental Protection Agency, Washington DC, EPA 822/R-93-001a.

[28] USEPA (1993) Standards for the Use or Disposal of Sewage Sludge; Final Rules. United States Environmental Protection Agency, Washington DC, 40 CFR Part 257 and 503.

[29] Haney, R.L., Haney, E.B., Hossner, L.R. and Arnold, J.G. (2006) A New Soil Extractant for Simultaneous Phosphorus, Ammonium, and Nitrate Analysis. *Communications in Soil Science and Plant Analysis*, 37, 1511-1523. http://dx.doi.org/10.1080/00103620600709977

[30] Haney, R.L., Brinton, W.F. and Evans, E. (2008) Soil CO2 Respiration: Comparison of Chemical Titration, CO_2 IRGA Analysis and the Solvita Gel System. *Renewable Agriculture and Food Systems*, 23, 171-176. http://dx.doi.org/10.1017/S174217050800224X

[31] Fierer, N. and Schimel, J.P. (2003) A Proposed Mechanism for the Pulse in Carbon Dioxide Production Commonly Observed Following the Rapid Rewetting of a Dry Soil. *Soil Science Society of America Journal*, 67, 798-805. http://dx.doi.org/10.2136/sssaj2003.0798

[32] Franzluebbers, A.J., Haney, R.L., Honeycutt, C.W., Schomberg, H.H. and Hons, F.M. (2000) Flush of CO_2 Following Rewetting of Dried Soil Relates to Active Organic Pools. *Soil Science Society of America Journal*, 64, 613-623. http://dx.doi.org/10.2136/sssaj2000.642613x

[33] Haney, R.L., Franzluebbers, A.J. and Hons, F.M. (2001) A Rapid Procedure for Prediction of N Mineralization. *Biology and Fertility of Soils*, 33, 100-104. http://dx.doi.org/10.1007/s003740000294

[34] Spohn, M. and Kuzyakov, Y. (2013) Phosphorus Mineralization Can Be Driven by Microbial Need for Carbon. *Soil Biology Biochemistry*, 61, 69-75. http://dx.doi.org/10.1016/j.soilbio.2013.02.013

This page left intentionally blank.

INDEX

A

Abiotic Soil Components, 244
Actinomycetes, 49, 223
Activation, 230
Aerobic Microbial of Soils, 266
Albumen, 83
Alfisols, 99
Algae, 225
Alluvial Soils, 142
Ammonia and Nitric Acid, 76
Anaerobiosis, 35
Andisols, 102
Animal Caseine, 84
Animal Management and Health, 277
Applying Soil Taxonomy, 108
Arid Soils, 97
Arthropod Fauna in Arable Crops, 255
Arthropod Fauna in Orchards, 256

B

Bacteria, 10
Basic Characteristics of Soils, 122
Beneficial Arthropods, 264
Biodegradation of Pesticides in Soil, 229
Biology of Ectomycorrhizal Fungi, 25
Biomass Measurement of Soil, 270
Black Soils, 143

C

Cane Sugar, 80
Carbon, 247
Carbon Cycle, 36
Carbonic Acid, 61
Carburetted Hydrogen, 66
Care and Feeding of Soil Organisms, 14
Category of the Soil Series, 112
Cellulose, 78
Changing the Spectrum of Toxicity, 230
Chemical Changes, 129
Chemical Weathering, 126
Chlorine, 93
Ciliophora, 227
Class Mastigophora, 227
Classified Soils in the Tropics, 133
Class-Rhizopoda, 226
Clay Minerals, 127
Climate Changes, 129
Collembola, 261
Components of Soil, 215
Compost Preparations, 285
Crop Rotations, 240

D

Decomposition of Ammonia, 89
Decomposition of Carbonic Acid, 87

Decomposition of Nitric Acid, 89
Decomposition of Water in the Plant, 88
Deposition, 128
Development of the Soil Survey, 166
Dextrine, 79
Distributing Amendments, 112
Drainage History, 129
Dry Deposition, 258

E

Early Concepts and Study of Soil, 158
Earthworm Field Tests, 256
Earthworms, 15
Ectomycorrhizal Fungi, 24
Effects of Cultivation on the Habitat, 234
Effects of Pesticides, 228
Enchytraeids, 262
Entisols, 99
Environmental Factors, 272
Ergosterol, 271
Ericoid Mycorrhizas, 26
Erosion, 168
Estimating the Degree of Erosion, 171
Evaluating Proposed Amendments, 112

F

Factors Affecting Soil Organism Populations, 17
Fallow and Top-soil Storage, 241
Families of Soil, 133
Farm Management, 277
Farm Participatory Research Model, 282
Farmyard Manure, 186
Field Tests, 253
Foliar Fertilizer, 206
Food Quality, 281
Forest Soils, 145
Frequency of Applications, 284
Fruits, 90

Functions / Role of Bacteria, 50
Fungi, 10
Fungicides, 239

G

Gelisols, 104
Glucosamine, 271
Goethean Phenomenology, 280
Grape Sugar, 80
Great Groups, 131

H

Haematite, 127
Harvesting, 192
Healthy Soil–Healthy Plant, 201
Herbicides, 239
Heterotrophs, 222
Higher Plants, 260
Histosols, 101
Honey Bee Field Test, 254
Hydric Soils, 179
Hydric Soils and Soil Survey, 173

I

Igneous Rocks, 4
Inceptisols, 99
Incrusting Matter, 78
Initial Loading, 128
Inorganic Constituents, 77
Insecticides, 239
Insects, 9
Integrated Strategies, 206
Investigation Results, 155
Irrigation, 240

L

Landslip Erosion, 169
Laterite Soils, 144

Leys, 275
Loading and Drainage History, 128
Lsopods and Millipedes, 261
Lumbricids, 263

M

Major Elements, 257
Management of Low-Activity Clays (LAC) Soils, 138
Margaric Acid, 81
Material and Methods of Soil, 154
Metamorphic Rocks, 5
Methodology, 210
Micas, 127
Microbial Activity, 39
Microbial Cultivation, 44
Microcosm Tests, 248
Milk Production Potential, 277
Mineral Weathering, 259
Mites, 8
Modern Concept of Soil, 163
Mollisols, 100
Molluscs, 264
Mycorrhizal Associations, 20

N

National Soil Classification System, 109
Nature of Soil, 142
Nematodes, 11
Nitric Acid, 66
Nitrogen, 244
Nitrogen Fixation, 258
Normal Errors of Observation, 114
Nutrient Uptake by Crops, 191

O

Official Soil Series Descriptions, 118
Oleic Acid, 82
Orders, 131

Organic Chemicals and Metals, 259
Organic Constituents of Plants, 60
Organic Pasture Systems of Soil, 274
Oribatid Mites, 261
Origins of Soils from Rocks, 126
Origins, Formation and Mineralogy, 126
Oxygen Uptake, 270

P

Pastoral Fertilizer Strategies, 205
Pasture and Grasing Management, 276
Peaty and Organic Soils, 145
Pectine and Pectic Acid, 81
Permanent Pasture Composition, 274
Persistence of Pesticides in Soil, 228
Pest and Diseases, 192
Pesticides, 238
Phosphoric Acid, 94
Phosphorus, 246
Physical Measurement of Biomass, 271
Physical Weathering, 126
Pictorial Imaging Methods, 280
Plant Disease, 33
Plant Roots and Soil Biota, 18
Policy and Responsibilities, 108
Principles, 279
Printed Soil Surveys, 174
Proposing and Naming a Soil Series, 115
Protein Content, 271
Protozoa, 226
Protozoans and Nematodes, 260
Purely Mineral Soil, 198

Q

Quantification of Input of Chemicals in the Soil, 257
Quartz, 127

R

Recommendations, 278
Red and Yellow Soils, 143
Redox Reactions, 37
Research Approaches, 280
Residual Effect of Fertilizers, 191
Respiratory Metabolism, 270

S

Saline Soils, 145
Sedimentary Rocks, 4
Series of Soil, 133
Shape of Grains, 124
Soil as an Engineering Material, 123
Soil Classification, 120
Soil Enzyme : Dehydrogenase, 252
Soil Enzyme : Phosphatase, 253
Soil Enzyme : Urease, 252
Soil Fauna, 55
Soil Fertility, 181
Soil Flora, 55
Soil Fungi, 16
Soil Genesis, 1
Soil Microbiology, 34
Soil Orders Found in Missouri, 99
Soil Organisms, 13
Soil Protozoa, 16
Soil Supplying Power, 191
Soil Survey and the Soil Map, 167
Soil Survey, 158
Soil Taxonomy 96
Soil-Borne Plant Pathogens, 32
Soil-Landscape Relationships, 165
Soils and their Classification, 129
Sporophora, 227
Starch Contains, 79
Stearic Acid, 82
Stirring, 285
Strategies for Bioremediation, 59
Structure or Fabric, 125
Submitting Proposed Amendments, 111
Sub-orders, 131
Sugar, 80
Sulphur, 245
Sulphuric Acid, 93

T

Temperature and Heat Transfer, 273
Test for Denitrification, 252
Test for Nitrogen Fixation, 252
Tests on Enzyme Activity in Soil, 252
Tests on Soil Microbial Processes, 250
Transportation, 128
Transportation and Deposition, 127
Tree Clearing, 242
Types of Manures and Fertilizers, 185
Types of Micro-organisms in Soil, 55
Typical Characteristics, 97

U

Ultisols, 100
Ultisols and Oxisols, 136
Unique Characteristics of Arid Soils, 98
Unloading, 128
Use of Horns, 284

V

Vegetable Fibrine, 83
Vertisols, 100

W

Water, 75
Water Erosion, 170
Water Management, 192
Weathering of Rocks, 126
Web-based Soil Survey, 175

Weed Management, 276
Wet Deposition, 258
What is the Origin of Root Exudates?, 18
Where do Soil Biota Live?, 242
Wider Environmental Issues, 277
Wind Erosion, 171

This page left intentionally blank.